Freddy Dumortier, Jaume Llibre, Joan C. Artés

Qualitative Theory of Planar Differential Systems

With 123 Figures and 10 Tables

 Springer

Freddy Dumortier
Hasselt University
Campus Diepenbeek
Agoralaan-Gebouw D
3590 Diepenbeek, Belgium
e-mail: freddy.dumortier@uhasselt.be

Jaume Llibre
Universitat Autònoma de
Barcelona
Dept. Matemátiques
08193 Cerdanyola
Barcelona, Spain
e-mail: jllibre@mat.uab.es

Joan C. Artés
Universitat Autònoma de
Barcelona
Dept. Matemátiques
08193 Cerdanyola
Barcelona, Spain
e-mail: artes@mat.uab.es

Mathematics Subject Classification (2000): 34Cxx (34C05, 34C07, 34C08, 34C14, 34C20, 34C25, 34C37, 34C41), 37Cxx (37C10, 37C15, 37C20, 37C25, 37C27, 37C29)

Library of Congress Control Number: 2006924563

1 00S 4S4 20S

ISBN-10 3-540-32893-9 Springer Berlin Heidelberg New York
ISBN-13 3-540-32902-1 Springer Berlin Heidelberg New York

Springer is a part of Springer Science+Business Media
springer.com
© Springer-Verlag Berlin Heidelberg 2006

Cover design: Erich Kirchner, Heidelberg
Typesetting by the authors and SPi using a Springer LaTeX macro package

Printed on acid-free paper   SPIN: 11371328   41/3100/SPi   5 4 3 2 1 0

Preface

Our aim is to study *ordinary differential equations* or simply *differential systems* in two real variables

$$\dot{x} = P(x, y),$$
$$\dot{y} = Q(x, y), \qquad (0.1)$$

where P and Q are C^r functions defined on an open subset U of $\mathbb{R}^2$, with $r = 1, 2, \ldots, \infty, \omega$. As usual C^ω stands for analyticity. We put special emphasis onto *polynomial differential systems*, i.e., on systems (0.1) where P and Q are polynomials.

Instead of talking about the differential system (0.1), we frequently talk about its associated *vector field*

$$X = P(x, y)\frac{\partial}{\partial x} + Q(x, y)\frac{\partial}{\partial y} \qquad (0.2)$$

on $U \subset \mathbb{R}^2$. This will enable a coordinate-free approach, which is typical in the theory of dynamical systems. Another way expressing the vector field is by writing it as $X = (P, Q)$. In fact, we do not distinguish between the differential system (0.1) and its vector field (0.2).

Almost all the notions and results that we present for two-dimensional differential systems can be generalized to higher dimensions and manifolds; but our goal is not to present them in general, we want to develop all these notions and results in dimension 2. We would like this book to be a nice introduction to the qualitative theory of differential equations in the plane, providing simultaneously the major part of concepts and ideas for developing a similar theory on more general surfaces and in higher dimensions. Except in very limited cases we do not deal with bifurcations, but focus on the study of individual systems.

Our goal is certainly not to look for an analytic expression of the global solutions of (0.1). Not only would it be an impossible task for most differential systems, but even in the few cases where a precise analytic expression can be found it is not always clear what it really represents. Numerical analysis of a differential system (0.1) together with graphical representation are essential ingredients in the description of the phase portrait of a system (0.1) on U; that is, the description of U as union of all the orbits of the system. Of course, we do not limit our study to mere numerical integration. In fact in trying to do this one often encounters serious problems; calculations can take an enormous amount of time or even lead to erroneous results. Based however on a priori knowledge of some essential results on differential systems (0.1), these problems can often be avoided.

Qualitative techniques are very appropriate to get such an overall understanding of a differential system (0.1). A clear picture is achieved by drawing a phase portrait in which the relevant qualitative features are represented; it often suffices to draw the "extended separatrix skeleton." Of course, for practical reasons, the representation must not be too far from reality and has to respect some numerical accuracy. These are, in a nutshell, the main ingredients in our approach.

The basic results on differential systems and their qualitative theory are introduced in Chap. 1. There we present the fundamental theorems of existence, uniqueness, and continuity of the solutions of a differential system with respect the initial conditions, the notions of α- and ω-limit sets of an orbit, the Poincaré–Bendixson theorem characterizing these limit sets and the use of Lyapunov functions in studying stability and asymptotic stability. We analyze the local behavior of the orbits near singular points and periodic orbits. We introduce the notions of separatrix, separatrix skeleton, extended (and completed) separatrix skeleton, and canonical region that are basic ingredients for the characterization of a phase portrait.

The study of the singular points is the main objective of Chaps. 2, 3, 4, and 6, and partially of Chap. 5. In Chap. 2 we mainly study the elementary singular points, i.e., the hyperbolic and semi-hyperbolic singular points. We also provide the normal forms for such singularities providing complete proofs based on an appropriate two-dimensional approach and with full attention to the best regularity properties of the invariant curves. In Chap. 3, we provide the basic tool for studying all singularities of a differential system in the plane, this tool being based on convenient changes of variables called blow-ups. We use this technique for classifying the nilpotent singularities.

A serious problem consists in distinguishing between a focus and a center. This problem is unsolved in general, but in the case where the singular point is a linear center there are algorithms for solving it. In Chap. 4 we present the best of these algorithms currently available.

Polynomial differential systems are defined in the whole plane $\mathbb{R}^2$. These systems can be extended to infinity, compactifying $\mathbb{R}^2$ by adding a circle, and extending analytically the flow to this boundary. This is done by the so-called "Poincaré compactification," and also by the more general "Poincaré–Lyapunov compactification." In both cases we get an extended analytic differential system on the closed disk. In this way, we can study the behavior of the orbits near infinity. The singular points that are on the circle at infinity are called the infinite singular points of the initial polynomial differential system. Suitably gluing together two copies of the extended system, we get an analytic differential system on the two-dimensional sphere.

In Chap. 6 we associate an integer to every isolated singular point of a two-dimensional differential system, called its index. We prove the Poincaré–Hopf theorem for vector fields on the sphere that have finitely many singularities: the sum of the indices is 2. We also present the Poincaré formula for computing the index of an isolated singular point.

After singular points the main subjects of two-dimensional differential systems are limit cycles, i.e., periodic orbits that are isolated in the set of all periodic orbits of a differential system. In Chap. 7 we present the more basic results on limit cycles. In particular, we show that any topological configuration of limit cycles is realizable by a convenient polynomial differential system. We define the multiplicity of a limit cycle, and we study the bifurcations of limit cycles for rotated families of vector fields. We discuss structural stability, presenting a number of results and some open problems. We do not provide complete proofs but explain some steps in the exercises.

For a two-dimensional vector field the existence of a first integral completely determines its phase portrait. Since for such vector fields the notion of integrability is based on the existence of a first integral the following natural question arises: *Given a vector field on $\mathbb{R}^2$, how can one determine if this vector field has a first integral?* The easiest planar vector fields having a first integral are the Hamiltonian ones. The integrable planar vector fields that are not Hamiltonian are, in general, very difficult to detect. In Chap. 8 we study the existence of first integrals for planar polynomial vector fields through the Darbouxian theory of integrability. This kind of integrability provides a link between the integrability of polynomial vector fields and the number of invariant algebraic curves that they have.

In Chap. 9 we present a computer program based on the tools introduced in the previous chapters. The program is an extension of previous work due to J. C. Artés and J. Llibre and strongly relies on ideas of F. Dumortier and the thesis of C. Herssens. Recently, P. De Maesschalck had made substantial adaptations. The program is called "Polynomial Planar Phase Portraits," abbreviated as P4 [9]. This program is designed to draw the phase portrait of any polynomial differential system on the compactified plane obtained by Poincaré or Poincaré–Lyapunov compactification; local phase portraits, e.g., near singularities in the finite plane or at infinity, can also be obtained. Of course, there are always some computational limitations that are described in Chaps. 9 and 10. This last chapter is dedicated to illustrating the use of the program P4.

Almost all chapters end with a series of appropriate exercises and some bibliographic comments.

The program P4 is freeware and the reader may download it at will from http://mat.uab.es/~artes/p4/p4.htm at no cost. The program does not include either MAPLE or REDUCE, which are registered programs and must be acquired separately from P4. The authors have checked it to be bug free, but nevertheless the reader may eventually run into a problem that P4 (or the symbolic program) cannot deal with, not even by modifying the working parameters.

To end this preface we would like to thank Douglas Shafer from the University of North Carolina at Charlotte for improving the presentation, especially the use of the English language, in a previous version of the book.

Contents

List of Figures

1

Basic Results on the Qualitative Theory of Differential Equations

In this chapter we introduce the basic results on the qualitative theory of differential equations with special emphasis on planar differential equations, the main topic of this book.

In the first section we recall the basic results on existence, uniqueness, and continuous dependence on initial conditions, as well as the basic notions of maximal solution and periodic solution. The basic notions of phase portrait, topological equivalence and conjugacy, and α- and ω-limit sets of an orbit of a differential equation are introduced in Sects. 2–4, respectively.

The local phase portrait at singular points and periodic orbits are studied in Sects. 5 and 6, respectively. The beautiful Poincaré–Bendixson Theorem, characterizing the α- and ω-limit sets of bounded orbits, is stated in Sect. 7. Finally, in Sect. 8 the notions of separatrix, separatrix skeleton, extended (and completed) separatrix skeleton and canonical region are given. These notions are fundamental for understanding the phase portrait of a planar system of differential equations.

1.1 Vector Fields and Flows

Let Δ be an open subset of the euclidean plane $\mathbb{R}^2$. We define a *vector field of class C^r* on Δ as a C^r map $X : \Delta \to \mathbb{R}^2$ where $X(x)$ is meant to represent the free part of a vector attached at the point $x \in \Delta$. Here the r of C^r denotes a positive integer, $+\infty$ or ω, where C^ω stands for an analytic function. The graphical representation of a vector field on the plane consists in drawing a number of well chosen vectors $(x, X(x))$ as in Fig. 1.1. Integrating a vector field means that we look for curves $x(t)$, with t belonging to some interval in $\mathbb{R}$, that are solutions of the *differential equation*

$$\dot{x} = X(x), \tag{1.1}$$

where $x \in \Delta$, and $\dot{x}$ denotes dx/dt (one can also write x' instead of $\dot{x}$). The variables x and t are called the *dependent variable* and the *independent*

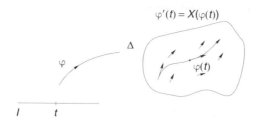

$$\varphi'(t) = X(\varphi(t))$$

Fig. 1.1. An integral curve

variable of the differential equation, respectively. Usually t is also called the *time*.

Since $X = X(x)$ does not depend on t, we say that the differential equation (1.1) is *autonomous*.

We recall that solutions of this differential equation are differentiable maps $\varphi : I \to \Delta$ (I being an interval on which the solution is defined) such that

$$\frac{d\varphi}{dt}(t) = X(\varphi(t)),$$

for every $t \in I$.

The vector field X is often represented by a differential operator

$$X = X_1 \frac{\partial}{\partial x_1} + X_2 \frac{\partial}{\partial x_2},$$

operating on functions that are at least C^1. For such a function f, the image

$$Xf = X_1 \frac{\partial f}{\partial x_1} + X_2 \frac{\partial f}{\partial x_2},$$

represents at x the derivative of $f \circ \varphi$, for any solution φ at t with $\varphi(t) = x$.

Associated to the vector field $X = (X_1, X_2)$ or to the differential equation (1.1) there is the 1–form

$$\omega = X_1(x_1, x_2)dx_2 - X_2(x_1, x_2)dx_1.$$

In this book we mainly talk about vector fields or differential equations, but we will see that it is sometimes useful or more appropriate to use the language of 1–forms, as we will for instance do in Chap. 4.

A point $x \in \Delta$ such that $X(x) = 0$ (respectively $\neq 0$) is called a *singular point* (respectively *regular point*) of X. Often the word *critical* is used instead of singular, but as critical may have different meanings depending on the context, we prefer the word singular.

Let x be a singular point of X. Then $\varphi(t) = x$, with $-\infty < t < \infty$, is a solution of (1.1), i.e., $0 = \varphi'(t) = X(\varphi(t)) = X(x)$.

Let $x_0 \in \Delta$ and $\varphi : I \to \Delta$ be a solution of (1.1) such that $\varphi(0) = x_0$. The solution $\varphi : I \to \Delta$ is called *maximal* if for every solution $\psi : J \to \Delta$ such that $I \subset J$ and $\varphi = \psi|_I$ then $I = J$ and, consequently $\varphi = \psi$. In this case we write $I = I_{x_0}$ and call it the *maximal interval*.

Let $\varphi : I_{x_0} \to \Delta$ be a maximal solution; it can be regular or constant. Its image $\gamma_\varphi = \{\varphi(t) : t \in I_{x_0}\} \subset \Delta$ endowed with the orientation induced by φ, in case φ is regular, is called the *trajectory*, *orbit* or *(maximal) integral curve* associated to the maximal solution φ.

We recall that for a solution defining an integral curve the tangent vector $\varphi'(t)$ at $\varphi(t)$ coincides with the value of the vector field X at the point $\varphi(t)$; see Fig. 1.1.

Theorem 1.1 *Let X be a vector field of class C^r with $1 \leq r \leq +\infty$ or $r = \omega$. Then the following statements hold.*

(i) *(Existence and uniqueness of maximal solutions). For every $x \in \Delta$ there exists an open interval I_x on which a unique maximal solution φ_x of (1.1) is defined and satisfies the condition $\varphi_x(0) = x$.*

(ii) *(Flow properties). If $y = \varphi_x(t)$ and $t \in I_x$, then $I_y = I_x - t = \{r - t : r \in I_x\}$ and $\varphi_y(s) = \varphi_x(t + s)$ for every $s \in I_y$.*

(iii) *(Continuity with respect to initial conditions). Let $\Omega = \{(t, x) : x \in \Delta, t \in I_x\}$. Then Ω is an open set in $\mathbb{R}^3$ and $\varphi : \Omega \to \mathbb{R}^2$ given by $\varphi(t, x) = \varphi_x(t)$ is a map of class C^r. Moreover, φ satisfies*

$$D_1 D_2 \varphi(t, x) = DX(\varphi(t, x)) D_2 \varphi(t, x)$$

for every $(t, x) \in \Omega$ where D_1 denotes the derivative with respect to time, D_2 denotes the derivative with respect to x, and DX denotes the linear part of the vector field.

The proof of this theorem (and the others in this chapter) is given in [152] and [151]. We can also refer to [44].

We denote by $\varphi : \Omega \to \mathbb{R}^2$ the *flow generated by the vector field X*.

It is clear that if $I_x = \mathbb{R}$ for every x, the flow generated by X is a flow defined on $\Omega = \mathbb{R} \times \Delta$. But many times one has $I_x \neq \mathbb{R}$. For this reason the flow generated by X is often called the *local flow generated by X*. In case $\Omega = \mathbb{R} \times \mathbb{R}^2$, condition *(ii)* of Theorem 1.1 defines a group homomorphism $t \to \varphi_t$ from the additive group of the reals to the group of C^r diffeomorphisms from $\mathbb{R}^2$ to $\mathbb{R}^2$, endowed with the operation of composition. In case $\Delta \neq \mathbb{R}^2$ or $I_x \neq \mathbb{R}$ the homomorphism property, expressed by condition *(ii)*, holds only when the composition makes sense, inducing the word "local" in the denomination. The name "flow" comes from the fact that points following trajectories of X resemble liquid particles following a laminar motion.

Theorem 1.2 *Let X be a vector field of class C^r with $1 \leq r \leq +\infty$ or $r = \omega$, and $\Delta \subset \mathbb{R}^2$. Let $x \in \Delta$ and $I_x = (\omega_-(x), \omega_+(x))$ be such that $\omega_+(x) < \infty$ (respectively $\omega_-(x) > -\infty$). Then $\varphi_x(t)$ tends to $\partial \Delta$ (the boundary of Δ) as $t \to \omega_+(x)$ (respectively $t \to \omega_-(x)$), that is, for every compact $K \subset \Delta$ there exists $\varepsilon = \varepsilon(K) > 0$ such that if $t \in [\omega_+(x) - \varepsilon, \omega_+(x))$ (respectively $t \in (\omega_-(x), \omega_-(x) + \varepsilon])$, then $\varphi_x(t) \notin K$.*

Proof. Contrary to what we wish to show we suppose that there exists a compact set $K \subseteq \Delta$ and a sequence $t_n \to w_+(x) < \infty$ such that $\varphi_x(t_n) \in K$ for all n. Taking a subsequence if necessary, we may assume that $\varphi_x(t_n)$ converges to a point $x_0 \in K$. Let $b > 0$ and $\alpha > 0$ such that $B_b \times I_\alpha \subseteq \Omega$, where $B_b = \{y \in \mathbb{R}^2 : |y - x_0| \leq b\} \subseteq \Delta$ and $I_\alpha = \{t \in \mathbb{R} : |t| < \alpha\}$. From statement *(iii)* of Theorem 1.1, Ω is open. From statement *(ii)*, $\varphi_x(t_n + s)$ is defined for $s < \alpha$ and coincides with $\varphi_y(s)$ for n sufficiently large, where, $y = \varphi_x(t_n)$. But then $t_n + s > w_+(x)$, producing the contradiction.   □

From Theorem 1.2 it follows that $w_+(x) = \infty$ (respectively $w_-(x) = -\infty$) if the orbit $\varphi_x(t)$ stays in some compact set K as $t \to w_+(x)$ (respectively $t \to w_-(x)$).

Let $\varphi_x(t)$ be an integral curve of X. We say that it is *periodic* if there exists a real number $c > 0$ such that $\varphi_x(t + c) = \varphi_x(t)$ for every $t \in \mathbb{R}$.

Proposition 1.3 *Let $\varphi_x(t)$ be a solution of X defined on the maximal interval I_x. If $\varphi_x(t_1) = \varphi_x(t_2)$ with $t_1 \neq t_2$, $t_1, t_2 \in I_x$ then $I_x = \mathbb{R}$ and $\varphi_x(t + c) = \varphi_x(t)$ for every $t \in \mathbb{R}$ with $c = t_2 - t_1$. Therefore, φ_x is a periodic solution of period c.*

Proof. If we define $\psi : [t_2, t_2 + c] \to \mathbb{R}^2$ by $\psi(t) = \varphi_x(t - c)$, we have $\psi'(t) = \varphi'_x(t - c) = X(\varphi_x(t - c)) = X(\psi(t))$ and $\psi(t_2) = \varphi_x(t_1) = \varphi_x(t_2)$. From the uniqueness of the solutions, we have $[t_2, t_2 + c] \subseteq I$ and $\varphi_x(t) = \varphi_x(t + c)$ if $t \in [t_2, t_2 + c]$. Proceeding in the same way, we have $I = \mathbb{R}$ and $\varphi_x(t + c) = \varphi_x(t)$ for all $t \in \mathbb{R}$.   □

1.2 Phase Portrait of a Vector Field

We recall that the orbit γ_p of a vector field $X : \Delta \to \mathbb{R}^2$ through the point p is the image of the maximal solution $\varphi_p : I_p \to \Delta$ endowed with an orientation if the solution is regular.

Note that if $q \in \gamma_p$ then $\gamma_p = \gamma_q$. Even more, if $q \in \gamma_p$, it means that exists $t_1 \in I_p$ such that $q = \varphi(t_1, p)$, $\varphi(t, q) = \varphi(t + t_1, p)$ and $I_p - t_1 = I_q$. In other words, given two orbits of X either they coincide or they are disjoint.

Theorem 1.4 *If φ is a maximal solution of a C^r differential system (1.1), then one of the following statements holds.*

(i) φ *is a bijection onto its image.*
(ii) $I = \mathbb{R}$, φ *is a constant function, and γ_φ is a point.*
(iii) $I = \mathbb{R}$, φ *is a periodic function of minimal period τ (that is, there exists a value $\tau > 0$ such that $\varphi(t + \tau) = \varphi(t)$ for every $t \in \mathbb{R}$, and $\varphi(t_1) \neq \varphi(t_2)$ if $|t_1 - t_2| < \tau$).*

Proof. If φ is not bijective, $\varphi(t_1) = \varphi(t_2)$ for some $t_1 \neq t_2$. Then by Proposition 1.3, $I = \mathbb{R}$ and $\varphi(t + c) = \varphi(t)$ for every $t \in \mathbb{R}$ and $c = t_2 - t_1 \neq 0$.

We will prove that the set $C = \{c \in \mathbb{R} : \varphi(t + c) = \varphi(t) \text{ for every } t \in \mathbb{R}\}$ is an additive subgroup of $\mathbb{R}$ which is closed in $\mathbb{R}$. In fact, if $c, d \in C$, then $c+d, -c \in C$, because $\varphi(t+c+d) = \varphi(t+c) = \varphi(t)$ and $\varphi(t-c) = \varphi(t-c+c) = \varphi(t)$. So C is an additive subgroup of $\mathbb{R}$.

But we also have that, if $c_n \in C$ and $c_n \to c$ then $c \in C$, because

$$\varphi(t + c) = \varphi(t + \lim_{n \to \infty} c_n) = \varphi(\lim_{n \to \infty} (t + c_n))$$
$$= \lim_{n \to \infty} \varphi(t + c_n) = \lim_{n \to \infty} \varphi(t) = \varphi(t).$$

As we will prove in the next lemma, any additive subgroup C of $\mathbb{R}$ is of the form $\tau \mathbb{Z}$ with $\tau \geq 0$, or C is dense in $\mathbb{R}$.

Since $C \neq \{0\}$ is closed, it follows that $C = \mathbb{R}$ or $C = \tau \mathbb{Z}$ with $\tau > 0$. Each of these possibilities corresponds respectively to the cases *(ii)* and *(iii)* of the theorem. $\square$

Remark 1.5 We will say period, instead of minimal period, if no confusion is possible.

Lemma 1.6 *Any additive subgroup $C \neq \{0\}$ of $\mathbb{R}$ is either of the form $C = \tau \mathbb{Z}$ where $\tau > 0$, or is dense in $\mathbb{R}$.*

Proof. Suppose that $C \neq \{0\}$. Then $C \cap \mathbb{R}_+ \neq \emptyset$, where $\mathbb{R}_+$ denotes the positive real numbers, since there exists $c \in C$, $c \neq 0$, which implies that c or $-c$ belongs to $C \cap \mathbb{R}_+$.

Let $\tau = \inf(C \cap \mathbb{R}_+)$. If $\tau > 0$, $C = \tau \mathbb{Z}$, because if $c \in C - \tau \mathbb{Z}$, there exists a unique $K \in \mathbb{Z}$ such that $K\tau < c < (K + 1)\tau$ and so, $0 < c - K\tau < \tau$ and $c - K\tau \in C \cap \mathbb{R}_+$. This contradicts the fact that $\tau = \inf(C \cap \mathbb{R}_+)$.

If $\tau = 0$, we verify that C is dense in $\mathbb{R}$. In fact, given $\varepsilon > 0$ and $t \in \mathbb{R}$, there exists $c \in C$ such that $|c - t| < \varepsilon$. To see this, it is enough to take $c_0 \in C \cap \mathbb{R}_+$ such that $0 < c_0 < \varepsilon$. Then the distance of any real number t to a point of $c_0 \mathbb{Z} \subseteq C$ is less than ε, because this set divides $\mathbb{R}$ in intervals of length $c_0 < \varepsilon$ with endpoints in $c_0 \mathbb{Z}$. $\square$

We note that in statements *(i)* and *(iii)* of Theorem 1.4 we can add that γ_φ is C^r–diffeomorphic to $\mathbb{R}$ and that γ_φ is C^r–diffeomorphic to a circle $\mathbb{S}^1$. For a proof see Corollary 1.14.

Let P and Q be two complex polynomials in the variables x and y of degrees m and n, respectively. Suppose that the two algebraic curves $P(x, y) = 0$ and $Q(x, y) = 0$ intersect in finitely many points; i.e., that the polynomials P and Q have no common factor in the ring of complex polynomials. Then the two algebraic curves $P(x, y) = 0$ and $Q(x, y) = 0$ intersect in at most mn points of the complex plane $\mathbb{C}^2$, and exactly in mn points of the complex projective plane $\mathbb{CP}^2$, if we take into account the multiplicity of the intersection points. This result is called *Bezout's Theorem*; for more details see page 10 of [43].

A differential system of the form

$$\dot{x} = P(x, y), \quad \dot{y} = Q(x, y),$$

where P and Q are polynomials in the real variables x and y is called a *polynomial differential system* of *degree m* if m is the maximum degree of the polynomials P and Q.

From Bezout's Theorem it follows that a polynomial differential system of degree m has either infinitely many singular points (i.e., a continuum of singularities), or at most m^2 singular points in $\mathbb{R}^2$.

By a *phase portrait of the vector field* $X : \Delta \rightarrow \mathbb{R}^2$ we mean the set of (oriented) orbits of X. It consists of singularities and regular orbits, oriented according to the maximal solutions describing them, hence in the sense of increasing t. In general, the phase portrait is represented by drawing a number of significant orbits, representing the orientation (in case of regular orbits) by arrows. In Sect. 1.9 we will see how to look for a set of significant orbits.

Now we consider some examples.

Example 1.7 We describe the phase portrait of a vector field $X = (P, Q)$ on $\mathbb{R}^2$ where $P(x, y) = P(x)$ and has a finite numbers of zeros and for which $Q(x, y) = -y$. Let $a_1 < a_2 < \cdots < a_n$ be the zeros of $P(x)$. We write $a_0 = -\infty$ and $a_{n+1} = \infty$.

First it is easy to check that the straight line $y = 0$ is invariant under the flow (i.e., is a union of orbits), as are all the vertical straight lines $x = a_i$ for $i = 1, \ldots, n$. Then for $i = 0, 1, \ldots, n$, on each interval (a_i, a_{i+1}) of the straight line $y = 0$, P has constant sign. We fix an interval (a_i, a_{i+1}) in which P is positive. Then for $x \in (a_i, a_{i+1})$ we have that if $\varphi(t, x)$ is a solution of $\dot{x} = P(x)$ passing through x, it has positive derivative in its entire maximal interval $I_x = (\omega_-(x), \omega_+(x))$.

So the following statements hold:
(i) When $t \rightarrow \omega_-(x)$, $\varphi(t, x) \rightarrow a_i$ and when $t \rightarrow \omega_+(x)$, $\varphi(t, x) \rightarrow a_{i+1}$. The reason is that if $\varphi(t, x) \rightarrow b > a_i$ as $t \rightarrow \omega_-(x)$, then because $\varphi(t, b)$ has positive derivative, the orbits γ_x and γ_b must intersect; but this implies $\gamma_x = \gamma_b$ which is a contradiction. In the same way we see that $\varphi(t, x) \rightarrow a_{i+1}$ when $t \rightarrow \omega_+(x)$.
(ii) If $i \geq 1$ we have that $\omega_-(x) = -\infty$, because for every $t \in I_x$ we have that $\varphi(t, x) > a_1 > -\infty$ and this implies, by Theorem 1.2, that $\omega_-(x) = -\infty$.
(iii) If $i < n$ we have that $\omega_+(x) = \infty$. The proof is identical to *(ii)*.

An equivalent result may be proved in an interval (a_i, a_{i+1}) on which P is negative.

The phase portrait of the vector field $X = (P, Q)$ is given in Fig. 1.2 which follows easily from the fact (taking into account the form of $Q(x, y)$) that the solution through the point (x_0, y_0) is given by $(\varphi(t, x_0), y_0 e^{-t})$. □

Example 1.8 Linear planar systems. The phase portraits of systems $\dot{x} = Ax$, where A is a 2×2 matrix with $\delta = \det A \neq 0$ are well–known (see [98]). If

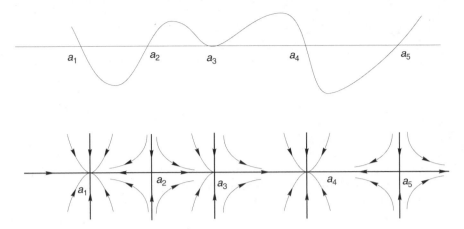

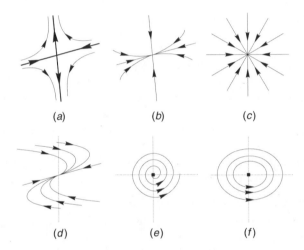

Fig. 1.2. Phase portrait of Example 1.7

Fig. 1.3. Phase portraits of Example 1.8

$\delta < 0$ we have a *saddle*; if $\delta > 0$ and $\rho = $ trace $(A) = 0$ we have a *linear center*; if $\delta > 0$ and $\rho^2 - 4\delta < 0$ we have a *focus*; and if $\delta > 0$ and $\rho^2 - 4\delta > 0$ we have a *node*. The corresponding phase portraits are given in Fig. 1.3.

The eigenvalues of A are

$$\lambda_1, \lambda_2 = \frac{\rho \pm \sqrt{\rho^2 - 4\delta}}{2}.$$

The corresponding eigenspaces are called E_1 and E_2, respectively. In the case of the saddle, the orbits of the linear system corresponding to the four orbits contained in $E_1 - \{0\}$ and $E_2 - \{0\}$, are called the *saddle separatrices* of the linear system. $\qquad\square$

1.3 Topological Equivalence and Conjugacy

We need to introduce several notions of equivalence between two vector fields which will allow us to compare their phase portraits.

Let X_1 and X_2 be two vector fields defined on open subsets Δ_1 and Δ_2 of $\mathbb{R}^2$, respectively. We say that X_1 is *topologically equivalent* (respectively C^r–*equivalent) to X_2 when there exists a homeomorphism (respectively a diffeomorphism of class C^r) $h : \Delta_1 \to \Delta_2$ which sends orbits of X_1 to orbits of X_2 preserving the orientation. More precisely, let $p \in \Delta_1$ and γ_p^1 be the oriented orbit of X_1 passing through p; then $h(\gamma_p^1)$ is an oriented orbit of X_2 passing through $h(p)$. Such a homeomorphism h is called a *topological equivalence* (respectively, C^r–*equivalence*) between X_1 and X_2.

Let $\varphi_1 : \Omega_1 \to \mathbb{R}^2$ and $\varphi_2 : \Omega_2 \to \mathbb{R}^2$ be the flows generated by the vector fields $X_1 : \Delta_1 \to \mathbb{R}^2$ and $X_2 : \Delta_2 \to \mathbb{R}^2$ respectively. We say that X_1 is *topologically conjugate* (respectively C^r–*conjugate) to X_2 when there exists a homeomorphism (respectively a diffeomorphism of class C^r) $h : \Delta_1 \to \Delta_2$ such that $h(\varphi_1(t,x)) = \varphi_2(t,h(x))$ for every $(t,x) \in \Omega_1$. In this case, it is necessary that the maximal intervals I_x for φ_1 and $I_{h(x)}$ for φ_2 be equal. Such a homeomorphism (or diffeomorphism) h is called a *topological conjugacy* (respectively C^r–*conjugacy*) between X_1 and X_2. Any conjugacy is clearly also an equivalence. One also uses "C^0–equivalent" and "C^0–conjugate" instead of respectively topological equivalent and topological conjugate.

A topological equivalence h defines an equivalence relation between vector fields defined on open sets Δ_1 and $\Delta_2 = h(\Delta_1)$ of $\mathbb{R}^2$. A topological equivalence h between X_1 and X_2 maps singular points to singular points, and periodic orbits to periodic orbits. If h is a conjugacy, the period of the periodic orbits is also preserved.

Example 1.9 The function $h : \mathbb{R}^2 \to \mathbb{R}^2$ defined by $h(x,y) = (x, y + x^3/4)$ is a C^r–conjugacy between $X(x,y) = (x,-y)$ and $Y(x,y) = (x, -y + x^3)$ as $\psi(t,(a,b)) = (ae^t, be^{-t})$ is a trajectory for X, $\varphi(t,(a,b)) = (ae^t, (b-a^3/4)e^{-t} + a^3 e^{3t}/4)$ is a trajectory for Y and $h(\psi(t,p)) = \varphi(t, h(p))$. $\square$

Example 1.10 Let $A = \begin{pmatrix} 0 & a \\ -a & 0 \end{pmatrix}$ and $B = \begin{pmatrix} 0 & b \\ -b & 0 \end{pmatrix}$ be matrices on $\mathbb{R}^2$ with $ab > 0$. All orbits of the systems $\dot{x} = Ax$ and $\dot{x} = Bx$ are periodic having period $2\pi/a$ and $2\pi/b$, respectively, with the exception of the origin which is a singular point. If $a \neq b$, these systems cannot be conjugate. But $h = \text{Identity}$ on $\mathbb{R}^2$ is a C^ω–equivalence (even a linear equivalence). $\square$

The next lemma gives a characterization for a C^r–conjugacy with $r \geq 1$.

Lemma 1.11 *Let $X_1 : \Delta_1 \to \mathbb{R}^2$ and $X_2 : \Delta_2 \to \mathbb{R}^2$ be vector fields of class C^r and $h : \Delta_1 \to \Delta_2$ a diffeomorphism of class C^r with $r \geq 1$. Then h is a conjugacy between X_1 and X_2 if and only if*

$$Dh_p X_1(p) = X_2(h(p)) \text{ for every } p \in \Delta_1. \tag{1.2}$$

Proof. Let $\varphi_1 : \Omega_1 \to \Delta_1$ and $\varphi_2 : \Omega_2 \to \Delta_2$ be the flows of X_1 and X_2, respectively. Assume that h satisfies (1.2). Given $p \in \Delta_1$, let $\psi(t) = h(\varphi_1(t, p))$ with $t \in I_1(p)$. Then ψ is a solution of $\dot{x} = X_2(x)$, $x(0) = h(p)$, because

$$\dot{\psi}(t) = Dh(\varphi_1(t,p)) \frac{d}{dt} \varphi_1(t,p) = Dh(\varphi_1(t,p))X_1(\varphi_1(t,p)) =$$
$$= X_2(h(\varphi_1(t,p))) = X_2(\psi(t)).$$

So $h(\varphi_1(t, p)) = \varphi_2(t, h(p))$. Conversely, assume that h is a conjugacy. Given $p \in \Delta_1$, we have that $h(\varphi_1(t,p)) = \varphi_2(t, h(p))$, $t \in I_1(p) = I_2(h(p))$. If we differentiate this relation with respect to t and evaluate at $t = 0$, we get (1.2). $\qquad\square$

Let $X : \Delta \to \mathbb{R}^2$ be a vector field of class C^r and $\Delta \subset \mathbb{R}^2$ and $A \subset \mathbb{R}$ open subsets. A C^r map $f : A \to \Delta$ is called a *transverse local section of X* when for every $a \in A$, $f'(a)$ and $X(f(a))$ are linearly independent. Take $\Sigma = f(A)$ with the induced topology. If $f : A \to \Sigma$ is a homeomorphism (meaning that f is an embedding) we say that Σ is a *transverse section of X*.

Theorem 1.12 (Flow Box Theorem) *Let p be a regular point of a C^r vector field $X : \Delta \to \mathbb{R}^2$ with $1 \le r \le +\infty$ or $r = \omega$, and let $f : A \to \Sigma$ be a transverse section of X of class C^r with $f(0) = p$. Then there exists a neighborhood V of p in Δ and a diffeomorphism $h : V \to (-\varepsilon, \varepsilon) \times B$ of class C^r, where $\varepsilon > 0$ and B is an open interval with center at the origin such that*

(i) $h(\Sigma \cap V) = \{0\} \times B$;
(ii) h *is a C^r-conjugacy between $X|_V$ and the constant vector field $Y : (-\varepsilon, \varepsilon) \times B \to \mathbb{R}^2$ defined by $Y = (1, 0)$. See Fig. 1.4.*

Proof. Let $\varphi : \Omega \to \Delta$ be the flow of X. Let $F : \Omega_A = \{(t, u) : (t, f(u)) \in \Omega\} \to \Delta$ be defined by $F(t, u) = \varphi(t, f(u))$. F maps parallel lines into integral curves of X. We will prove that F is a local diffeomorphism in $0 = (0,0) \in \mathbb{R} \times \mathbb{R}$. By the Inverse Function Theorem, it is enough to prove that $DF(0)$ is an isomorphism.

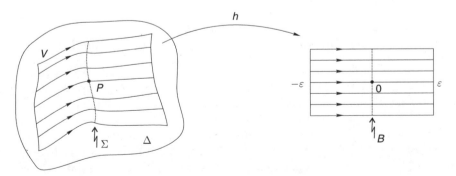

Fig. 1.4. The Flow Box Theorem

We have that

$$D_1 F(0) = \frac{d}{dt}\varphi(t, f(0))\,|_{t=0} = X(\varphi(0, p)) = X(p),$$

and $D_2 F(0) = D_1 f(0)$ because $\varphi(0, f(u)) = f(u)$ for all $u \in A$. So the vectors $D_1 F(0)$ and $D_2 F(0)$ generate $\mathbb{R}^2$ and $DF(0)$ is an isomorphism.

By the Inverse Function Theorem, there exists $\varepsilon > 0$ and a neighborhood B in $\mathbb{R}$ around the origin such that $F|(-\varepsilon, \varepsilon) \times B$ is a diffeomorphism on an open set $V = F((-\varepsilon, \varepsilon) \times B)$. Let $h = (F|(-\varepsilon, \varepsilon) \times B)^{-1}$. Then $h(\Sigma \cap V) = \{0\} \times B$, since $F(0, u) = f(u) \in \Sigma$ for all $u \in B$. This proves (i). On the other hand, h^{-1} conjugates Y and X:

$$
\begin{aligned}
Dh^{-1}(t, u)Y(t, u) &= DF(t, u)(1, 0) \\
&= D_1 F(t, u)) \\
&= X(\varphi(t, f(u))) \\
&= X(F(t, u)) \\
&= X(h^{-1}(t, u)),
\end{aligned}
$$

for every $(t, u) \in (-\varepsilon, \varepsilon) \times B$. This ends the proof.   □

Corollary 1.13 *Let Σ be a transverse section of X. For every point $p \in \Sigma$, there exist $\varepsilon = \varepsilon(p) > 0$, a neighborhood V of p in $\mathbb{R}^2$ and a function $\tau : V \to \mathbb{R}$ of class C^r such that $\tau(V \cap \Sigma) = 0$ and:*

(i) *for every $q \in V$, an integral curve $\varphi(t, q)$ of $X|V$ is defined and bijective in $J_q = (-\varepsilon + \tau(q), \varepsilon + \tau(q))$.*
(ii) *$\xi(q) = \varphi(\tau(q), q) \in \Sigma$ is the only point where $\varphi(\cdot, q)|J_q$ intersects Σ. In particular, $q \in \Sigma \cap V$ if and only if $\tau(q) = 0$.*
(iii) *$\xi : V \to \Sigma$ is of class C^r and $D\xi(q)$ is surjective for every $q \in V$. Even more, $D\xi(q)v = 0$ if and only if $v = \alpha X(q)$ for some $\alpha \in \mathbb{R}$.*

Proof. Let h, V and ε be as in the Flow Box Theorem. We write $h = (-\tau, \eta)$. The vector field Y of that theorem satisfies all the statements of the corollary. Since h is a C^r–conjugacy, it follows that X also satisfies the statements.   □

Corollary 1.14 *If γ is a maximal solution of a C^r differential system (1.1) and γ is not a singular point, then γ is C^r diffeomorphic to $\mathbb{R}$ or $\mathbb{S}^1$.*

Proof. Let p be a point of γ. Let Σ be a transverse section of (1.1) such that $p \in \Sigma \cap \gamma$. We define $D = \{t \in I_p : t \geq 0, \varphi(t, p) \in \Sigma\}$. We claim that taking Σ sufficiently small $\varphi(t, p)$ with $t \geq 0$ has a unique point on Σ. Indeed, because of the Flow Box Theorem, we know that D consists of isolated points. If $D \neq \{0\}$, let 0 and t_0 be two consecutive elements of D. Now $\{\varphi(t, p) : t \in [0, t_0]\}$, together with the segment of Σ in between $\varphi(0) = p$ and $\varphi(t_0) = q$, form a topological circle C, which by the Jordan's Curve Theorem divides the plane in two connected components, like we represent in Fig. 1.5(a) or (b). In both cases it is clear that the orbit through p cannot have

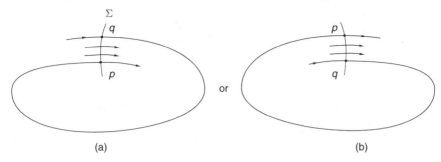

Fig. 1.5. The arc $\{\varphi(t,p) : t \in [0, t_0]\}$

other intersections with Σ, besides p, for Σ sufficiently small. So the claim is proved.

If in the arguments above $p = q$, then γ is a periodic orbit C^r diffeomorphic to $\mathbb{S}^1$. If $p \neq q$, by the Flow Box Theorem applied again to this Σ, there is a neighborhood of p in γ which is C^r diffeomorphic to an open interval. Since p is an arbitrary point of γ, it follows that γ is C^r diffeomorphic to $\mathbb{R}$.   □

Remark 1.15 The statement concerning regular periodic orbits is also valid on surfaces in general and even in any dimension, but this is not the case for the statement concerning regular non–periodic orbits.

1.4 α- and ω-limit Sets of an Orbit

Let Δ be an open subset of $\mathbb{R}^2$ and let $X : \Delta \to \mathbb{R}^2$ be a vector field of class C^r where $1 \leq r \leq \infty$ or $r = \omega$.

Let $\varphi(t) = \varphi(t, p) = \varphi_p(t)$ be the integral curve of X passing through the point p, defined on its maximal interval $I_p = (\omega_-(p), \omega_+(p))$. If $\omega_+(p) = \infty$ we define the set

$$\omega(p) = \{q \in \Delta : \text{there exist } \{t_n\} \text{ with } t_n \to \infty$$
$$\text{and } \varphi(t_n) \to q \text{ when } n \to \infty\}.$$

In the same way, if $\omega_-(p) = -\infty$ we define the set

$$\alpha(p) = \{q \in \Delta : \text{there exist } \{t_n\} \text{ with } t_n \to -\infty$$
$$\text{and } \varphi(t_n) \to q \text{ when } n \to \infty\}.$$

The sets $\omega(p)$ and $\alpha(p)$ are called the *ω-limit set* (or simply *ω-limit*) and the *α-limit set* (or *α-limit*) of p, respectively.

We begin with some examples:

Example 1.16 Let $X : \mathbb{R}^2 \to \mathbb{R}^2$ be a vector field given by $X(x, y) = (x, -y)$. Then:

(i) If $p = (0, 0)$, $\alpha(p) = \omega(p) = \{(0, 0)\}$.
(ii) If $p \in \{(x, 0) : x \neq 0\}$, $\alpha(p) = \{(0, 0)\}$ and $\omega(p) = \emptyset$.

Fig. 1.6. A limit cycle and some orbits spiralling to it

(iii) If $p \in \{(0, y) : y \neq 0\}$, $\omega(p) = \{(0, 0)\}$ and $\alpha(p) = \emptyset$.
(iv) If $p \in \{(x, y) : xy \neq 0\}$, $\alpha(p) = \omega(p) = \emptyset$. □

Example 1.17 Let $\varphi(t) = \varphi(t, p)$ be a periodic orbit of period τ. Then $\omega(p) = \{\varphi(t) : t \in \mathbb{R}\} = \alpha(p)$. □

Example 1.18 Let $X : \mathbb{R}^2 \to \mathbb{R}^2$ be the vector field given by

$$X(x, y) = (y + x(1 - x^2 - y^2), -x + y(1 - x^2 - y^2)).$$

Its phase portrait is given in Fig. 1.6. It is easy to check that $C = \{(x, y) : x^2 + y^2 = 1\}$ is the only periodic orbit of this vector field. Then

(i) $\alpha(p) = \{(0, 0)\}$ if $p \in \text{Int}(C) = \{(x, y) : x^2 + y^2 < 1\}$.
(ii) $\alpha(p) = \emptyset$ if $p \in \text{Ext}(C) = \{(x, y) : x^2 + y^2 > 1\}$.
(iii) $\alpha(p) = C$ if $p \in C$.
(iv) $\omega(p) = C$ for any p different from the origin. □

Let γ_p be the orbit of X through the point p and $q \in \gamma_p$; then $\omega(p) = \omega(q)$. This is due to the fact that if $q \in \gamma_p$, then there exists $c \in \mathbb{R}$ such that $\varphi(t, p) = \varphi(t + c, q)$. In the same way $\alpha(p) = \alpha(q)$.

We define the *α-limit set* of an orbit γ as the set $\alpha(p)$ for some $p \in \gamma$.

We define the *ω-limit set* of an orbit γ as the set $\omega(p)$ for some $p \in \gamma$.

Let $\varphi(t) = \varphi(t, p)$ be an integral curve of a vector field X through the point p, and $\psi(t) = \psi(t, p)$ an integral curve of the vector field $-X$ through the point p; then $\psi(t, p) = \varphi(-t, p)$. From this, it follows that the ω-limit set of $\psi(t)$ is equal to the α-limit set of $\varphi(t)$ and conversely, the α-limit set of $\psi(t)$ is equal to the ω-limit set of $\varphi(t)$. For this reason, in order to study the general properties of the α-limit and ω-limit sets of orbits, it is enough to study the ω-limit sets.

Theorem 1.19 *Let $X : \Delta \to \mathbb{R}^2$ be a vector field of class C^r defined on an open set $\Delta \subset \mathbb{R}^2$ and $\gamma^+(p) = \{\varphi(t, p) : t \geq 0\}$ (respectively $\gamma^-(p) = \{\varphi(t, p) : t \leq 0\}$) a positive semi-orbit (respectively negative semi-orbit) of the vector field X through the point p. If $\gamma^+(p)$ (respectively $\gamma^-(p)$) is contained in a compact subset $K \subset \Delta$, then:*

(i) $\omega(p) \neq \emptyset$ (respectively $\alpha(p)$);
(ii) $\omega(p)$ is compact (respectively $\alpha(p)$);

(iii) $\omega(p)$ is invariant for X (respectively $\alpha(p)$), that is, if $q \in \omega(p)$, then an integral curve passing through q is contained in $\omega(p)$;

(iv) $\omega(p)$ is connected (respectively $\alpha(p)$).

(v) If $\omega(\gamma) \subset \gamma$ (respectively $\alpha(\gamma) \subset \gamma$) then $\omega(\gamma) = \gamma$ (respectively $\alpha(\gamma) = \gamma$), and γ is either a singular point or a periodic orbit.

Proof. From the previous observation it is sufficient to prove the theorem for a set $\omega(p)$.

(i) $\omega(p) \neq \emptyset$.

Let $t_n = n \in \mathbb{N}$. From the assumptions we have that $\{\varphi(t_n)\} \subset K$ with K compact. Then there exists a subsequence $\{\varphi(t_{n_k})\}$ which converges to a point $q \in K$. We have then that $t_{n_k} \to \infty$ as $n_k \to \infty$ and $\varphi(t_{n_k}) \to q$. Then by definition $q \in \omega(p)$.

(ii) $\omega(p)$ is compact.

We have that $\omega(p) \subset \overline{\gamma^+(p)} \subset K$, so it is sufficient to prove that $\omega(p)$ is closed. Let $q_n \to q$, with $q_n \in \omega(p)$. We will prove that $q \in \omega(p)$. Since $q_n \in \omega(p)$, there exists for every q_n a sequence $\{t_m^{(n)}\}$ such that $t_m^{(n)} \to \infty$ and $\varphi(t_m^{(n)}, p) \to q_n$ as $m \to \infty$.

For every sequence $\{t_m^{(n)}\}$ we choose a point $t_n = t_{m(n)}^{(n)} > n$ such that $d(\varphi(t_n, p), q_n) < 1/n$. Then we have that:

$$d(\varphi(t_n, p), q) \leq d(\varphi(t_n, p), q_n) + d(q_n, q) < \frac{1}{n} + d(q_n, q).$$

Therefore, it follows that $d(\varphi(t_n, p), q) \to 0$ as $n \to \infty$, that is, $\varphi(t_n, p) \to q$. Since $t_n \to \infty$ as $n \to \infty$, we get that $q \in \omega(p)$.

(iii) $\omega(p)$ is invariant under X.

Let $q \in \omega(p)$ and let $\psi : I(q) \to \Delta$ be an integral curve of X passing through the point q. Let $q_1 = \varphi(t_0, q) = \psi(t_0)$. We will prove that $q_1 \in \omega(p)$. As $q \in \omega(p)$, there exists a sequence $\{t_n\}$ such that $t_n \to \infty$ and $\varphi(t_n, p) \to q$ as $n \to \infty$. Since φ is continuous, it follows that:

$$q_1 = \varphi(t_0, q) = \varphi(t_0, \lim_{n\to\infty} \varphi(t_n, p))$$
$$= \lim_{n\to\infty} \varphi(t_0, \varphi(t_n, p)) = \lim_{n\to\infty} \varphi(t_0 + t_n, p).$$

We have then a sequence $\{s_n\} = \{t_0 + t_n\}$ such that $s_n \to \infty$ and $\varphi(s_n, p) \to q_1$ as $n \to \infty$, that is $q_1 \in \omega(p)$; see Fig. 1.7.

(iv) $\omega(p)$ is connected.

Assume that $\omega(p)$ is not connected. Then $\omega(p) = A \cup B$, where A and B are closed, non-empty and $A \cap B = \emptyset$. As $A \neq \emptyset$, there exists a subsequence $\{t'_n\}$ such that $t'_n \to \infty$ and $\varphi(t'_n) \to a \in A$ as $n \to \infty$. In the same way, there exists a sequence $\{t''_n\}$ such that $t''_n \to \infty$ and $\varphi(t''_n) \to b \in B$ as $n \to \infty$. Then we can construct a sequence $\{t_n\}$ such that $t_n \to \infty$ as $n \to \infty$ and such that $d(\varphi(t_n), A) < d/2$ and $d(\varphi(t_{n+1}), A) > d/2$, where $d = d(A, B) > 0$ for every n odd.

Fig. 1.7. A subsequence converging to a point of a limit cycle

Since the function $g(t) = d(\varphi(t), A)$, for $t_n \leq t \leq t_{n+1}$ for all n odd, is continuous, and $g(t_n) < d/2$ and $g(t_{n+1}) > d/2$, it follows (from the Intermediate Value Theorem) that there exists t_n^*, with $t_n < t_n^* < t_{n+1}$ such that

$$g(t_n^*) = d(\varphi(t_n^*), A) = d/2.$$

Since the sequence $\{\varphi(t_n^*)\}$ is contained in the compact set $Q = \{x \in \Delta : d(x, A) = d/2\}$, it has a convergent subsequence which we also denote by $\{\varphi(t_n^*)\}$. Let $p^* = \lim_{n\to\infty} \varphi(t_n^*)$. Then $p^* \in \omega(p)$. But $p^* \notin A$, because $d(p^*, A) = d/2 > 0$; also $p^* \notin B$, because $d(p^*, B) \geq d(A, B) - d(p^*, A) = d/2 > 0$. This is a contradiction.

(v) If $\omega(\gamma) \subset \gamma$ then $\omega(\gamma) = \gamma$. We claim that γ is either a singular point or a periodic orbit.

By Theorem 1.4 and Corollary 1.14 an orbit γ is homeomorphic to a point, to a circle or to $\mathbb{R}$. So *(v)* will be proved if, assuming that γ is homeomorphic to $\mathbb{R}$ and $\omega(\gamma) \subset \gamma$, we arrive at a contradiction. Clearly $\omega(\gamma) \neq \gamma$ because γ is not compact and $\omega(\gamma)$ is compact by *(ii)*. Then $\omega(\gamma)$ is not invariant, in contradiction to statement *(iii)*.   □

We can also remark, from the proof of Theorem 1.19, that even if $\gamma^+(p)$ (respectively $\gamma^-(p)$) is not contained in a compact subset, nevertheless $\omega(p)$ (respectively $\alpha(p)$) is invariant under X and is closed. The closure of γ is $\bar{\gamma} = \gamma \cup \omega(\gamma) \cup \alpha(\gamma)$. Of course, both $\omega(\gamma)$ and $\alpha(\gamma)$ can be empty and do not need to be connected.

We consider the vector field X of Fig. 1.6 restricted to the open set $\Delta = \mathbb{R}^2 \setminus \{p_1, p_2\}$, where p_1 and p_2 are different points of the circle of radius 1 centered at the origin. If $p \neq (0,0)$ and $p \notin C \setminus \{p_1, p_2\}$, then $\omega(p) = C \setminus \{p_1, p_2\}$, which shows that $\omega(p)$ is not connected. Consequently the compactness of K cannot be removed from the hypotheses of Theorem 1.19

1.5 Local Structure of Singular Points

Let p be a regular point of a planar C^r vector field X with $1 \leq r \leq \infty$ or $r = \omega$. By the Flow Box Theorem, we know that there exists a C^r diffeomorphism which conjugates X in a neighborhood of p with the constant flow $Y = (1,0)$. Then both vector fields X and Y are locally C^r-conjugate near the regular points. Near regular points there is a unique model for C^r-conjugacy.

Let p be a singular point of a planar C^r vector field $X = (P, Q)$. In general the study of the local behavior of the flow near p is quite complicated. Already the linear systems show different classes, even for local topological equivalence. We say that

$$DX(p) = \begin{pmatrix} \dfrac{\partial P}{\partial x}(p) & \dfrac{\partial P}{\partial y}(p) \\[2ex] \dfrac{\partial Q}{\partial x}(p) & \dfrac{\partial Q}{\partial y}(p) \end{pmatrix}$$

is the *linear part* of the vector field X at the singular point p.

The singular point p is called *non–degenerate* if 0 is not an eigenvalue.

The singular point p is called *hyperbolic* if the two eigenvalues of $DX(p)$ have real part different from 0.

The singular point p is called *semi-hyperbolic* if exactly one eigenvalue of $DX(p)$ is equal to 0. Hyperbolic and semi-hyperbolic singularities are also said to be *elementary singular points*.

The singular point p is called *nilpotent* if both eigenvalues of $DX(p)$ are equal to 0 but $DX(p) \neq 0$.

The singular point p is called *linearly zero* if $DX(p) \equiv 0$.

The singular point p is called a *center* if there is an open neighborhood consisting, besides the singularity, of periodic orbits. The singularity is said to be *linearly a center* if the eigenvalues of $DX(p)$ are purely imaginary without being zero. In that case, and if we suppose the vector field X to be analytic (see Chap. 4), the vector field X can have either a center or a focus at p. To distinguish between a center and a focus is a difficult problem in the qualitative theory of planar differential equations; see Chap. 4. We note that a center-focus problem also exists for nilpotent or linearly zero singular points.

In order to study the local phase portrait at the singular point p we define the determinant, the trace and the discriminant at p as

$$\det(p) = \begin{vmatrix} \dfrac{\partial P}{\partial x}(p) & \dfrac{\partial P}{\partial y}(p) \\[2ex] \dfrac{\partial Q}{\partial x}(p) & \dfrac{\partial Q}{\partial y}(p) \end{vmatrix},$$

$$\operatorname{tr}(p) = \frac{\partial P}{\partial x}(p) + \frac{\partial Q}{\partial y}(p),$$

$$\Delta(p) = \operatorname{tr}(p)^2 - 4\det(p),$$

respectively. It is easy to check that

(i) if $\det(p) \neq 0$, then the singular point is non–degenerate and it is either hyperbolic, or linearly a center;

(ii) if $\det(p) = 0$ but $\text{tr}(p) \neq 0$, then the singular point is semi-hyperbolic;

(iii) if $\det(p) = 0$ and $\text{tr}(p) = 0$, then the singular point is linearly zero or nilpotent depending on whether $DX(p)$ is the zero matrix or not.

It is obvious that if $p = (x_0, y_0)$ is a singular point of the differential system

$$\dot{x} = P(x, y),$$
$$\dot{y} = Q(x, y),$$

then the point $(0, 0)$ is a singular point of the system

$$\dot{\bar{x}} = P(\bar{x}, \bar{y}),$$
$$\dot{\bar{y}} = Q(\bar{x}, \bar{y}), \tag{1.3}$$

where $x = \bar{x} + x_0$, and $y = \bar{y} + y_0$, and now the functions $P(\bar{x}, \bar{y})$ and $Q(\bar{x}, \bar{y})$ start with terms of order 1 in $\bar{x}$ and $\bar{y}$. In other words, we can always move a singular point to the origin of coordinates in which case system (1.3) becomes (dropping the bars over x and y)

$$\dot{x} = ax + by + F(x, y),$$
$$\dot{y} = cx + dy + G(x, y),$$

where F and G vanish together with their first partial derivatives at $(0, 0)$. By a linear change of coordinates the linearization $DX(0, 0)$ regarded as the matrix

$$\begin{pmatrix} a & b \\ c & d \end{pmatrix},$$

can be placed in real Jordan canonical form.

If the singularity is hyperbolic, the Jordan form is

$$\begin{pmatrix} \lambda_1 & 0 \\ 0 & \lambda_2 \end{pmatrix} \quad \text{or} \quad \begin{pmatrix} \lambda_1 & 1 \\ 0 & \lambda_1 \end{pmatrix} \quad \text{or} \quad \begin{pmatrix} \alpha & \beta \\ -\beta & \alpha \end{pmatrix}$$

with $\lambda_1 \lambda_2 \neq 0$, $\alpha \neq 0$ and $\beta > 0$.

In the semi-hyperbolic case and the linearly center case, we obtain, respectively

$$\begin{pmatrix} \lambda & 0 \\ 0 & 0 \end{pmatrix} \quad \text{and} \quad \begin{pmatrix} 0 & \beta \\ -\beta & 0 \end{pmatrix}$$

with $\lambda \neq 0$ and $\beta > 0$, while we obtain

$$\begin{pmatrix} 0 & 1 \\ 0 & 0 \end{pmatrix} \quad \text{and} \quad \begin{pmatrix} 0 & 0 \\ 0 & 0 \end{pmatrix}$$

in the nilpotent case and the linearly zero case, respectively.

If moreover we allow a time rescaling, introducing a new time $u = \gamma t$ for some $\gamma > 0$, as is usual when working with equivalences, then we can also suppose that in the hyperbolic case one of the numbers λ_1 or λ_2 is equal to ± 1 and either $\alpha = \pm 1$ or $\beta = 1$, while in the semi-hyperbolic case $\lambda = \pm 1$ and in the linearly center case $\beta = 1$.

We are now going to study these different cases systematically.

Let p be a singular point. A *characteristic orbit* $\gamma(t)$ at p is an orbit tending to p in positive time (respectively in negative time) with a well defined slope, i.e., $\gamma(t) \to p$ for $t \to \infty$ (respectively $t \to -\infty$) and the limit $\lim_{t\to\infty}(\gamma(t) - p)/\|\gamma(t) - p\|$ (respectively $\lim_{t\to-\infty}(\gamma(t) - p)/\|\gamma(t) - p\|$) exists.

In Chap. 6 we will work along circles that can be obtained as the image of an orientation preserving injective regular parametrization

$$\rho : \mathbb{S}^1 \to \mathbb{R}^2$$
$$e^{2\pi i t} \to \rho(e^{2\pi i t})$$

with ρ of class C^1, $\rho' \neq 0$ everywhere and ρ injective. We mean that there exists some C^1 functions $\Psi : \mathbb{R} \to \mathbb{R}^2$ with $\Psi(t) = \rho(e^{2\pi i t})$, $\Psi|_{[0,1)}$ injective, and such that $\Psi'(t) = (\Psi_1'(t), \Psi_2'(t)) \neq 0$ and $n(t) = (-\Psi_2'(t), \Psi_1'(t))$ points out of the (topological) disk encircled by $\rho(\mathbb{S}^1)$. We can also write $\Psi'(t) = \rho'(e^{2\pi i t})$ and $N(e^{2\pi i t}) = n(t)$; ρ' and N are C^0 functions on $\mathbb{S}^1$. We will call ρ a *permissible circle parametrization*.

Let X be a C^1 vector field defined in a compact neighborhood V of p, for which ∂V is the image of a C^2 permissible circle parametrization $\rho : \mathbb{S}^1 \to \partial V$, and suppose that $X(p) = 0$ and $X(q) \neq 0$ for all $q \in V \setminus \{p\}$.

(i) We say that $X|_V$ is a *center* if ∂V is a periodic orbit and all orbits in $V \setminus \{p\}$ are periodic.

(ii) We say that $X|_V$ is an *attracting focus/node* if at all points of ∂V the vector field points inward and for all $q \in V \setminus \{p\}$, $\omega(q) = \{p\}$ and $\gamma^-(q) \cap \partial V \neq \emptyset$.

(iii) We say that $X|_V$ is a *repelling focus/node* if at all points of ∂V the vector field points outward and for all $q \in V \setminus \{p\}$, $\alpha(q) = \{p\}$ and $\gamma^+(q) \cap \partial V \neq \emptyset$.

(iv) We say that $X|_V$ has a *non-trivial finite sectorial decomposition* if we are not in the case (i), (ii) or (iii) and if there exist a finite number of characteristic orbits $c_0, \ldots, c_{n-1}$, each cutting ∂V transversely at one point p_i, in the sense that ∂V is a transverse section near p_i, and with the property that between c_i and c_{i+1} (with $c_n = c_0$ and ordered in such a way that $p_0, \ldots, p_{n-1}$ follows the cyclic order of ρ), we have one of the following situations with respect to the sector S_i, defined as the compact region bounded by $\{p\}$, c_i, c_{i+1} and the piece of ∂V between p_i and p_{i+1}:

    (1) *Attracting parabolic sector.* At all points of $[p_i, p_{i+1}] \subset \partial V$ the vector field points inward, and for all $q \in S_i \setminus \{p\}$, $\omega(q) = \{p\}$ and $\gamma^-(q) \cap \partial V \neq \emptyset$.

(2) Repelling parabolic sector. At all points of $[p_i, p_{i+1}] \subset \partial V$ the vector field points outward, and for all $q \in S_i \setminus \{p\}$, $\alpha(q) = \{p\}$ and $\gamma^+(q) \cap \partial V \neq \emptyset$.

(3) Hyperbolic sector. There exists a point $q_i \in (p_i, p_{i+1}) \subset \partial V$ with the property that at all points of $[p_i, q_i)$ the vector field points inward (respectively outward) while at all points of $(q_i, p_{i+1}]$ the vector field points outward (respectively inward); at q_i the vector field is tangent at ∂V and the tangency is external in the sense that the x–orbit of q_i stays outside V; and for all $q \in S_i \setminus \overline{c_i \cup c_{i+1} \cup q_i}$ we have $\gamma^+(q) \cap \partial V \neq \emptyset$ and $\gamma^-(q) \cap \partial V \neq \emptyset$.

(4) Elliptic sector. There exists a point $q_i \in (p_i, p_{i+1}) \subset \partial V$ with the property that $\gamma(q_i) \subset V$ with $\omega(q_i) = \alpha(q_i) = \{p\}$; at all points $q \in [p_i, q_i)$ the vector field points inward, $\gamma^+(q) \subset V$ and $\omega(q) = p$. We denote by $S_{[p_i, q_i]} = \bigcup\limits_{q \in [p_i, q_i]} \gamma^+(q)$; at all points of $q \in (q_i, p_{i+1}]$ the vector field points outward, $\gamma^-(q) \subset V$ and $\alpha(q) = p$. We denote by $S_{[q_i, p_{i+1}]} = \bigcup\limits_{q \in [q_i, p_{i+1}]} \gamma^-(q)$; at all points q of $S \setminus (S_{[p_i, q_i]} \cup S_{[q_i, p_{i+1}]} \cup \{p\})$ we have $\gamma(q) \subset V$ with $\omega(q) = \alpha(q) = p$.

The same may also be true for $[p_i, q_i]$ and $[q_i, p_{i+1}]$ interchanged.

See Figure 1.8 for a picture of the different sectors.

Let X be a C^1 vector field defined in a neighborhood W of some singularity p. We say that X has the *finite sectorial decomposition property* at p if there exists some neighborhood $V \subset W$ of p such that $X|_V$ satisfies one of the conditions *(i)*, *(ii)*, *(iii)* or *(iv)*.

In the first three cases we speak about a trivial sectorial decomposition, since there is but one sector. We remark that the distinction between a focus and a node is not topological but differentiable. We will deal with it in Chap. 2. In the last case (case *(iv)*), we denote respectively by e, h and p the number of elliptic, hyperbolic and parabolic sectors. Since we are not in the cases *(i)*, *(ii)* or *(iii)*, we clearly need that e or h, or both, are different from zero. We try to keep p as small as possible, both by joining two adjacent parabolic sectors, in other words, not accepting two adjacent parabolic sectors, and by adding a parabolic sector to an elliptic one if it is adjacent to it. Hence the remaining

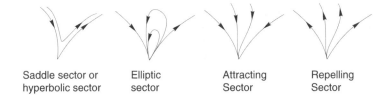

Saddle sector or   Elliptic   Attracting   Repelling
hyperbolic sector   sector   Sector   Sector

Fig. 1.8. Sectors near a singular point

parabolic sectors can only be the ones lying between two hyperbolic sectors. We call this a *minimal sectorial decomposition.*

Since $X(p_i)$ and $X(p_{i+1})$ cannot both be pointing inward (or outward) if S_i is a hyperbolic or an elliptic sector, it is clear that $e + h$ is always even. It is also clear that in a minimal non-trivial sectorial decomposition we have $p \leq h$. For the sake of completeness we define $(e, h) = (0, 0)$ in the cases *(i)*, *(ii)* or *(iii)*.

It is possible to define a more general notion of sectorial decomposition that could be called "C^0 finite sectorial decomposition" and that avoids the notion of tangency. However, since we are interested in analytic vector fields, we do not need to do this. In the use of the finite sectorial decomposition, we will therefore not try to be as general as possible, but we will require the differentiability (the specific C^1-class) which permits us to provide the simplest proof.

In Chap. 3 we will show how to prove that every analytic vector field has the finite sectorial decomposition property at every isolated zero. The same does not necessarily hold for C^∞ vector fields, but it does hold if we add the so called Łojasiewicz condition (see Chap. 3). In both cases we will see that the permissible parametrization of the boundary ∂V can be taken as a C^∞ mapping $\rho : \mathbb{S}^1 \to \partial V$.

For an oriented regular simple closed curve surrounding a point p, like a closed orbit of X or an oriented boundary of a neighborhood V of p, we will sometimes say that it is "clockwise" or "counter-clockwise." Such a notion does not seem to need a definition, at least physically, but mathematically it does.

To avoid ambiguity, we first adopt the convention that we call the standard basis (e_1, e_2), with $e_1 = (1, 0)$ and $e_2 = (0, 1)$, positive and that we always represent it with e_1 pointing to the right and e_2 pointing upward, like in Fig. 1.9. If we then consider an oriented regular simple closed C^1 curve γ surrounding p, then we call it *clockwise* (respectively *counter-clockwise*), if at each t, the basis $(\gamma'(t), n_e(t))$ is positive (respectively negative), where $n_e(t)$ stands for the normal at $\gamma(t)$ pointing into the exterior of γ, as in Fig. 1.9.

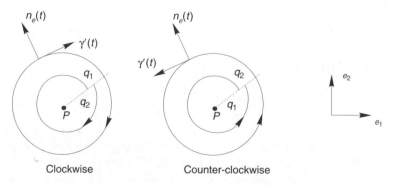

Fig. 1.9. Curves surrounding a point p

Both notions easily extend to piecewise C^1 oriented simple closed curves as well as to curves γ that are not closed, but move from a point $q_1 \neq p$ to some q_2 with $\overrightarrow{pq_2} = r(\overrightarrow{pq_1})$, with $r > 0$, in a way that the closed curve formed by γ and $\overrightarrow{q_2q_1}$ is a piecewise C^1 oriented simple closed curve surrounding p.

The permissible circle parametrizations, that we have used in the definition of finite sectorial decomposition, turn in a clockwise way.

It can be proven that the topological equivalence classes for a given isolated singular point are characterized by the number of elliptic, hyperbolic and parabolic sectors (denoted by e, h, p, respectively) and the arrangement of these sectors. There also exist results describing which triples of non-negative integers (e, h, p) are possible and which arrangement can be realized for each of these triples, depending on the *minimal degree* m of the singularity, which is the lowest degree of the non-zero terms in the Taylor expansion of the vector field at the singular point. We state a theorem on the number of elliptic and hyperbolic sectors without giving a proof.

Theorem 1.20 *Suppose that (e, h) is a couple of non-negative integers such that $e + h > 0$. Then there is a singular point of minimal degree m whose local phase portrait has e elliptic sectors and h hyperbolic sectors if and only if*

(i) $e + h \equiv 0 \pmod 2$;
(ii) $e \leq 2m - 1$ *and* $e + h \leq 2m + 2$;
(iii) *if* $e \neq 0$, *then* $e + h \leq 2m$.

There also exist results on the maximum number of parabolic sectors p depending on m, e and h. For a precise description of such results and their proofs see the bibliographical comments in Sect. 1.11.

We remark that Theorem 1.20 does not give any information about centers or foci, since then $e + h = 0$, but we have the well-known fact that the center-focus class is non-empty if and only if m is odd.

By using Theorem 1.20, we can easily get a finite list of pictures for any given m such that the local phase portrait of a singularity of a vector field of minimal degree m is topologically equivalent to one of them. To do this, it suffices to draw all possible minimal sectorial decompositions for any triple (e, h, p) with (e, h) satisfying Theorem 1.20 and $p \leq h$. Notice that this list may contain some pictures which cannot be realized.

1.6 Local Structure Near Periodic Orbits

Let $\gamma = \{\varphi_p(t) : t \in \mathbb{R}\}$ be a periodic orbit of period τ_0 of a vector field X of class C^r defined in an open subset $\Delta \subset \mathbb{R}^2$, where r denotes a positive integer, $+\infty$ or ω. Let Σ be a transverse section to X at p. Due to the continuity of the flow φ of X, for every point $q \in \Sigma$ close to p, the trajectory $\varphi_q(t)$ remains close to γ, with t in a given compact interval, for example, $[0, 2\tau_0]$. We define $f(q)$ as the first point where $\varphi_q(t)$ intersects Σ. Let Σ_0 be the domain of f. Of course we have that $p \in \Sigma_0$ and $f(p) = p$; see Fig. 1.10.

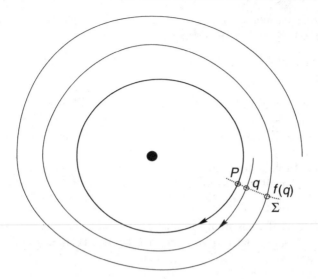

Fig. 1.10. Local behavior near a periodic orbit

The properties of X close to γ are reflected in f. For example, the periodic orbits of X close to γ correspond to fixed points of f, which are points $q \in \Sigma$ for which $f(q) = q$. The asymptotic behavior of the orbits of X close to γ is also described by f. So $\lim_{n \to \infty} f^n(q) = p$ implies that $\lim_{t \to \infty} d(\varphi_q(t), \gamma) = 0$, where d denotes the Euclidean distance on $\mathbb{R}^2$.

We want to observe that a section Σ as we use here is a differentiable curve contained in Δ. We may assume that Σ is an "open" segment of $\mathbb{R}^2$ in the sense that it is the regular image of an open interval in $\mathbb{R}$.

Given Σ we will define the *Poincaré map* $f : \Sigma_0 \to \Sigma$ as the return map of the flow on Σ, i.e., for each point of Σ belonging to a specific orbit, the map will give us the first point where the orbit intersects Σ in positive time. We suppose that Σ_0 is sufficiently small such that f is defined for all points in Σ_0,

Proposition 1.21 *Let φ be a C^r-flow with $1 \le r \le \infty$ or $r = \omega$. Then the Poincaré map $f : \Sigma_0 \to \Sigma$ is a diffeomorphism of class C^r onto its image Σ_1.*

A periodic orbit γ of X is called a *limit cycle* if there exists a neighborhood V of γ such that γ is the only periodic orbit contained in V.

Let γ be a periodic orbit of $\mathbb{R}^2$. We denote by $mathrmExt(\gamma)$ the set of points which belong to the unbounded component of $\mathbb{R}^2 \setminus \gamma$ (the exterior of γ), and by $\mathrm{Int}(\gamma)$ the set of points which belong to the bounded component of $\mathbb{R}^2 \setminus \gamma$ (the interior of γ).

Proposition 1.22 *There exist only three different types of limit cycles in $\mathbb{R}^2$:*

(i) Stable, when $\omega(q) = \gamma$ for every $q \in V$;
(ii) Unstable, when $\alpha(q) = \gamma$ for every $q \in V$;

(iii) *Semi-stable, when* $\omega(q) = \gamma$ *for every* $q \in V \cap Ext(\gamma)$ *and* $\alpha(q) = \gamma$ *for every* $q \in V \cap Int(\gamma)$, *or conversely.*

We see that γ is a limit cycle if and only if p is an isolated fixed point of f. Moreover,

(i) γ is stable if and only if $|f(x) - p| < |x - p|$ for every $x \neq p$ sufficiently close to p;

(ii) γ is unstable if and only if $|f(x) - p| > |x - p|$ for every $x \neq p$ sufficiently close to p;

(iii) γ is semi-stable if and only if $|f(x) - p| < |x - p|$ for every $x \in \Sigma \cap$ Ext (γ), $x \neq p$ sufficiently close to p, and $|f(x) - p| < |x - p|$ for every $x \in \Sigma \cap$ Int (γ), $x \neq p$ sufficiently close to p, or conversely.

In particular, if X is analytic and $f(x)$ is not the identity, then $f(x) = x + a_k(x - p)^k + \cdots$ with $a_k \neq 0$. So if k is odd, then γ is stable if $a_k < 0$, and unstable if $a_k > 0$. If k is even, then γ is semi-stable. If $f(x) \equiv x$, then all a_k are zero and γ lies in the interior of an annulus of periodic orbits of X, that is, it is not a limit cycle.

If $f'(p) < 1$ then we can apply the Intermediate Value Theorem and deduce that γ is stable. Equivalently, γ is unstable if $f'(p) > 1$; see Fig. 1.11. We remark that $f'(p) > 0$, because a Poincaré map on the plane is orientation preserving. It is easy to see that $f'(p)$ is independent of the chosen regular parameter x on Σ as well as on $p \in \gamma$; it is called the *characteristic exponent* of γ.

In what follows we shall study the Poincaré map for a differential equation in the plane. We consider a differential equation in the plane given by

$$\dot{x} = P(x, y),$$
$$\dot{y} = Q(x, y).$$

Let $X(x, y) = (P(x, y), Q(x, y))$ be the corresponding vector field. Denote the divergence of X at q by $(\text{div}X)(q)$. Assume that Σ is a transverse section and $\Sigma' \subset \Sigma$ is an open subset on which the Poincaré map is defined, $f : \Sigma' \to \Sigma$.

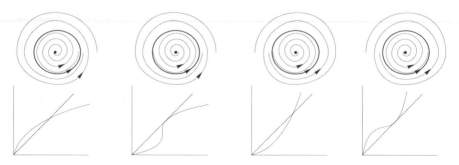

Fig. 1.11. Different classes of limit cycles and their Poincaré maps

Since the differential equations are in the plane, Σ is a curve which can be parametrized by $\gamma : I \to \Sigma$ with $\gamma(I') = \Sigma'$ and $|\gamma'(s)| = 1$. Let $X_\perp(q)$ be the scalar component to the tangent line to Σ at q given by

$$X_\perp \circ \gamma(s) = \det(\gamma'(s), X \circ \gamma(s)).$$

In the case that Σ is a horizontal line, $\{(x, y^*) : x_1 < x < x_2\}$, $X_\perp(q) = Q(q)$. In the case that Σ is an vertical line, $\{(x^*, y) : y_1 < y < y_2\}$, $X_\perp(q) = P(q)$.

Let $\tau(q)$ be the return time for $q \in \Sigma'$, so $f(p) = \varphi^{\tau(q)}(q) = \varphi_q(\tau(q))$.

The next theorem gives a sufficient condition for determining the stability of a limit cycle.

Theorem 1.23 *Let $\gamma : I' \to \Sigma'$ be a parametrization of the transverse section Σ' as above with $|\gamma'(s)| = 1$. Then for $s \in I'$,*

$$(\gamma^{-1} \circ f \circ \gamma)'(s) = \frac{X_\perp \circ \gamma(s)}{X_\perp \circ f \circ \gamma(s)} \exp\left(\int_0^{\tau \circ \gamma(s)} (divX) \circ \varphi^t \circ \gamma(s) \, dt\right).$$

Thus if $f(p) = p$ and $\gamma(s_0) = p$, then the characteristic exponent of γ is given by

$$f'(p) = (\gamma^{-1} \circ f \circ \gamma)'(s_0) = \exp\left(\int_0^{\tau(p)} (divX) \circ \varphi^t(p) \, dt\right).$$

Proof. The first variational equation states that

$$\frac{d}{dt} D\varphi_q^t = DX_{\varphi^t(q)} D\varphi_q^t.$$

Since $\det(D\varphi_q^0) = \det(id) = 1$, Liouville's formula for time-dependent linear equations gives

$$\det(D\varphi_q^{\tau(q)}) = \exp\left(\int_0^{\tau(q)} (divX) \circ \varphi^t(q) \, dt\right).$$

Notice that the right-hand side of this equality is the exponential in the formula for $(\gamma^{-1} \circ f \circ \gamma)'(s)$ as stated in the theorem. Therefore to complete the proof, we must relate $(\gamma^{-1} \circ f \circ \gamma)'(s)$ to $\det(D\varphi_{\gamma(s)}^{\tau \circ \gamma(s)})$.

Taking the derivative of $f \circ \gamma(s) = \varphi^{\tau \circ \gamma(s)}(\gamma(s))$ with respect to s yields

$$(f \circ \gamma)'(s) = (D\varphi_{\gamma(s)}^{\tau \circ \gamma(s)})\gamma'(s) + (\tau \circ \gamma)'(s)[X \circ \varphi^{\tau \circ \gamma(s)}(\gamma(s))]$$

$$= (D\varphi_{\gamma(s)}^{\tau \circ \gamma(s)})\gamma'(s) + (\tau \circ \gamma)'(s)[X \circ f \circ \gamma(s)].$$

Then

$$
\begin{aligned}
(\gamma^{-1} \circ f \circ \gamma)'(s)[X_\perp \circ f \circ \gamma(s)] &= \det((f \circ \gamma)'(s), X \circ f \circ \gamma(s)) \\
&= \det((D\varphi_{\gamma(s)}^{\tau \circ \gamma(s)})\gamma'(s), X \circ f \circ \gamma(s)) \\
&\quad + \det((\tau \circ \gamma)'(s)[X \circ f \circ \gamma(s)], X \circ f \circ \gamma(s)) \\
&= \det((D\varphi_{\gamma(s)}^{\tau \circ \gamma(s)})\gamma'(s), (D\varphi_{\gamma(s)}^{\tau \circ \gamma(s)})X \circ \gamma(s)) \\
&= \det((D\varphi_{\gamma(s)}^{\tau \circ \gamma(s)})\det(\gamma'(s), X \circ \gamma(s)) \\
&= \exp\left(\int_0^{\tau \circ \gamma(s)} (\operatorname{div}X) \circ \varphi^t \circ \gamma(s)\, dt\right) X_\perp \circ \gamma(s).
\end{aligned}
$$

Dividing by $X_\perp \circ f \circ \gamma(s)$, the theorem is proved.   □

Theorem 1.23 is contained in Sect. 28 of [4] or in Sect. 5.8.3 of [136].

The following theorem provides another way to compute the stability of a limit cycle. This result is due independently to Freire, Gasull and Guillamon [68] and to Chicone and Liu [34].

Theorem 1.24 *Let γ be a τ_0-periodic orbit of a C^2 planar vector field X. Assume that in a neighborhood V of γ, X admits a transverse Lie symmetry given by the vector field Y, i.e., $[Y, X] = \mu Y$. Let $\Sigma = \{\psi(p, s) : s \in \mathbb{R}\} \cap V$, be a transverse section to γ, where $\psi(p, s)$ is the solution of $\dot{x} = Y(x)$ with $\psi(p, 0) = p$. Then the characteristic exponent of γ is given by*

$$
f'(p) = \exp\left(\int_0^{\tau_0} \mu(\gamma(s))\, ds\right),
$$

where f is the Poincaré map on Σ.

1.7 The Poincaré–Bendixson Theorem

In what follows, we are going to assume that Δ is an open subset of $\mathbb{R}^2$ and X is a vector field of class C^r with $r \geq 1$. Also, in Δ, γ_p^+ denotes a positive semi–orbit passing through the point p.

Theorem 1.25 (Poincaré–Bendixson Theorem I) *Let $\varphi(t) = \varphi(t, p)$ be an integral curve of X defined for all $t \geq 0$, such that γ_p^+ is contained in a compact set $K \subset \Delta$. Assume that the vector field X has at most a finite number of singularities in K. Then one of the following statements holds.*

(i) If $\omega(p)$ contains only regular points, then $\omega(p)$ is a periodic orbit.

(ii) If $\omega(p)$ contains both regular and singular points, then $\omega(p)$ is formed by a set of orbits, every one of which tends to one of the singular points in $\omega(p)$ as $t \to \pm\infty$.

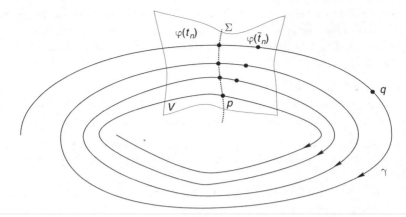

Fig. 1.12. Scheme of the section

(iii) If $\omega(p)$ does not contain regular points, then $\omega(p)$ is a unique singular point.

The next lemmas will facilitate the proof of this theorem.

Lemma 1.26 *If $p \in \Sigma \cap \omega(\gamma)$ where Σ is a transverse section of X and $\gamma = \{\varphi(t)\}$ is an orbit of X, then p is the limit of a sequence of points $\varphi(t_n)$ in Σ as $t_n \to \infty$*

Proof. We assume that $\gamma = \{\varphi(t)\} = \{\varphi(t, q)\}$ and $p \in \Sigma \cap \omega(\gamma)$, as in Fig. 1.12.

We consider a neighborhood V of p, and a map $\tau : V \to \mathbb{R}$, as given by Corollary 1.13.

Since $p \in \omega(\gamma)$ there exists a sequence $\{\tilde{t}_n\}$ such that $\tilde{t}_n \to \infty$ and $\varphi(\tilde{t}_n) \to p$ as $n \to \infty$.

Now there exists $n_0 \in \mathbb{N}$ such that $\varphi(\tilde{t}_n) \in V$ for every $n \geq n_0$. Let $t_n = \tilde{t}_n + \tau(\varphi(\tilde{t}_n))$ for $n \geq n_0$. We have that

$$\varphi(t_n) = \varphi(\tilde{t}_n + \tau(\varphi(\tilde{t}_n)), q) = \varphi(\tau(\varphi(\tilde{t}_n)), \varphi(\tilde{t}_n)),$$

and by the definition of τ we get that $\varphi(t_n) \in \Sigma$.

Since τ is continuous, it follows that

$$\lim_{n \to \infty} \varphi(t_n) = \lim_{n \to \infty} \varphi(\tau(\varphi(\tilde{t}_n)), \varphi(\tilde{t}_n)) = \varphi(0, p) = p,$$

because $\varphi(\tilde{t}_n) \to p$ and $\tau(\varphi(\tilde{t}_n)) \to \tau(p) = 0$ as $n \to \infty$. $\square$

We remark that a transverse section Σ for a vector field X has dimension 1, because we are considering a vector field on $\mathbb{R}^2$. Then locally, Σ is a diffeomorphic image of an interval of the straight line. From now on, we will suppose that every transverse section Σ is a diffeomorphic image of an interval. So Σ has a total order "$\leq$" induced by the total order of the interval. We can then talk about monotonic sequences in Σ. By an argument similar to the one used in Corollary 1.14 one can prove the next lemma. We repeat the argument, since it is very crucial in the proof of Theorem 1.25.

Lemma 1.27 *Let Σ be a transverse section for X contained in Δ. If γ is an orbit of X and $p \in \Sigma \cap \gamma$, then $\gamma_p^+ = \{\varphi(t,p) : t \geq 0\}$ intersects Σ in a (finite or infinite) monotonic sequence $p_1, p_2, \ldots, p_n, \ldots$.*

Proof. Let $D = \{t \in \mathbb{R}^+ : \varphi(t,p) \in \Sigma\}$. From the Flow Box Theorem, we get that D is discrete. Then we can order the set $D = \{0 < t_1 < t_2 < \cdots < t_n < \cdots\}$.

Let $p_1 = p$. We define, in the case that it exists, $p_2 = \varphi(t_1,p)$. By induction, we define $p_n = \varphi(t_{n-1},p)$. If $p_1 = p_2$, then γ is a periodic orbit of period $\tau = t_1$, and $p = p_n$ for all n. If $p_1 \neq p_2$, say $p_1 < p_2$ and if p_3 exists, we shall prove that $p_3 > p_2$.

We take an orientation on the section Σ as in Fig. 1.13(a), and we see that due to the fact that Σ is connected, and from the continuity of the vector field, the orbits of X always cross the section in the same sense, we say, from left to right, as it is shown in Fig. 1.13(b).

We recall that in $\mathbb{R}^2$ the Jordan Curve Theorem holds, which says that if J is a continuous and simple closed curve (J is a homeomorphic image of a circle), then $\mathbb{R}^2 \setminus J$ has two connected components: S_i (bounded) and S_e (unbounded) having J as their common boundary.

We consider now a Jordan curve formed by the union of the segment $\overline{p_1 p_2} \subset \Sigma$ with the arc $\widehat{p_1 p_2}$ of the orbit $\widehat{p_1 p_2} = \{\varphi(t,p) : 0 \leq t \leq t_1\}$, as shown in Fig. 1.14.

In particular, the orbit γ, starting at p_2 (that is, for values of $t > t_1$) is contained in S_i. In fact, it cannot intersect the arc $\widehat{p_1 p_2}$ due to the uniqueness of the orbits (see Fig. 1.15(a)) and it cannot intersect the segment $\overline{p_1 p_2}$ because it would go in the direction opposite to the flow; see Fig. 1.15(b).

In short, in the case that t_3 exists, we must have $p_1 < p_2 < p_3$ as shown in Fig. 1.16. Repeating these arguments, we have that $p_1 < p_2 < p_3 < \cdots < p_n < \cdots$. So $\{p_n\}$ is a monotonic sequence. If $p_2 < p_1$ the proof is analogous. $\qquad\square$

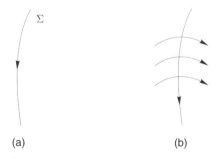

(a)   (b)

Fig. 1.13. Scheme of flow across the section

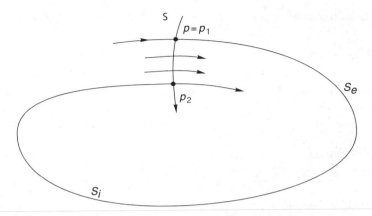

Fig. 1.14. Definition of Jordan's curve

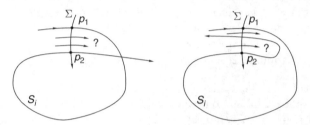

Fig. 1.15. Impossible configurations

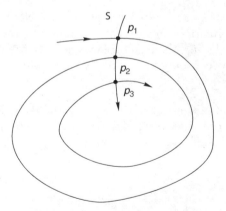

Fig. 1.16. Possible configuration

Lemma 1.28 *If Σ is a transverse section of a vector field X and $p \in \Delta$, then Σ intersects $\omega(p)$ at most in one point.*

Proof. From Lemma 1.27, a set of points of γ_p^+ in Σ has at most one limiting point, because this set is a monotonic sequence. Therefore, from Lemma 1.26 the result follows.   □

Lemma 1.29 *Let $p \in \Delta$, with γ_p^+ contained in a compact set, and γ an orbit of X with $\gamma \subset \omega(p)$. If $\omega(\gamma)$ contains regular points, then γ is a closed orbit and $\omega(p) = \gamma$.*

Proof. Let $q \in \omega(\gamma)$ be a regular point, and let V be a neighborhood of q given by Corollary 1.13 and Σ_q be a corresponding transverse section. From Lemma 1.26 there exists a sequence $t_n \to \infty$ such that $\gamma(t_n) \in \Sigma_q$. As $\gamma(t_n) \in \omega(p)$, the sequence $\{\gamma(t_n)\}$ reduces to a point, by Lemma 1.28. This proves that γ is periodic.

Now we prove that $\gamma = \omega(p)$. Since $\omega(p)$ is connected and γ is closed and non-empty, it is enough to prove that γ is open in $\omega(p)$.

Let $\bar{p} \in \gamma$, $V_{\bar{p}}$ a neighborhood of $\bar{p}$ given by Corollary 1.13, and $\Sigma_{\bar{p}}$ a corresponding transverse section. We will prove that $V_{\bar{p}} \cap \gamma = V_{\bar{p}} \cap \omega(p)$, inducing that γ is open in $\omega(p)$.

Obviously, $V_{\bar{p}} \cap \gamma \subset V_{\bar{p}} \cap \omega(p)$. Assume, contrary to what we wish to show, that there exists $\bar{q} \in V_{\bar{p}} \cap \omega(p)$ such that $\bar{q} \notin \gamma$. From the Flow Box Theorem, and the invariance of $\omega(p)$, there exists $t \in \mathbb{R}$ such that $\varphi(t, \bar{q}) \in \omega(p) \cap \Sigma_{\bar{p}}$ and $\varphi(t, \bar{q}) \neq \bar{p}$. So there exist two different points of $\omega(p)$ in $\Sigma_{\bar{p}}$, which is impossible by Lemma 1.28. Thus $V_{\bar{p}} \cap \gamma = V_{\bar{p}} \cap \omega(p)$.   □

Proof of Theorem 1.25. Under assumption *(i)* take $q \in \omega(p)$, then the orbit $\gamma_q \subset \omega(p)$. Since $\omega(p)$ is compact, we get that $\omega(\gamma_q) \neq \emptyset$. It immediately follows from Lemma 1.29 that $\omega(p) = \gamma_q$, a periodic orbit; see Fig. 1.17.

Under assumption *(ii)*, if γ is a regular orbit contained in $\omega(p)$, then from Lemma 1.29, and since $\alpha(\gamma)$ and $\omega(\gamma)$ are connected, it follows that both $\alpha(\gamma)$ and $\omega(\gamma)$ are singular points of the vector field X. Note that X has only a finite number of singularities in $\omega(p)$; see the different examples of Fig. 1.18.

Under assumption *(iii)*, the result follows directly from the fact that $\omega(p)$ is connected and X can only have finitely many singularities in $\omega(p)$; see Fig. 1.19.   □

Using the same arguments as in the proof of the Poincaré–Bendixson Theorem I, we obtain the next result for C^r vector fields with $r \geq 1$ whose singularities have a "finite sectorial decomposition."

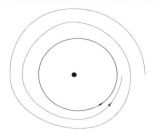

Fig. 1.17. Periodic orbit as ω-limit

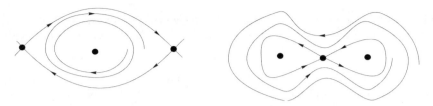

Fig. 1.18. Possible ω-limit sets

Fig. 1.19. A singular point as ω-limit

Corollary 1.30 (Poincaré–Bendixson Theorem II) *Let* $\varphi(t) = \varphi(t,p)$ *be an integral curve of* X *defined for all* $t \geq 0$, *such that* γ_p^+ *is contained in a compact set* $K \subset \Delta$. *Assume that the vector field* X *has a finite number of singularities in* $\omega(p)$, *each having a "finite sectorial decomposition." Then one of the following statements holds.*

(i) If $\omega(p)$ *contains only regular points, then* $\omega(p)$ *is a periodic orbit.*

(ii) If $\omega(p)$ *contains regular and singular points, then* $\omega(p)$ *is formed by a finite number of orbits* $\gamma_1, \ldots, \gamma_n$ *and a finite number of singular points* $p_1, \ldots, p_n$ *such that* $\alpha(\gamma_i) = p_i$, $\omega(\gamma_i) = p_{i+1}$ *for* $i = 1, \ldots, n-1$, $\alpha(\gamma_n) = p_n$ *and* $\omega(\gamma_n) = p_1$. *Possibly, some of the singular points* p_i *are identified.*

(iii) If $\omega(p)$ *does not contain regular points, then* $\omega(p)$ *is a singular point.*

As we have observed (without proof) in Sect. 1.5, analytic vector fields have the finite sectorial decomposition property at their singularities, and under mild conditions (to be specified in Sect. 2.7) C^∞ vector fields also have it. However, there exist C^∞ vector fields having singularities without the finite sectorial decomposition property, and for which it is possible to encounter an ω-limit set $\omega(q)$ consisting of one singularity p together with an infinity of regular orbits having p as their ω-limit and α-limit set.

The Poincaré–Bendixson Theorem also holds for vector fields on the two-dimensional sphere $\mathbb{S}^2$. If X has a finite number of singular points in $\mathbb{S}^2$, then the ω-limit set of an orbit passing through $p \in \mathbb{S}^2$ has the same possibilities *(i)*, *(ii)* and *(iii)* as given in the Poincaré–Bendixson Theorem of $\mathbb{R}^2$.

As an application of the Poincaré–Bendixson Theorem we present the next result.

Theorem 1.31 *Let X be a vector field of class C^1 on an open set $\Delta \subset \mathbb{R}^2$. If γ is a periodic orbit of X such that $Int(\gamma) \subset \Delta$, then there exists a singular point of X contained in $Int(\gamma)$.*

Proof. We will suppose that there are no singular points in $Int(\gamma)$. We consider the set Γ of closed orbits of X contained in $\overline{Int(\gamma)}$, ordered with the following partial order: $\gamma_1 \leq \gamma_2$ if and only if $\overline{Int(\gamma_1)} \supseteq \overline{Int(\gamma_2)}$.

We will show that any completely ordered subset S of Γ (i.e., $\gamma_1 \neq \gamma_2$ in S implies that either $\gamma_1 < \gamma_2$ or $\gamma_1 > \gamma_2$), admits an upper bound; that is, an element greater than or equal to any other element of S.

Let $\sigma = \cap\{\overline{Int(\gamma_i)} : \gamma_i \in S\}$. We see that $\sigma \neq \emptyset$, since every $\overline{Int(\gamma_i)}$ is compact and the family $\{\overline{Int(\gamma_i)} : \gamma_i \in S\}$ has the finite intersection property. That is, any finite intersection of elements of a family cannot be empty. Let $q \in \sigma$. From the Poincaré–Bendixson Theorem, $\omega(q)$ is a closed orbit contained in σ, since this set is invariant under X and does not contain singular points. This orbit is an upper bound of S.

From Zorn's Lemma, Γ has a maximal element. So there does not exist any element of Γ contained in $Int(\mu)$. But if $p \in Int(\mu)$, $\alpha(p)$ and $\omega(p)$ are closed orbits by the Poincaré–Bendixson Theorem (since there do not exist singular points). As $\alpha(p)$ and $\omega(p)$ cannot be both equal to μ, one of them will be contained in $Int(\mu)$. This contradiction proves that there must be singular points in $Int(\mu)$. □

1.8 Lyapunov Functions

In trying to follow orbits in order to detect their asymptotic behavior, it is often interesting to use the method of Lyapunov functions.

Therefore, throughout this section, we will consider a C^1 vector field $X : \Delta \subset \mathbb{R}^2 \to \mathbb{R}^2$ defined on some open set $\Delta \subset \mathbb{R}^2$ (it in fact suffices to require that X be locally Lipschitz) and we denote by $\phi(p,t)$ its flow, defined on some open domain $D \subset \mathbb{R}^2 \times \mathbb{R}$. Let $f : \Delta \subset \mathbb{R}^2 \to \mathbb{R}$ be a C^1 function; then for $p \in \Delta$, the derivative of the function f along the solution $\phi(p,t)$ is given by

$$Xf(p) = \tfrac{d}{dt}f(\phi(p,t)) = Df_p(X(p)) = <\nabla f_p, X(p)>$$
$$= \tfrac{\partial f}{\partial x}(p)X_1(p) + \tfrac{\partial f}{\partial y}(p)X_2(p), \quad (1.4)$$

where $\nabla f_p = \nabla f(p)$ denotes the gradient of f at p. It is common to also denote $Xf(p)$ by $\dot{f}(p)$.

Definition 1.32 Under the condition expressed earlier, f is called a *Lyapunov function* for X if $\dot{f}(p) = Xf(p) < 0$ on Δ.

By definition a Lyapunov function is strictly decreasing along the orbits of f, meaning, among other things, that X is everywhere transverse to the

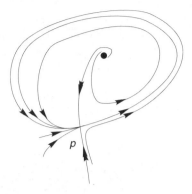

Fig. 1.20. A saddle-node loop

level curves of f at regular points of f. We will now see in a few examples how the use of Lyapunov functions can be very helpful. We will also see that it is often not necessary to require that $\dot{f}(p) < 0$ everywhere, but that one may allow that on certain closed subsets $\dot{f}(p) = 0$ or, in other words, that f is a Lyapunov function on a smaller open domain than Δ. We now first consider the stability of the singularities, always under the conditions expressed earlier.

Definition 1.33 *(i)* A singularity p of X is called *Lyapunov stable* if for each neighborhood $V \subset \Delta$ of p there exists a neighborhood $W \subset V$ of p such that for all $q \in W$ and for all $t \in [0, \infty)$, $\phi(q, t) \in V$.
(ii) If moreover $\lim_{t \to \infty} \phi(q, t) = p$ for all $q \in W$, then p is called *asymptotically stable*.

Remark 1.34 The example given in Fig. 1.20 shows that the limit condition in *(ii)* does not imply the stability, hence not what we defined as "asymptotic stability."

The following criterion for stability relies on the use of Lyapunov functions.

Theorem 1.35 *Let X be a C^1 vector fields on an open set $\Delta \subset \mathbb{R}^2$, let p be a singularity of X and V a neighborhood of p such that f is continuous on V and C^1 on $V \setminus \{p\}$, $f(p) = 0$ and for all $q \in V \setminus \{p\}$: $f(q) > 0$ and $\dot{f}(q) \le 0$ (respectively $\dot{f}(q) < 0$). Then p is stable (respectively asymptotically stable).*

Proof. For every neighborhood W of p, choose $\varepsilon > 0$ so small that $\overline{B}_\varepsilon(p) \subset W$. Let $m_\varepsilon = \min\{f(q) | q \in S_\varepsilon(p)\}$, where $S_\varepsilon(p) = \partial B_\varepsilon(p)$; by supposition $m_\varepsilon > 0$. We now choose $\delta > 0$ such that $f(q) < m_\varepsilon$ for all $q \in B_\delta(p)$. It is now clear, from the fact that $\dot{f}(q) \le 0$, that $\phi(q, t)$ remains in $B_\varepsilon(p)$ for all $t \in (0, \infty)$ since $f(\phi(q, t))$ can never reach the value m_ε. This implies the stability of p.

If moreover we have the condition $\dot{f}(q) < 0$, then first we can observe that X has no other singularities in V but p, so by the Poincaré–Bendixson Theorem, either $\phi(q, t) \to p$, for all $q \in B_\delta(p)$, or for some q, $\phi(q, t)$ tends to an ω-limit set that contains some point p_1 at which $f(p_1) \neq 0$. We leave it as an exercise to show that this is not possible since $f(\phi(q,t))$ has to be strictly decreasing.   □

Remark 1.36 In the second part of the previous proof it is not necessary to rely on the Poincaré–Bendixson Theorem. Simple reasoning that can be found in many text books (e.g., [128]) applies. We leave the alternate proof as an exercise.

Remark 1.37 In the case that $\dot{f} \equiv 0$, f is called a first integral of X (see Chap. 8). By Theorem 1.35 it is clear that if this condition is satisfied in a neighborhood V of p on which $f(q) > 0$ for all $q \in V \setminus \{p\}$, then p is a stable, although not asymptotically stable, singularity of X.

Example 1.38 *(i)* The system

$$\begin{aligned} \dot{x} &= -y^3, \\ \dot{y} &= \ \ x^3, \end{aligned} \tag{1.5}$$

has $x^4 + y^4$ as a first integral and hence has a stable but not asymptotically stable singularity at the origin.
(ii) The system

$$\begin{aligned} \dot{x} &= -y^3 - x^3, \\ \dot{y} &= \ \ x^3 - y^3, \end{aligned} \tag{1.6}$$

has $x^4 + y^4$ as a Lyapunov function on $\mathbb{R}^2 \setminus \{(0,0)\}$ and hence has a (globally) asymptotically stable singularity at the origin.

Lyapunov functions are also useful for detecting the stability and asymptotic stability of periodic orbits. We will show this with an example.

Definition 1.39 *(i)* Under the conditions expressed earlier, a periodic orbit γ of X is called *Lyapunov stable* if for each neighborhood V of γ there exists a neighborhood $W \subset V$ of γ such that for all $q \in W$ and for all $t \in (0, \infty)$, $\phi(q,t) \in V$.
(ii) If moreover $\phi(q,t)$ has γ as ω-limit for all $q \in W$, then γ is called *asymptotically stable*.

We leave it to the reader to prove the following theorem. The proof is similar to the proof of Theorem 1.35.

Theorem 1.40 *Under the conditions expressed earlier, let γ be a periodic orbit of X and V a neighborhood of γ such that f is continuous on V and C^1 on $V \setminus \gamma$, $f|_\gamma = 0$ and for all $q \in V \setminus \gamma$ we have $f(q) > 0$ and $\dot{f}(q) \leq 0$ (respectively $\dot{f}(q) < 0$). Then γ is stable (respectively asymptotically stable)*

Example 1.41 Consider the system

$$\begin{aligned} \dot{x} &= y + (1 - x^2 - y^2)x, \\ \dot{y} &= -x + (1 - x^2 - y^2)y. \end{aligned} \qquad (1.7)$$

We see that $\gamma = \mathbb{S}^1 = \{(x, y) | \, x^2 + y^2 = 1\}$ is a periodic orbit, while f defined by

$$f(x, y) = \begin{cases} 1 - (x^2 + y^2) & \text{when } x^2 + y^2 < 1 \\ (x^2 + y^2) - 1 & \text{when } x^2 + y^2 > 1, \end{cases}$$

satisfies the conditions of Theorem 1.40 on $\mathbb{R}^2 \setminus \{(0,0)\}$. As such, γ is stable and moreover all orbits in $\mathbb{R}^2 \setminus \{(0,0)\}$ have γ as their ω-limit set.

1.9 Essential Ingredients of Phase Portraits

For linear vector fields it is possible to describe all conjugacy classes, but this is not possible when we study non-linear vector fields. However, there exists a general characterization of the topological equivalence classes for vector fields on the plane. To present it we will introduce some definitions. These definitions together with Proposition 1.42 and Theorem 1.43 can be found in [107] and [116]. In these works much more general situations are considered.

In this section we do not need X to be C^1; it need only be sufficiently regular to admit the existence of a continuous flow (local Lipchitz continuity suffices). We consider a differential equation $\dot{x} = X(x)$ where X is a locally Lipschitz function on $\mathbb{R}^2$ and let $\phi(s, x)$ be the flow defined by the differential equation. Following the notation of the works of Markus and Neumann, we denote by $(\mathbb{R}^2, \phi)$ the flow defined by the differential equation. By the theorem of continuous dependence on initial conditions, the flow $(\mathbb{R}^2, \phi)$ is continuous.

We say that a flow $(\mathbb{R}^2, \phi)$ is a *parallel* flow if it is topologically equivalent to one of the following flows:

(i) The flow defined on $\mathbb{R}^2$ by the differential system $\dot{x} = 1$, $\dot{y} = 0$, which we denote by *strip flow*.
(ii) The flow defined on $\mathbb{R}^2 \setminus \{0\}$ by the differential system given in polar coordinates $r' = 0$, $\theta' = 1$, which we denote by *annulus flow*.
(iii) The flow defined on $\mathbb{R}^2 \setminus \{0\}$ by the differential system given in polar coordinates $r' = r$, $\theta' = 0$, which we denote by *spiral* or *nodal flow*.

Given a maximal open region on $\mathbb{R}^2$ on which the flow is parallel, it is interesting to know the orbit-structure of its boundary. Clearly following types of orbits can be present:

(i) a singular point,

(ii) a periodic orbit for which there does not exist a neighborhood entirely consisting of periodic orbits,

(iii) an orbit $\gamma(p)$, homeomorphic to $\mathbb{R}$ for which there does not exist a neighborhood N of $\gamma(p)$ such that

(1) For all $q \in N$ $\alpha(q) = \alpha(p)$ and $\omega(q) = \omega(p)$,

(2) The boundary ∂N of N, that is $\partial N = \bar{N} \setminus N$, is formed by $\alpha(p)$, $\omega(p)$ and two orbits $\gamma(q_1)$ and $\gamma(q_2)$ such that $\alpha(p) = \alpha(q_1) = \alpha(q_2)$ and $\omega(p) = \omega(q_1) = \omega(q_2)$.

Orbits satisfying *(i)*, *(ii)* or *(iii)* are called "separatrices" in [107] and [116]. In this book we prefer to keep the name *separatrix* for an orbit satisfying *(iii)*, meaning for instance that a linear saddle has four separatrices. Indeed, orbits limiting on a singularity and bordering a hyperbolic sector are always separatrices. We also remark that for analytic flows, the orbits satisfying *(ii)* are the limit cycles. In view of presenting the next result we introduce the *separatrix skeleton* consisting of the set of all orbits satisfying either *(i)* or *(iii)*; if we also add the orbits satisfying *(ii)* we speak about the *extended separatrix skeleton*. It is easy to see that the union S of all orbits in the extended separatrix skeleton is a closed set invariant under the flow. Let V be a (maximal) connected component of $\mathbb{R}^2 \setminus S$. It necessarily is invariant under the flow. We call V a *canonical region*.

Proposition 1.42 *Every canonical region of $(\mathbb{R}^2, \phi)$ is parallel, given by either a strip, an annular or a spiral flow.*

Proof. Let U be a canonical region of the C^0 flow $(\mathbb{R}^2, \phi)$. We denote the flow ϕ on U by (U, ϕ') with $\phi' = \phi|_U$ where U does not contain singularities. Since there are no separatrices in U nor closed orbits satisfying *(ii)*, the set consisting of orbits homeomorphic to $\mathbb{S}^1$ is open, and similarly the set consisting of orbits homeomorphic to $\mathbb{R}$ (which we term *line orbits*) is open. Hence U consists entirely of closed orbits or entirely of line orbits.

We claim that two orbits of ϕ' can be separated by disjoint parallel neighborhoods. To prove this, we suppose to the contrary that $\gamma(p)$ and $\gamma(q)$ are distinct orbits (closed or not) which cannot be separated. Then for any parallel neighborhood N_p of p, we have $q \in \mathrm{cl}(N_p)$; i.e.,

$$q \in \bigcap_{N_p} \mathrm{cl}(N_p) = \alpha(p) \cup \gamma(p) \cup \omega(p).$$

This means that $q \in \alpha(p)$ (or $q \in \omega(p)$). But this is impossible because $q \in N_q \subset U$ and $\alpha(p) \cup \omega(p) \subset \mathrm{cl}(N_q) \setminus N_q \not\subset U$.

It follows that the quotient space U/ϕ', obtained by collapsing orbits of (U, ϕ') to points, is a (Hausdorff) connected one dimensional manifold. Hence the natural projection $\pi : U \to U/\phi'$ is a locally trivial fibering of which there are only three possibilities, namely the three classes of parallel flows described earlier. This proves the proposition. □

Given a flow $(\mathbb{R}^2, \phi)$ by the *completed separatrix skeleton* we mean the union of the extended separatrix skeleton of the flow together with one orbit from each one of the canonical regions (it is also called *separatrix configuration* by other authors). Let C_1 and C_2 be the completed separatrix skeletons of the flows $(\mathbb{R}^2, \phi_1)$ and $(\mathbb{R}^2, \phi_2)$, respectively. We say that C_1 and C_2 are *topologically equivalent* if there exists a homeomorphism from $\mathbb{R}^2$ to $\mathbb{R}^2$ that maps the orbits of C_1 to the orbits of C_2 preserving the orientation.

Theorem 1.43 (Markus–Neumann–Peixoto Theorem) *Assume that* $(\mathbb{R}^2, \phi_1)$ *and* $(\mathbb{R}^2, \phi_2)$ *are two continuous flows with only isolated singular points. Then these flows are topologically equivalent if and only if their completed separatrix skeletons are equivalent.*

From this result, it is enough to describe the completed separatrix skeleton in order to determine the topological equivalence class of a differential system.

1.10 Exercises

Exercise 1.1 Let X be a linear vector field on a neighborhood of $0 \in \mathbb{R}^2$, with $X(0) = 0$, and denote by $\varphi(t, x)$ the flow of X. Show that $D_x\varphi(t, 0) = e^{At}$, with $A = DX(0)$.

Exercise 1.2 Consider the equation $x'' = F(x)$ defined on an interval of $\mathbb{R}$, which corresponds to an one dimensional conservative system. Clearly this is equivalent to the system

$$\begin{aligned} \dot{x} &= v, \\ \dot{v} &= F(x). \end{aligned} \tag{1.8}$$

(i) Prove that the total energy $E = T + U$ is a first integral of (1.8) where $T(v) = v^2/2$ is the *kinetic energy* and $U(x) = -\int_{x_0}^{x} F(z)dz$ is the *potential energy*.

(ii) Prove that all the singular points of (1.8) are on the x-axis. Prove also that all the periodic orbits of (1.8) intersect the x-axis and are symmetric with respect to it.

(iii) Prove that if $U(x_1) = U(x_2) = c$, $U'(x_j) \neq 0$ for $j = 1, 2$ and $U(x) < c$ for all $x_1 < x < x_2$, then (1.8) has a periodic orbit passing through the points $(x_1, 0)$ and $(x_2, 0)$.
   Hint: An orbit that passes through $(x_0, 0)$ is given by $v^2/2 + V(x) = E$ where E is its energy. Use the fact that $dv/dx = F(x)/v$ to conclude that this orbit must again cut the x-axis and that this must happen at $(x_2, 0)$. Then use (ii).

(iv) Assume that $F(x) \neq 0$ for $0 < |x - x_0| < a$. Prove that (1.8) has a center or a saddle at $(x_0, 0)$ when $U(x_0)$ is a relative minimum or maximum.

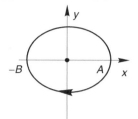

Fig. 1.21. Figure for Exercise 1.5

Exercise 1.3 Following the previous exercise, determine the phase portrait of the following equations:

(i) $x'' = -x$ (spring),
(ii) $x'' = -\sin x$ (pendulum),
(iii) $x'' = -1/x^2$ (gravitation).

Exercise 1.4 Consider the equation $x''+q(x) = 0$ where $q \in C^1(\mathbb{R})$, $q(0) = 0$ and $xq(x) > 0$ if $x \neq 0$. This is the equation of the motion of a unit mass attached to an elastic spring which reaches a displacement x with a force $q(x)$. We define the rigidity of the spring by $h(x) = q(x)/x$. From *(iv)* in Exercise 1.2 we know that $(0,0)$ is a center of the phase portrait (x, v).

(i) Given an orbit close to the origin, with energy E and limits of oscillation $-B$ and A (see Fig. 1.21), show that its period is

$$T = 2 \int_{-B}^{A} \frac{dx}{\sqrt{2(E - U(x))}}.$$

　　Hint: Note that $x' = v = \sqrt{2(E - U(x))}$.
(ii) Consider two springs with $h_1(x) \geq h(x)$ which oscillate inside the same limits (see *(i)*). Let T_1 and T be their periods. Then $T \geq T_1$.
　　Hint: Note that at the point A, $E = U(A) = \int_0^A q(u)du$ and so $E-U(x) = \int_x^A q(u)du$. Use this to prove that $E - U(x) \leq E - U_1(x)$. Then apply *(i)*.
(iii) A spring for which $h(x) = h(-x)$ is called *symmetric*. In this case, $U(x) = U(-x)$ and $B = A$ in *(i)*. The number A is called the *amplitude of the oscillation*. Then we will say that a symmetric spring is *strong* if $h''(0) > 0$, and *weak* if $h''(0) < 0$. Prove that the period of a strong (respectively weak) spring decreases (respectively increases) when the amplitude of the oscillations increases.
　　Hint: Let $A_1 = cA$ with $c > 1$. By symmetry it is necessary to consider only the time that the spring needs to oscillate between 0 and A (respectively 0 and A_1). Take $x = cy$ and obtain the equation $y'' + yh(cy) = 0$. Note that the oscillation of the amplitude A for this equation corresponds to the oscillation of the amplitude $A_1 = cA$ for the original equation, both with the same period. Then apply *(ii)*.

Exercise 1.5 Let a, b, c, d be real numbers and $f, g : B \to \mathbb{R}$ functions of class C^1 defined in a ball B with center at the origin $(0,0)$ of $\mathbb{R}^2$ and radius r. The system

$$\begin{aligned} \dot{x} &= ax + by + f(x, y), \\ \dot{y} &= cx + dy + g(x, y), \end{aligned} \qquad (1.9)$$

is called a perturbed system of the linear system

$$\begin{aligned} \dot{x} &= ax + by, \\ \dot{y} &= cx + dy. \end{aligned}$$

(i) Prove that if $f = O(r)$, $g = O(r)$ and $ad - bc \neq 0$, then the origin $(0,0)$ is an isolated singular point of (1.9). As usual, $f = O(r)$ means that $f = rF(x, y)$ with $F(x, y)$ bounded in a neighborhood of the origin.

(ii) Suppose that $f = g = 0$. Determine the conditions on a, b, c, d such that $(0,0)$ is a hyperbolic singularity of (1.9). In this case, describe the phase portrait of (1.9) in a neighborhood of the origin. There are three topological types.

(iii) Describe the phase portraits of the systems later. Show that they are not topologically equivalent each other or to the types found in *(ii)*.

$$\dot{x} = x^2, \qquad \dot{y} = -y,$$

$$\dot{x} = e^{-1/x^2} \sin \frac{1}{x}, \qquad \dot{y} = -y.$$

Exercise 1.6 Let X be a C^1 vector field on $\Delta \subset \mathbb{R}^2$. Prove that if $\varphi(t)$ is a trajectory of X defined on the maximal interval (ω_-, ω_+) and such that $\lim_{t \to \omega_+} \varphi(t) = p \in \Delta$, then $\omega_+ = \infty$ and p is a singular point of X.

Exercise 1.7 Let $\varphi(t, x)$ be the flow generated by a C^1 vector field X on $\mathbb{R}^2$. A non-empty subset $S \subset \mathbb{R}^2$ is called *minimal* (for X) if it is *invariant* (i.e., if $x \in S$ then $\varphi(t, x) \in S$ for all $t \in \mathbb{R}$), compact, and does not contain proper subsets with these properties.

Prove that in $\mathbb{R}^2$, the only minimal subsets for X are the singular points and the periodic orbits of X.

Exercise 1.8 Determine $\omega(p)$ and $\alpha(p)$ for $p \in \mathbb{R}^2$ and for the vector field $Z = (X, Y)$ given by

$$X = -y + x(x^2 + y^2) \sin \left(\frac{\pi}{\sqrt{x^2 + y^2}} \right),$$

$$Y = x + y(x^2 + y^2) \sin \left(\frac{\pi}{\sqrt{x^2 + y^2}} \right).$$

Hint: Study the inner product $\langle (x, y), Z(x, y) \rangle = xX + yY$.

Exercise 1.9 Determine the set $\omega(p)$ for all $p \in \mathbb{R}^2$ for the system

$$\dot{x} = y[y^2 + (x^2 - 1)^2] + x(1 - x^2 - y^2),$$
$$\dot{y} = -y[y^2 + (x^2 - 1)^2] + y(1 - x^2 - y^2).$$

Hint: The same as in the previous exercise.

Exercise 1.10 Determine the singular points of the system

$$\dot{x} = y,$$
$$\dot{y} = -b\sin x - ay,$$

with $a, b > 0$. Assuming that it has no periodic orbits, make a sketch of the phase portrait for this system.

Exercise 1.11 Prove that

$$\dot{x} = 2x - x^5 - xy^4,$$
$$\dot{y} = y - y^3 - x^2 y,$$

has no periodic orbits.

Hint: Show that all singularities of this system are on the coordinate axes. Consider the phase portrait of this system restricted to the axes and prove that the existence of a closed orbit provides a contradiction.

Exercise 1.12 Let X be a C^1 vector field on $\mathbb{R}^2$. Let p be a regular point of X such that $p \in \omega(p)$. Then $\omega(p)$ is a periodic orbit.

Exercise 1.13 Let X be a C^1 vector field on $\mathbb{R}^2$ and γ an orbit of X. Prove that if γ is not a periodic orbit, then either it is a singular point or $\omega(\gamma) \cap \alpha(\gamma) = \emptyset$.

Exercise 1.14 Let X be a C^1 vector field on $\mathbb{R}^2$ such that there exists a neighborhood V of the origin on which $X|_V$ is the linear vector field

$$(x_1, x_2) \to (\lambda_1 x_1, \lambda_2 x_2)$$

with $\lambda_1 \lambda_2 < 0$ and $\lambda_1 + \lambda_2 < 0$.

Let $L = \gamma_p \cup \{(0, 0)\}$ be like in Fig. 1.22, and J_L be the bounded connected component of $\mathbb{R}^2 \setminus L$. Prove that there exists a neighborhood W_L of L such that for all $q \in W_L \cap J_L$ we have that $\omega(q) = L$.

Hint: Consider Fig. 1.22. Note that it is possible to define a Poincaré map π for the loop L using the transverse segment Σ which is contained in the upper side of the square. Prove that $\pi = f \circ g$ where g maps points from this segment to Σ_0 and $f : \Sigma_0 \to \Sigma$. Prove that $g(x) = x^\theta$ with $\theta > 1$ and conclude that $\pi'(x) < 1$.

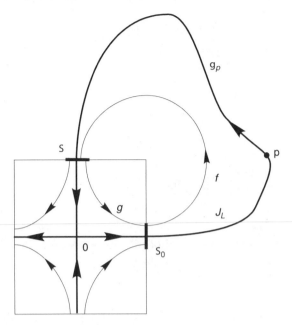

Fig. 1.22. Hint for Exercise 1.14

Exercise 1.15 Prove that the origin is an asymptotically stable singular point of the system

$$\dot{x} = -x - \frac{x^3}{3} - 2\sin y,$$

$$\dot{y} = -y - \frac{y^3}{3}.$$

That is, prove that the ω-limit of every solution in a neighborhood of the origin is the origin.

Exercise 1.16 Let $f : \mathbb{R}^2 \to \mathbb{R}^2$ be a C^1 function such that $f(0,0) = (0,0)$ and $\langle x, f(x)\rangle < 0$ for all $x \neq 0$. Prove that $x \to |x|^2$ is a Lyapunov function for the system $x' = f(x)$.

Exercise 1.17 Let X be a C^1 vector field on some open domain $\Delta \subset \mathbb{R}^2$. Let $S = \{p \in \Delta : X(p) = 0\}$ and let $f : \Delta \setminus S \to \mathbb{R}$ be a C^1 Lyapunov function for X on $\Delta \setminus S$. Show that for all $q \in \Delta \setminus S$, $\omega(q) \subset S$.

Exercise 1.18 Let $\Delta \subset \mathbb{R}^2$ be an open set and $V : \Delta \to \mathbb{R}$ a C^2 function. The *gradient vector field* associated to V is defined by $\dot{x} = -\mathrm{grad}V(x)$ for $x \in \Delta$ where

$$\mathrm{grad}V(x) = \left(\frac{\partial V}{\partial x_1}(x), \frac{\partial V}{\partial x_2}(x)\right).$$

Note that the C^1 vector field grad$V$ satisfies

$$DV_x \cdot y = < \text{grad}V(x), y >,$$

for all $x \in \Delta$ and $y \in \mathbb{R}^2$. Let $\dot{V}$ be a derivative of V along the trajectories of the field. Prove:

(i) $\dot{V}(x) \leq 0$ for all $x \in \Delta$, and $\dot{V}(x) = 0$ if and only if x is a singularity of grad$V$.

(ii) If x_0 is an isolated minimum of V, then x_0 is an asymptotically stable singularity of $-$grad$V$.

(iii) $-$grad$V$ does not have periodic solutions.

Exercise 1.19 Let $X = \nabla f = \text{grad} f$, where f is a C^r function, $r \geq 2$, defined on an open set $\Delta \subset \mathbb{R}^2$. Prove that X has no periodic orbits. If X has only isolated singular points, then for all $p \in \Delta$ the ω-limit set of p is either empty or is a singular point.

    Hint: If $\varphi(t)$ is a trajectory of X note that $df(\varphi(t))/dt > 0$; that is, $f \circ \varphi$ is an increasing function.

Exercise 1.20 Let $\Delta \subset \mathbb{R}^2$ be open and let $V : \Delta \to \mathbb{R}$ be of class C^2. Given $c \in \mathbb{R}$, the set $V^{-1}(c)$ is called a *level* of V. If $x \in V^{-1}(c)$ is a regular point (that is, $DV_x \neq 0$), then $V^{-1}(c)$ is a C^1 curve near x. Prove that in this case, grad$V(x)$ is perpendicular to $V^{-1}(c)$ at x. For each case later, make a sketch of the graph of V and of the phase portrait of $-$grad$V$.

(i) $V(x,y) = x^2 + y^2$;
(ii) $V(x,y) = x^2 - y^2$;
(iii) $V(x,y) = x^4 - x^2 + y^2$.

Exercise 1.21 Let $f : \mathbb{R}^2 \to \mathbb{R}^2$ be a C^1 function such that $f(0,0) = (0,0)$. The singular point $(0,0) \in \mathbb{R}^2$ is called *globally stable* when is stable and $\lim_{t \to \infty} \varphi(t) = (0,0)$ for every solution $\varphi(t)$ of

$$x' = f(x). \tag{1.10}$$

    Let $V : \mathbb{R}^2 \backslash \{(0,0)\} \to \mathbb{R}$ be a Lyapunov function of system (1.10). Suppose that for every $c > 0$ there exists $R > 0$ such that if $|x| > R$ then $V(x) > c$ for all $x \in \mathbb{R}^2$. Then $(0,0)$ is a globally stable solution of (1.10). Note that the condition $\{x \in \mathbb{R}^2 : V(x) = 0\} = \{(0,0)\}$ is not necessary. It is sufficient to suppose that this set does not contain a full trajectory different from $x(t) = (0,0)$.

Exercise 1.22 Let $f : \mathbb{R} \to \mathbb{R}$ be a C^1 function such that $f(0) = 0$. Consider the system

$$x'' + ax' + f(x) = 0, \tag{1.11}$$

for $x \in \mathbb{R}$. If $a > 0$ and $f(x)x > 0$ for all $x \neq 0$, then a *null solution* is an asymptotically stable solution for system (1.11) (that is, the corresponding solution for the associated first order system on $\mathbb{R}^2$ is an asymptotically stable solution). If $f(x)/x > \varepsilon > 0$ for all $x \neq 0$, then a null solution is globally stable (see Exercise 1.21).

Hint: Take $V(x,y) = y^2 + 2\int_0^x f(x)dx$.

Exercise 1.23 Under the assumptions that $a < 0 < b$, $F, g \in C^1(a,b)$, $xg(x) > 0$ for $x \neq 0$, $G(x) = \int_0^x g(s)ds \to \infty$ as $x \to a$ if $a = -\infty$ and $G(x) \to \infty$ as $x \to b$ if $b = \infty$, $f(x)/g(x)$ is monotone increasing on $(a,0) \cup (0,b)$ and is not constant on any neighborhood of $x = 0$, it follows that the system

$$\dot{x} = y - F(x),$$
$$\dot{y} = -g(x),$$

has at most one limit cycle in the region $a < x < b$ and if exists it is stable.

Exercise 1.24 Under the assumptions that $g(x) = x$, $F \in C^1(\mathbb{R})$, $f(x) = F'(x)$ is an even function with exactly two positive zeros $a_1 < a_2$ with $F(a_1) > 0$ and $F(a_2) < 0$, and $f(x)$ is monotone increasing for $x > a_2$, it follows that the system from previous exercise has at most two limit cycles.

1.11 Bibliographical Comments

For a deeper introduction to the topics of this chapter, the reader can look at the following books: Chow and Hale [35], Guckenheimer and Holmes [77], Hirsch and Smale [84], Hartman [80], Sotomayor [151], Zhang, Ding, Huang and Dong [169] and mainly the book of Sotomayor [152]. Many of the proofs of the results of this chapter are translations of the corresponding proofs from the book of Sotomayor [152].

The results related to Theorem 1.20 are due to several authors. Condition (i) was obtained by Bendixson [15]; and (ii) and (iii) were proved by Berlinskii [16] and [17]. For results on the number of parabolic sectors p we refer to Sagalovich [138] and [139].

The proof of Theorem 1.43 is due to Markus [108], to Neumann [116] and to Peixoto [123].

2

Normal Forms and Elementary Singularities

In this chapter we study the elementary singular points; i.e., the hyperbolic and semi-hyperbolic singular points. The first are those singular points having two eigenvalues with non-zero real part, the second have a unique non-zero eigenvalue. We also provide complete information about invariant manifolds as well as good normal forms for such singularities.

2.1 Formal Normal Form Theorem

Since it is no simpler to limit to $\mathbb{R}^2$, we present the general normal form theorem for C^∞ vector fields on $\mathbb{R}^n$.

Let $X = A + f$ be a vector field on $\mathbb{R}^n$ with A linear, and f a C^∞ function such that $f(0) = 0$ and $Df(0) = 0$. The aim of a normal form theory is to determine for each given linear vector field A a restricted class of non-linearities F_n as small and as simple as possible, and such that for each f, the equation can, by appropriate C^∞ coordinate change, be brought to the form $X' = A + \tilde{f}$ with $\tilde{f} \in F_n$. We also recall that a C^r coordinate change h, with $r \geq 1$, operates on a vector field X_1 to change it into X_2 as given by expression (1.2), namely

$$Dh_{h^{-1}(p)}X_1(h^{-1}(p)) = X_2(p);$$

we also write $X_2 = h_* X_1$.

In view of finding appropriate coordinate changes, consider for each $m \in \mathbb{N}$ with $m \geq 1$ the so called "adjoint action":

$$ad_m A : H^m(\mathbb{R}^n) \to H^m(\mathbb{R}^n)$$
$$X \to [A, X],$$

where $H^m(\mathbb{R}^n)$ denotes the set of polynomial vector fields on $\mathbb{R}^n$ which are homogeneous of degree m. Here, $[A, X] = A \circ X - X \circ A$, seen as differential operators. Let $B^m = ad_m A(H^m(\mathbb{R}^n))$ and let G^m be any complement of B^m, i.e., $B^m \oplus G^m = H^m(\mathbb{R}^n)$.

Theorem 2.1 (Formal Normal Form Theorem) *Let X be a C^r vector field defined in a neighborhood of 0 with $X(0) = 0$ and $DX(0) = A$, $r \in \mathbb{N}$ with $r \geq 1$. Let B^k and G^k be as above. Then there is an analytic change of coordinates $\phi : (\mathbb{R}^n, 0) \to (\mathbb{R}^n, 0)$ in a neighborhood of 0 such that $X' = \phi_*(X)$ is of the form*

$$X'(y) = Ay + g_2(y) + \cdots + g_r(y) + o(\|y\|^r)$$

with $g_i \in G^i$, for all $i = 2, \ldots, r$.

Proof. The proof proceeds by induction on $2 \leq s < r$. We assume that

$$X(x) = A(x) + g_2(x) + \cdots + g_{s-1}(x) + f_s(x) + o(\|x\|^s),$$

with $g_i \in G^i$, for all $i = 2, \ldots, s - 1$ and f_s is homogeneous of degree s. We try a coordinate transformation of the form

$$x = h(y) = y + P(y),$$

where P is a homogeneous (vector valued) polynomial of degree s, which needs to be determined. Substitution gives

$$(I + DP(y))\dot{y} = A(y + P(y)) + g_2(y) + \cdots + g_{s-1}(y) + f_s(y) + o(\|y\|^s)$$

or

$$\begin{aligned}
\dot{y} &= (I + DP(y))^{-1}[Ay + g_2(y) + \cdots + g_{s-1}(y) + f_s(y) + AP(y) + o(\|y\|^s)] \\
&= (I - DP(y) + O(\|y\|^s))[Ay + \cdots] \\
&= Ay + g_2(y) + \cdots + g_{s-1}(y) + f_s(y) + AP(y) - DP(y)Ay + o(\|y\|^s)].
\end{aligned}$$

The $(s-1)$-jet of X remains unchanged, while for the terms of order s we get

$$f_s(y) - ad_s A(P)(y).$$

Clearly a suitable choice of P will make this term lie in G_s.   □

Remark 2.2 The proof we use can be found in [77]. It is constructive, and it can be used to implement the calculation in specific examples.

The original proof in [156] is quite similar but uses the time 1-mapping of the flow P_t instead of $I + P$.

Remark 2.3 Using Theorem 2.1 and Borel's Theorem (see [19] and [115]) on the representation of ∞-jets, one can show the following.

If X is a C^∞ vector field defined in a neighborhood of $0 \in \mathbb{R}^n$, with $X(0) = 0$, then there is a C^∞ coordinate change $\phi : (\mathbb{R}^n, 0) \to (\mathbb{R}^n, 0)$ in a neighborhood of 0 such that $X' = \phi_*(X)$ has an ∞-jet:

$$j_\infty X'(0) = A + \sum_{i=2}^{\infty} g_i,$$

with $g_i \in G^i$, for all $i = 2, 3, \ldots$. Note that this result is purely formal. It provides a simple form for the ∞-jet but not for the vector field itself. Later on we will see essential differences.

We recall that Borel's Theorem states that for every $p \in \mathbb{R}^n$ and for every ∞-jet (i.e., formal series in n variables), there exists a C^∞ function f defined on a neighborhood of p whose ∞-jet $j_\infty f(p)$ (i.e., its Taylor series at p) is equal to the given ∞-jet.

Remark 2.4 A great deal of information concerning normal forms can be found in [20]. Especially interesting is the treatment of singular vector fields belonging to special Lie subalgebras like, for instance, Hamiltonian vector fields, volume preserving vector fields or reversible vector fields ($L_*X = -X$, with L a linear involution). In these cases one can find normal forms belonging to the same Lie subalgebra, and this for a coordinate change lying in the Lie subgroup belonging to the Lie subalgebra. The presentation of the results is done in the framework of filtered Lie algebras having the Borel property.

Choice of the G_m From a computational point of view, the normal form theory presented in this way is not yet satisfactory. An improvement was made in [23] and [61]. It concerns the choice of the complementary spaces G^m.

The idea is that one can always use

$$G^m = \mathrm{Ker}(ad_m A^T),$$

where A^T denotes the transpose of A, defined by $\langle A^T x, y \rangle = \langle x, Ay \rangle$, or equivalently $(A^T)_{ij} = A_{ji}$. We give a sketch of the proof, referring to [162] for a detailed elaboration. We consider an inner product $\langle \, , \, \rangle_m$ on $H^m(\mathbb{R}^n)$ defined as

$$\left\langle \sum_{|\sigma|=m} a_\sigma x^\sigma, \sum_{|\sigma|=m} b_\sigma x^\sigma \right\rangle_m = \sum_{|\sigma|=m} \sigma! \langle a_\sigma, b_\sigma \rangle, \tag{2.1}$$

for all $a_\sigma, b_\sigma \in \mathbb{R}^m$ with $|\sigma| = m$.

For this inner product $\langle \, , \, \rangle_m$ we have

$$(ad_m A^T) = (ad_m A)^T,$$

and hence $H^m(\mathbb{R}^n) = \mathrm{Im}(ad_m A) \oplus \mathrm{Ker}(ad_m A)^T$.

We show by an example that $\mathrm{Ker}(ad A^T)$ can serve as a complement G^m, but not $\mathrm{Ker}(ad A)$. For this take

$$A = \begin{pmatrix} 0 & 1 \\ 0 & 0 \end{pmatrix}.$$

Look at the 2-jet

$$\left[y\frac{\partial}{\partial x}, x^i y^{2-i}\frac{\partial}{\partial x}\right] = i x^{i-1} y^{3-i}\frac{\partial}{\partial x},$$

$$\left[y\frac{\partial}{\partial x}, x^i y^{2-i}\frac{\partial}{\partial y}\right] = i x^{i-1} y^{3-i}\frac{\partial}{\partial y} - x^i y^{2-i}\frac{\partial}{\partial x}.$$

The image contains $xy(\partial/\partial x), y^2(\partial/\partial x), y^2(\partial/\partial y) - xy(\partial/\partial x), 2xy(\partial/\partial y) - x^2(\partial/\partial x)$. It is four dimensional such that $\mathrm{Ker}(ad_2 A)$ is two-dimensional.

As $y^2(\partial/\partial x) \in \mathrm{Ker}(ad_2 A)$, we cannot have $\mathrm{Im}(ad_2 A) \oplus \mathrm{Ker}(ad_2 A) = H^2(\mathbb{R}^2)$.

For $ad_2 A^T$ we have

$$\left[x\frac{\partial}{\partial y}, x^i y^{2-i}\frac{\partial}{\partial x}\right] = (2-i)x^{i+1} y^{1-i}\frac{\partial}{\partial x} - x^i y^{2-i}\frac{\partial}{\partial y},$$

$$\left[x\frac{\partial}{\partial y}, x^i y^{2-i}\frac{\partial}{\partial y}\right] = (2-i)x^{i+1} y^{1-i}\frac{\partial}{\partial y},$$

and $\mathrm{Ker}(ad_2 A^T)$ is spanned by $x^2(\partial/\partial y)$ and $x^2(\partial/\partial x) + xy(\partial/\partial y)$.

This gives one possible choice of the normal form; in certain proofs it is better to take $x^2(\partial/\partial y), xy(\partial/\partial y)$ as a basis for G^2, because then the 2-jet takes the form $y(\partial/\partial x) + (ax^2 + bxy)(\partial/\partial y)$ which, for instance, corresponds to a second order scalar differential equation. We will come back to these nilpotent singularities in Sect. 3.4.

We now start a systematic study of the singularities, starting with the elementary ones.

2.2 Attracting (Repelling) Hyperbolic Singularities

Throughout this section we deal with vector fields that are at least C^∞, and have an attracting or repelling hyperbolic singularity. We always position the singularity of interest at the origin.

The differential equation X near such a singularity can be written as

$$\dot{x} = ax + by + P(x,y),$$
$$\dot{y} = cx + dy + Q(x,y),$$

with $P(0,0) = Q(0,0) = DP(0,0) = DQ(0,0) = 0$, and where the matrix of $DX(0) = \begin{pmatrix} a & b \\ c & d \end{pmatrix}$ can be written as

$$\begin{pmatrix} \lambda_1 & 0 \\ 0 & \lambda_2 \end{pmatrix}, \qquad \begin{pmatrix} \lambda_1 & 1 \\ 0 & \lambda_1 \end{pmatrix}, \qquad \text{or} \qquad \begin{pmatrix} \alpha & \beta \\ -\beta & \alpha \end{pmatrix}$$

with $\lambda_1 \lambda_2 > 0$, $\alpha \neq 0$ and $\beta > 0$.

We will now show that such a singularity is locally C^0-conjugate to its linear part. It suffices to give the proof for the attracting case, meaning that

Fig. 2.1. Transverse section around an attracting singular point

$\lambda_1 < 0$, $\lambda_2 < 0$ and $\alpha < 0$. The repelling case can be reduced to it by reversing time. We start with the third case, for which $S_r = \{(x,y) : x^2 + y^2 = r^2\}$ is a transverse section for both X and $L = DX(0)$, if $r > 0$ is taken sufficiently small (see Fig. 2.1). We fix such value $r = r_0 > 0$. It clearly follows that for all points $p \in \overline{B}_{r_0}(0) = \{(x,y) : x^2 + y^2 \leq r_0^2\}$, both the X-orbit and the L-orbit tend to 0 for $t \to \infty$ and leave $B_{r_0}(0) = \{(x,y) : x^2 + y^2 < r_0^2\}$ for $t \to -\infty$.

Let $\varphi(p,t)$ and $\psi(p,t)$ denote the respective flows of X and L. Taking any point $p \in B_{r_0}(0) \setminus \{0\}$, there exists a unique time $t_p > 0$ such that $\varphi(p, -t_p) \in S_{r_0}$. We define the local C^0-conjugacy $h : B_{r_0}(0) \to B_{r_0}(0)$ as follows:

$$h(0) = 0,$$
$$h(p) = \psi(\varphi(p, -t_p), t_p). \tag{2.2}$$

We leave it to the reader to check that h is indeed a homeomorphism conjugating the flows φ and ψ.

For the second case we first observe that it is possible to find a linear coordinate change bringing the linear part $DX(0)$ into the form

$$\begin{pmatrix} \lambda_1 & \varepsilon \\ 0 & \lambda_1 \end{pmatrix}$$

where $\varepsilon > 0$ is any given positive number. In taking $\varepsilon > 0$ sufficiently small, the same construction as in the former case shows that X is locally C^0-conjugate to $L = DX(0)$.

A similar method works in the first case, except that instead of working with circles S_r and a related ball $B_{r_0}(0)$ it is preferable to use $S_r' = \{(x,y) : |\lambda_2|x^2 + |\lambda_1|y^2 = r^2\}$ and $B_{r_0}'(0) = \{(x,y) : |\lambda_2|x^2 + |\lambda_1|y^2 < r_0^2\}$.

In any case the simple argumentation that we have presented proves this (very) special case of the *Hartman–Grobman Theorem* which states that in any dimension, a vector field at a hyperbolic singularity is locally C^0-conjugate to its linear part $DX(0)$. We recall that a singular point p is *hyperbolic* if the eigenvalues of the linear part of the system at p have non-zero real part. A similar method shows that all attracting (respectively repelling) hyperbolic singularities in $\mathbb{R}^n$ are mutually C^0-conjugate. We will also develop a simple, although more complicated, argumentation to prove the same statements for the saddle case in dimension two.

Before proving the Hartman–Grobman Theorem in dimension two, we however describe the C^∞-conjugacy classes of attracting and repelling hyperbolic singularities. To that purpose, we start by applying the Formal Normal Form Theorem presented in Sect. 2.1. It depends on the respective Jordan Normal Form for the linear part. We restrict to the repelling case.

Case 1: Let $DX(0) = \lambda_1 x(\partial/\partial x) + \lambda_2 y(\partial/\partial y)$ with $\lambda_2 \geq \lambda_1 > 0$. The Lie–bracket operation of Sect. 2.1 (dividing all calculations by λ_1) gives

$$\left[x\frac{\partial}{\partial x} + \lambda y\frac{\partial}{\partial y}, x^m y^n \frac{\partial}{\partial x} \right] = ((m-1) + n\lambda)x^m y^n \frac{\partial}{\partial x},$$

$$\left[x\frac{\partial}{\partial x} + \lambda y\frac{\partial}{\partial y}, x^m y^n \frac{\partial}{\partial y} \right] = (m + (n-1)\lambda)x^m y^n \frac{\partial}{\partial y},$$

with $\lambda = \lambda_2/\lambda_1 \geq 1$. Besides the linear terms we see that all terms in the Taylor development can be removed, except for $x^m(\partial/\partial y)$ in the case $\lambda = m$. Based on Borel's Theorem on the realization of formal power series we hence know the existence of a C^∞ coordinate change near 0 such that the vector field can be written as

$$\dot{x} = \lambda_1 x + f(x,y),$$
$$\dot{y} = \lambda_2 y + g(x,y),$$

where f and g are C^∞ functions and $j_\infty f(0,0) = j_\infty g(0,0) = 0$, in the case $\lambda_2/\lambda_1 \notin \mathbb{N}$.

If $\lambda_2 = m\lambda_1$ for some $m \in \mathbb{N}$ with $m \geq 1$, then a C^∞ normal form is given by

$$\dot{x} = \lambda_1 x + f(x,y),$$
$$\dot{y} = \lambda_2 y + ax^m + g(x,y),$$

for some $a \in \mathbb{R}$, where f and g are C^∞ functions and $j_\infty f(0,0) = j_\infty g(0,0) = 0$. By a linear change $(x,y) \longmapsto (bx, \delta y)$ for some $b \in \mathbb{R}$ and $\delta \in \{-1, 1\}$ we can reduce to $a = 1$.

In Sect. 2.7 we will show that it is possible to reduce to $f = g = 0$, which is a special case of *Sternberg's Theorem*; see [153]. Before proving this, we consider the other cases.

Case 2: Let $DX(0) = \lambda(x(\partial/\partial x) + (x+y)(\partial/\partial y))$ with $\lambda > 0$. We again divide by λ in the Lie-bracket calculations:

$$\left[x\frac{\partial}{\partial x} + (x+y)\frac{\partial}{\partial y}, x^m y^n \frac{\partial}{\partial x} \right] = (m+n-1)x^m y^n \frac{\partial}{\partial x} + nx^{m+1}y^{n-1}\frac{\partial}{\partial x}$$
$$- x^m y^n \frac{\partial}{\partial y},$$

$$\left[x\frac{\partial}{\partial x} + (x+y)\frac{\partial}{\partial y}, x^m y^n \frac{\partial}{\partial y} \right] = (m+n-1)x^m y^n \frac{\partial}{\partial y} + nx^{m+1}y^{n-1}\frac{\partial}{\partial y}.$$

By making the appropriate linear combinations it is clear that all terms of order at least two can be removed, inducing a C^∞ normal form

$$\dot{x} = \lambda x + f(x,y),$$
$$\dot{y} = \lambda y + x + g(x,y),$$

where f and g are C^∞ functions and $j_\infty f(0,0) = j_\infty g(0,0) = 0$.

Case 3: Let $DX(0) = \alpha((x+\gamma y)(\partial/\partial x) + (-\gamma x + y)(\partial/\partial y))$ with $\alpha\gamma \neq 0$. We divide by α in the normal form calculations and get

$$\left[(x+\gamma y)\frac{\partial}{\partial x} + (-\gamma x + y)\frac{\partial}{\partial y}, x^m y^n \frac{\partial}{\partial x} \right] = (m+n-1)x^m y^n \frac{\partial}{\partial x}$$
$$+ m\gamma x^{m-1}y^{n+1}\frac{\partial}{\partial x} - n\gamma x^{m+1}y^{n-1}\frac{\partial}{\partial x} + \gamma x^m y^n \frac{\partial}{\partial y},$$

$$\left[(x+\gamma y)\frac{\partial}{\partial x} + (-\gamma x + y)\frac{\partial}{\partial y}, x^m y^n \frac{\partial}{\partial y} \right] = (m+n-1)x^m y^n \frac{\partial}{\partial y}$$
$$+ m\gamma x^{m-1}y^{n+1}\frac{\partial}{\partial y} - n\gamma x^{m+1}y^{n-1}\frac{\partial}{\partial y} - \gamma x^m y^n \frac{\partial}{\partial x}.$$

Again appropriate linear combinations show that all terms of order at least two can be removed, inducing a C^∞ normal form

$$\dot{x} = \alpha x + \beta y + f(x,y),$$
$$\dot{y} = -\beta x + \alpha y + g(x,y),$$

where $\alpha\beta \neq 0$, f and g are C^∞ functions and $j_\infty f(0,0) = j_\infty g(0,0) = 0$.

In Sect. 2.7 we will show how in all three cases the "flat" terms f and g can be removed, providing linear normal forms for C^∞-conjugacy, except in the case $\lambda = m \in \mathbb{N}$ with $m \geq 1$, for which the normal form is

$$\dot{x} = \lambda x, \qquad \dot{y} = m\lambda y + \delta x^m, \tag{2.3}$$

for some $\lambda \neq 0$ and $\delta \in \{0,1\}$.

In the analytic case it is even possible to prove that the obtained normal forms can be obtained by an analytic change of coordinates, but we will not prove this. For the proof in the C^∞ case we refer to Sect. 2.7.

2.3 Hyperbolic Saddles

2.3.1 Analytic Results

The differential equation X near such a singularity can be written as

$$\dot{x} = \lambda_1 x + P(x,y),$$
$$\dot{y} = \lambda_2 y + Q(x,y), \tag{2.4}$$

where $P(0,0) = Q(0,0) = DP(0,0) = DQ(0,0) = 0$ with $\lambda_1 > 0$ and $\lambda_2 < 0$.

From the beginning we normalize to $\lambda_1 = 1$ and $\lambda_2 = -\lambda$ with $\lambda \geq 1$. All other cases can easily be reduced to this one, through multiplication by a non-zero real number. We hence study X given by

$$\begin{aligned} \dot{x} &= x + P(x,y), \\ \dot{y} &= -\lambda y + Q(x,y), \end{aligned} \qquad (2.5)$$

with $\lambda > 0$. We start by taking P and Q analytic. We want to prove the existence of an analytic function $y(x)$ with $y(0) = y'(0) = 0$ and with the property that its graph $(x, y(x))$ represents an invariant manifold of (2.5). This manifold is called the *unstable manifold* of (2.5).

It suffices to use this result (existence of unstable manifold) on $-X$ in order to find for (2.5) the existence of a function $x(y)$ with $x(0) = x'(0) = 0$, whose graph $(x(y), y)$ is invariant. This is called the *stable manifold* of (2.5). To prove the existence of $y(x)$ we first write $(x, y) = (x, xu)$. This changes (2.5) into:

$$\begin{aligned} \dot{x} &= x + P(x, xu), \\ u\dot{x} + x\dot{u} &= -\lambda xu + Q(x, xu), \end{aligned}$$

which we can write as

$$\begin{aligned} \dot{x} &= x(1 + R(x, u)), \\ \dot{u} &= -\mu u + S(x, u), \end{aligned} \qquad (2.6)$$

with $\mu = \lambda + 1 > 1$ and R and S analytic.

We again write u as y and μ as λ. In order to prove the existence of the unstable manifold of (2.5) as an expression $y = xu(x)$, it hence suffices to consider (2.6), multiplied by $(1 + R(x, u))^{-1}$, whose study can be reduced to the following lemma that we will prove using a simple majorization argument.

Lemma 2.5 *Consider the differential equation*

$$x\frac{dy}{dx} = -(\lambda + B(x,y))y + A(x), \qquad (2.7)$$

with A and B analytic, $A(0) = 0$, $B(0,0) = 0$ and $\lambda > 0$. Then there exists an analytic solution $y(x)$ defined on some $(-\varepsilon, \varepsilon)$ with $\varepsilon > 0$, such that $y(0) = 0$ and $y'(0) = A'(0)/(\lambda + 1)$.

Proof. We search a formal power series solution

$$y = y(x) = \sum_{n=1}^{\infty} y_n x^n, \qquad (2.8)$$

with $y(0) = 0$ of (2.7) and show uniqueness and convergence.

By substituting (2.8) into (2.7), one easily finds that $y_1 = A'(0)/(\lambda + 1)$. To simplify the study of the recursion, one introduces the change of variables $y = x(y_1 + \tilde{y})$:

$$x\frac{d\tilde{y}}{dx} + \mu\tilde{y} = xf(x, \tilde{y}), \qquad (2.9)$$

with f analytic near $(0,0)$ and with $\mu = \lambda + 1 > 1$. In the sequel we drop the tildes however and proceed as if (2.9) is the main equation to solve.

It is now easily seen that (2.9) is a recursive relation for the sequence $(y_n)_n$ after substitution of (2.8). Indeed, when evaluating (2.9) at order x^k, one sees in the left-hand side $(k+\mu)y_k$, and in the right-hand side one finds an expression $H_k(y_1, \ldots, y_{k-1})$ that involves only lower order coefficients. This argument shows the existence of (2.8) solving (2.7) formally, and it also shows the unicity.

We now take care of the convergence properties. If at any step of the recursion we replace the right-hand side by a larger (positive) value, we certainly get larger values of y_k, so one can find bounds for the coefficients by replacing the right-hand side by an expression that at every step yields a larger value. Since

$$f(x,y) = \sum_{m=0}^{\infty} \sum_{n=0}^{\infty} f_{mn} x^m y^n$$

is analytic, we know that $|f_{mn}| \leq M r^m s^n$ for $M > 0$, $r > 0$ and $s > 0$. In the light of what was said earlier, we consider (2.9), replacing f by

$$F(x,y) = \sum_{m=0}^{\infty} \sum_{n=0}^{\infty} M r^m s^n x^m y^n = \frac{M}{(1-rx)(1-sy)}.$$

The new equation certainly leads to a formal solution that majorizes the original one. We are hence interested in

$$x\frac{dy}{dx} + \mu y = \frac{Mx}{(1-rx)(1-sy)}. \tag{2.10}$$

The recursive relation that is generated by (2.10) can be written down explicitly, if we multiply both sides of the equation by $(1-sy)$:

$$(n+\mu)y_n - s \sum_{k+l=n} (k+\mu)y_k y_l = M r^{n-1}.$$

Keeping in mind that $y_0 = 0$ this yields

$$y_n = \frac{M r^{n-1}}{n+\mu} + s \sum_{k=1}^{n-1} \frac{k+\mu}{n+\mu} y_k y_{n-k}.$$

We again replace this recursion by a "larger" recursion, i.e., a recursion that has a solution for which all coefficients are larger: we consider

$$Y_n = \frac{M}{\mu+1} r^{n-1} + s \sum_{k=1}^{n-1} Y_k Y_{n-k}.$$

This is the recursion that is generated from the simple equation

$$Y(x) = \frac{M}{\mu+1}\frac{x}{1-rx} + sY(x)^2, \quad Y(0) = 0.$$

One can easily compute that there is a unique analytic solution Y with radius of convergence

$$\rho = \frac{\mu+1}{r(\mu+1)+4Ms}.$$

This is a lower bound for the radius of convergence of the solution (2.8) to our original equation (2.7).   □

2.3.2 Smooth Results

In the case that we start with an expression (2.5) in which P and Q are only C^∞, it is possible to prove the existence of a C^∞ unstable (respectively stable) manifold. In the same way as in the analytic case, it follows from the next lemma.

Lemma 2.6 *Consider the differential equation*

$$x\frac{dy}{dx} = -(\lambda + B(x,y))y + A(x), \tag{2.11}$$

where A and B are C^∞ functions, $A(0) = 0$, $B(0,0) = 0$ and $\lambda > 0$. Then there exists a C^∞ solution $y(x)$ defined on some $(-\varepsilon,\varepsilon)$ with $\varepsilon > 0$, such that $y(0) = 0$ and $y'(0) = A'(0)/(\lambda+1)$.

Proof. Exactly as in the proof of Lemma 2.5 we can prove the existence of a unique power series

$$y = \hat{g}(x) = \sum_{n=1}^{\infty} y_n x^n,$$

with $y_1 = A'(0)/(\lambda+1)$ such that $\hat{g}(x)$ is a formal solution of (2.11).

Let $g(x)$ be any C^∞ function with $j_\infty g(0) = \hat{g}$, of which the existence is guaranteed by Borel's Theorem, which permits one to write every formal power series as the Taylor development of a C^∞ function. If we now write

$$y = Y + g(x),$$

then (2.11) changes into

$$x\frac{dY}{dx} = -(\lambda + \overline{B}(x,Y))Y + \overline{A}(x),$$

where $\overline{A}$ and $\overline{B}$ are C^∞ functions, $j_\infty\overline{A}(0) = 0$, $\overline{B}(0,0) = 0$ and $\lambda > 0$. We hence continue working with expression (2.11), having the extra property that

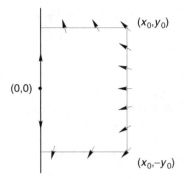

Fig. 2.2. The flow on the boundary of V_0

$j_\infty A(0) = 0$. Expression (2.11) can also be represented by the planar system

$$\dot{x} = -x,$$
$$\dot{y} = (\lambda + B(x,y))y - A(x).$$

We continue working on $x \geq 0$, proving the existence of a C^∞ solution $y(x)$ on $[0, x_0]$ with $x_0 > 0$ and $j_\infty y(0) = 0$. Similar reasoning will clearly be possible on $x \leq 0$, too.

On $V = [0, x_0] \times [-y_0, y_0]$, for $x_0 > 0$ and $y_0 > 0$ sufficiently small (see Fig. 2.2), we leave it as an exercise to prove that there exists at least one point on $\{x_0\} \times [-y_0, y_0]$ with the property that the ω-limit of its orbit is $(0,0)$. We observe that because $\dot{x} = -x \neq 0$ on V, this orbit, completed with the origin, is the graph $(x, f(x))$ of a continuous function $f : [0, x_0] \to \mathbb{R}$. In fact f is C^∞ for $x > 0$.

We call the graph of f a stable manifold and we now prove that such a stable manifold is unique.

We therefore introduce the new coordinates $(x, Y) = (x, y - f(x))$. Since the coordinate change is C^∞ for $x > 0$, it operates on the vector field X and changes it into Y given by

$$\dot{x} = -x,$$
$$\dot{Y} = Y(\lambda + F(x, Y + f(x))).$$

(2.12)

At $x = 0$, this vector field Y extends in a C^0 way to $Y|_{\{x=0\}} = X|_{\{x=0\}}$. It is clear that $\{Y = 0\}$ is the unique stable manifold for Y, implying the unicity of the stable manifold for X, too.

We hence know that an equation like in (2.11) has, for $x_0 > 0$ sufficiently small, a unique continuous solution $f : [0, x_0] \to \mathbb{R}$ with $f(0) = 0$; f is C^∞ on $(0, x_0)$.

We now prove that f must also be C^∞ at 0 and that $j_\infty f(0) = 0$. To that end we take an arbitrary $n \in \mathbb{N}$ with $n \geq 1$ and try a solution of the form

$f(x) = x^n f_n(x)$. Recalling that $j_\infty A(0) = 0$, we see that $f_n(x)$ has to be a solution of

$$x \frac{dy}{dx}(x) = -y(\lambda + n + B(x, x^n y)) + A_n(x),$$

with $A_n(x) = A(x)/x^n$. Since we know that this equation has a solution $f_n(x)$ where f_n is C^∞ on $(0, x_0]$, for some $x_0 > 0$, and with $f_n(0) = 0$, we hence obtain that necessarily the unique solution f of (2.11) has the property $f(x) = O(x^n)$ for every $n \in \mathbb{N}$ with $n \geq 1$.

By differentiating (2.11) it is now easy to inductively show that all derivatives of f tend to zero for $x \to 0$ and hence to obtain that f is C^∞ with $j_\infty f(0) = 0$. □

We can now summarize the results in the following theorem on *hyperbolic saddles*.

Theorem 2.7 *Take a differential equation*

$$\begin{aligned}
\dot{x} &= \lambda_1 x + P(x, y), \\
\dot{y} &= \lambda_2 y + Q(x, y),
\end{aligned} \tag{2.13}$$

where P and Q are C^∞ functions, $P(0,0) = Q(0,0) = 0$ and also $DP(0,0) = DQ(0,0) = 0$, $\lambda_1 > 0$ and $\lambda_2 < 0$. Then in a sufficiently small neighborhood of the origin there exist invariant manifolds W^u and W^s with $W^u = \{(x, \alpha(x))\}$, $W^s = \{(\beta(y), y)\}$, where α and β are both C^∞ with $\alpha(0) = \alpha'(0) = \beta(0) = \beta'(0) = 0$. Moreover, both W^u and W^s are uniquely defined. If P and Q are analytic, then α and β are analytic.

If we successively introduce the new coordinates

$$(x, Y) = (x, y - \alpha(x)),$$

and then

$$(X, Y) = (x - \overline{\beta}(Y), Y),$$

for $(\overline{\beta}(Y), Y)$ representing the stable manifold in the (x, Y)–coordinates, then we change (2.13) into

$$\begin{aligned}
\dot{x} &= x(\lambda_1 + R(x, y)), \\
\dot{y} &= y(\lambda_2 + S(x, y)),
\end{aligned} \tag{2.14}$$

for some C^∞ functions R and S. We can even take R and S to be analytic when starting with analytic P and Q.

The special form of expression (2.14) clearly shows the saddle like structure of the phase portrait, as represented in Fig. 2.3.

In Sect. 2.4 we prove that two hyperbolic saddles are C^0-conjugate near the origin, implying the *Hartman–Grobman Theorem* in this case. Before doing this, we will however first study the C^∞-conjugacy classes. We start by applying the Formal Normal Form Theorem. As in the attracting/repelling case, we divide by λ_1 in the calculations, writing $\lambda = \lambda_2/\lambda_1$. Then we have

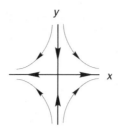

Fig. 2.3. A hyperbolic saddle

$$\left[x\frac{\partial}{\partial x} - \lambda y\frac{\partial}{\partial y}, x^m y^n \frac{\partial}{\partial x} \right] = (m - 1 - n\lambda)x^m y^n \frac{\partial}{\partial x},$$

$$\left[x\frac{\partial}{\partial x} - \lambda y\frac{\partial}{\partial y}, x^{\overline{m}} y^{\overline{n}} \frac{\partial}{\partial y} \right] = (\overline{m} - (\overline{n} - 1)\lambda)x^{\overline{m}} y^{\overline{n}} \frac{\partial}{\partial y}.$$

Clearly all terms of order at least two can be removed when λ is irrational. If $\lambda = k/l \in \mathbb{Q}$, then the normal form is given by

$$\left(1 + \sum_{i=1}^{\infty} a_i (x^k y^l)^i \right) x\frac{\partial}{\partial x} + \left(-\frac{k}{l} + \sum_{i=1}^{\infty} b_i (x^k y^l)^i \right) y\frac{\partial}{\partial y}.$$

By using Borel's Theorem on the realization of formal power series, together with Theorem 2.7, we obtain that a C^∞ hyperbolic saddle is C^∞-conjugate to one of the following C^∞ systems:

$$\begin{aligned} \dot{x} &= x(\lambda_1 + R(x,y)), \\ \dot{y} &= y(\lambda_2 + S(x,y)), \end{aligned} \tag{2.15}$$

with $\lambda_2/\lambda_1 \in \mathbb{R} \setminus \mathbb{Q}$ and $j_\infty R(0,0) = j_\infty S(0,0) = 0$; or

$$\begin{aligned} \dot{x} &= x(\lambda_1 + f(x^k y^l) + R(x,y)), \\ \dot{y} &= y(\lambda_2 + g(x^k y^l) + S(x,y)), \end{aligned} \tag{2.16}$$

with $\lambda_2/\lambda_1 = k/l \in \mathbb{Q}$ and $j_\infty R(0,0) = j_\infty S(0,0) = 0$.

In Sect. 2.7 we will prove that in both cases the flat terms R and S can be removed. Unlike the attracting/repelling case the theory of analytic normal forms is not that simple. In starting with an analytic hyperbolic saddle we cannot guarantee the existence of an analytic linearization when $\lambda \in \mathbb{R} \setminus \mathbb{Q}$; neither we can guarantee that f and g are analytic in the case that $\lambda \in \mathbb{Q}$. For more information on this very delicate matter we refer to [148].

2.4 Topological Study of Hyperbolic Saddles

Having the nice C^∞ normal forms (for C^∞-conjugacy) that one can find, as explained in Sect. 2.3, it is now easy to prove that two hyperbolic saddles are

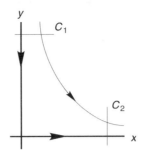

Fig. 2.4. The transition close to a saddle

mutually C^0-conjugate. The proof is essentially based on a time analysis along the X-orbits near the singularity. Because of its intrinsic interest we will start the presentation in a rather general context.

Take any hyperbolic saddle X, of which we suppose that the stable and unstable manifolds are represented, respectively, by the y-axis and the x-axis.

Let C_1 (respectively C_2) be a C^∞ segment, transversely cutting the y-axis (respectively, the x-axis) at $y_0 > 0$ (respectively, $x_0 > 0$). Let s be a regular C^∞ parameter on C_1 with $s > 0$ lying in the first quadrant $\{x > 0, y > 0\}$, as is represented in Fig. 2.4.

We denote by $T(s)$, with $s > 0$, the time an X-orbit needs to pass from C_1 to C_2. We intend to prove that $T(s)$ tends in a monotone way to ∞ for $s \to 0$, by showing that $T'(s) = (dT/ds)(s)$ tends to $-\infty$.

A first observation, which we leave as an exercise, consists in proving that this statement does not depend on the choice of C_1 and C_2, nor on the regular parametrization.

The second observation is that it suffices to consider the normal forms (2.15) and (2.16) taking the flat terms R and S identically zero. In Sect. 2.7 we will indeed remove these terms R and S by means of a C^∞ change of coordinates φ with the property $j_\infty\varphi(0,0) = Id$, a result that will not rely on the Hartman–Grobman Theorem that we are treating now. Because of our first observation, in the normal form we can even consider $C_1 = \{y = 1\}$, $C_2 = \{x = 1\}$ and $s = x$ as a regular parameter on C_1.

In the case that the normal form is linear, which is always the case if the *hyperbolicity ratio* λ_2/λ_1 is irrational, then the statement follows by a simple calculation. Indeed for $(\dot{x} = \lambda x, \dot{y} = -\mu y)$ one immediately gets $T(x) = -\ln x/\lambda$.

For a resonant normal form, which one obtains in case the hyperbolicity ratio is rational, we can proceed as follows.

Consider X given by

$$\begin{aligned} \dot{x} &= x(l + f(x^k y^l)), \\ \dot{y} &= y(-k + g(x^k y^l)), \end{aligned} \qquad (2.17)$$

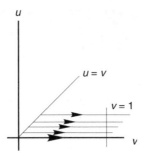

Fig. 2.5. Modified transition close to a saddle

with $l, k \in \mathbb{N}$ with $l, k \geq 1$, both f and g of class C^∞ and $f(0) = g(0) = 0$. If we introduce $v = x^k$ and $u = x^k y^l$, then (2.17) changes into Y given by

$$\dot{v} = v(kl + kf(u)) = v(kl + G(u)),$$
$$\dot{u} = u(kf(u) + lg(u)) = u^2 F(u)(kl + G(u)), \tag{2.18}$$

where F and G are C^∞ and $G(0) = 0$. The curves $\{x = 1\}$ and $\{y = 1\}$ are now given by $\{v = 1\}$ and $\{u = v\}$ respectively; see Fig. 2.5.

We continue by making the necessary calculations for system (2.18), writing x instead of v. We also introduce the system Z given by

$$\dot{x} = x, \qquad \dot{u} = u^2 F(u), \tag{2.19}$$

which, for $x > 0$, can also be expressed as the scalar differential equation

$$x \frac{du}{dx} = u^2 F(u). \tag{2.20}$$

Let the solution of (2.20) be denoted as $u(x_0, x)$, with as initial condition $u(x_0, x_0) = x_0$. The time which the flow of (2.19) spends along such a solution from (x_0, x_0) to $(1, u(x_0, 1))$ is given by $-\ln x_0$, fulfilling the requirements.

We now show that the requirements also hold for (2.18). The orbits of (2.18) are the same as those of (2.19), hence the time function along these orbits, going from (x_0, x_0) to $(1, u(x_0, 1))$ is now given by

$$T(x_0) = \int_{x_0}^{1} \frac{dv}{v(kl + G(u(x_0, v)))},$$

while

$$\frac{dT}{dx_0}(x_0) = -\frac{1}{x_0(kl + G(x_0))} - \int_{x_0}^{1} \frac{G'(u(x_0, v)) \dfrac{\partial u}{\partial x_0}(x_0, v)}{v(kl + G(u(x_0, v)))^2} dv. \tag{2.21}$$

From the variational equation of (2.20) it follows that

$$\frac{\partial u}{\partial x_0}(x_0, v) = \exp\left(\int_{x_0}^{v} \frac{u(x_0, s)}{s}(2F(u(x_0, s)) + u(x_0, s)F'(u(x_0, s)))ds\right).$$

Recall that we work on a region in the (x, u)-plane in which $u \leq x \leq 1$, with $u > 0$ and u sufficiently small. In this region it is easy to see that $u(x_0, x)/x$ stays bounded, and hence that $(\partial/\partial x_0)(u(x_0, x))$ stays bounded for $x_0 \approx 0$. For sufficiently small x_0, the absolute value of the integral in expression (2.21) is hence bounded by

$$C\int_{x_0}^{1} \frac{dv}{v} = -C\ln x_0 = o\left(\frac{1}{x_0}\right), \quad \text{as } x_0 \to 0,$$

for some $C > 0$; we may suppose that $1/2 < 1 + G(x_0) < 3/2$. It clearly follows that $(dT/dx_0)(x_0) \to -\infty$ for $x_0 \to 0$, which is what we wanted to prove.

Knowing the monotonicity of the time between C_1 and C_2 for a hyperbolic saddle, it is now easy to prove that not only a hyperbolic saddle is C^0-conjugate to its linear part, but also that any two hyperbolic linear saddles are mutually C^0-conjugate.

We show how this construction works by considering two hyperbolic saddles X and Y with related transverse sections C_1 and C_2 for X and D_1 and D_2 for Y; see Fig. 2.6. We suppose that $\{x = 0\}$ represents the stable manifold for both X and Y, while $\{y = 0\}$ represents the unstable manifold for both vector fields. Let $r_1 = C_1 \cap \{x = 0\}$, $r_2 = C_2 \cap \{y = 0\}$, $s_1 = D_1 \cap \{x = 0\}$ and $s_2 = D_2 \cap \{y = 0\}$.

The construction is such that we send C_1–D_1 and C_2–D_2, in a way that a piece γ of an X-orbit between $p_1 \in C_1$ and $p_2 \in C_2$ is sent to a piece δ of a Y-orbit between $q_1 \in D_1$ and $q_2 \in D_2$ (i.e., $h(\gamma) = \delta$, $h(p_1) = q_1$ and $h(p_2) = q_2$), taking care that the time to travel from q_1 to q_2 along δ is exactly equal to the time to travel from p_1 to p_2 along γ. Because both for X and Y, the time function tends in a monotone way to infinity, the choice for δ is uniquely determined by γ, at least for p_1 sufficiently close to r_1 (and hence p_2 sufficiently close to r_2). The other points of each γ are sent to points of the related δ by respecting time. This construction clearly leads

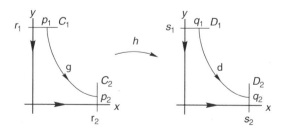

Fig. 2.6. Comparing transitions close to two saddles

to a homeomorphism which extends in a unique way to a homeomorphism on the closure (hence including both axes and the origin). We leave the details as an exercise. This construction clearly provides a C^0-conjugacy.

By this construction we have proven the Hartman–Grobman Theorem for two-dimensional hyperbolic saddles. In the literature different proofs exist for this result. We have chosen to give a proof based on the monotonicity of the time function and we have proven this monotonicity using suitable C^∞ normal forms.

2.5 Semi-Hyperbolic Singularities

2.5.1 Analytic and Smooth Results

As before we work with vector fields that are at least C^∞, and we position the singularity at the origin. By the Jordan Normal Form Theorem, we can write the vector field X as

$$\begin{aligned}
\dot{x} &= -\lambda x + F(x,y), \\
\dot{y} &= G(x,y),
\end{aligned} \qquad (2.22)$$

where $\lambda \neq 0$, F and G are C^∞ functions, and $F(0,0) = G(0,0) = DF(0,0) = DG(0,0) = 0$. We take $\lambda > 0$. A first observation is that there exists a unique invariant one dimensional manifold of the form $(x, y(x))$, where $y(x)$ is a C^∞ function and $y(0) = y'(0) = 0$, on which X is a hyperbolic contraction. We call it the *strong stable manifold*.

Theorem 2.8 *Let X be a C^∞ vector field as in (2.22). Then there exists near $x = 0$ a unique C^∞ function $y = \alpha(x)$, with $\alpha(0) = \alpha'(0) = 0$, whose graph $(x, \alpha(x))$ is invariant under X. If F and G are analytic, then $\alpha(x)$ is analytic.*

Proof. We introduce new variables $(x, y) = (x, xY)$. In these new variables the expression (2.22) changes into

$$\begin{aligned}
\dot{x} &= x(-\lambda + H(x,Y)), \\
\dot{Y} &= \lambda Y + K(x,Y),
\end{aligned}$$

with $H(x,Y) = F(x,xY)/x$ and $K(x,Y) = (G(x,xY) - YF(x,Y))/x$.

The result now immediately follows from the stable manifold theorem for hyperbolic saddles. $\qquad\square$

Depending on whether X is analytic or C^∞ we can find a coordinate change, which is respectively analytic or C^∞, straightening the strong stable manifold to $\{y = 0\}$, and hence changing expression (2.22) into

$$\begin{aligned}
\dot{x} &= -\lambda x + F(x,y), \\
\dot{y} &= yH(x,y).
\end{aligned} \qquad (2.23)$$

We can do even better, by linearizing $X|_{\{y=0\}}$ in a respectively analytic or C^∞ way. A similar result holds on the invariant manifolds of a hyperbolic saddle.

Proposition 2.9 *Let*

$$\dot{u} = \lambda u(1 + g(u)), \tag{2.24}$$

be a differential equation on $\mathbb{R}$, where $\lambda \neq 0$, g is a C^∞ function and $g(0) = 0$. Then there exists a unique change of coordinates

$$u = x(1 + \alpha(x)), \tag{2.25}$$

with α a C^∞ function and $\alpha(0) = 0$, changing (2.24) into $\dot{x} = \lambda x$. If g is analytic, then α is analytic, too.

Proof. The coordinate change (2.25) changes (2.24) into

$$(1 + \alpha(x) + x\frac{d\alpha}{dx}(x))\dot{x} = \lambda x(1 + \alpha(x))(1 + g(x(1 + \alpha(x)))),$$

for which we need

$$1 + \alpha(x) + x\frac{d\alpha}{dx}(x) = (1 + \alpha(x))(1 + g(x(1 + \alpha(x)))).$$

A straightforward calculation shows that $\alpha(x)$ needs to be solution of the differential equation

$$\frac{dy}{dx} = (1 + y)h(x, y), \tag{2.26}$$

with $h(x, y) = g(x(1 + y(x)))/x$.

Equation (2.26) clearly has a unique solution α with $\alpha(0) = 0$. It is C^∞, and is analytic if h is analytic. □

Remark 2.10 The unicity of the linearizing coordinate in Proposition 2.9 comes from the fact that we require $(du/dx)(0) = 1$. We leave it as an exercise to check what happens if we require $(du/dx)(0) = a$ with $a \neq 0$.

To continue the study of (2.23) we will now look for one dimensional invariant manifolds that are tangent to the y-axis, the axis representing the eigenspace of the zero eigenvalue. To better see the link with previous results, we prefer to switch coordinates and continue working with X given by

$$\begin{aligned} \dot{x} &= xG(x, y), \\ \dot{y} &= \lambda y + F(x, y), \end{aligned} \tag{2.27}$$

where $\lambda \neq 0$, F and G are C^∞ functions and $G(0, 0) = DF(0, 0) = 0$.

A first question to deal with concerns the singularities of X in a neighborhood of the origin. They lie on

$$\lambda y + F(x, y) = 0,$$

which by the Implicit Function Theorem is given by the graph of $y = f(x)$ for some C^∞ function f; f is analytic in case F is.

The x-coordinates of the singularities are hence solutions of

$$G(x, f(x)) = 0.$$

From now on we will deal only with the case that this function is not flat, i.e.,

$$G(x, f(x)) = x^n h(x)$$

for some C^∞ function h with $h(0) \neq 0$. We say that X has *non-flat center behavior*.

Flat center behavior for an analytic X implies that $G(x, f(x)) \equiv 0$, in which case X has an analytic one dimensional manifold of singularities. By an analytic change of coordinates, straightening the curve of singularities, system (2.27) can then be transformed into

$$\begin{aligned} \dot{x} &= xy\overline{G}(x,y), \\ \dot{y} &= y(\lambda + \overline{F}(x,y)). \end{aligned} \tag{2.28}$$

It is C^ω equivalent to

$$\begin{aligned} \dot{x} &= yB(x,y), \\ \dot{y} &= \delta y, \end{aligned}$$

for some analytic function B and $\delta = \pm 1$, which is nothing but a regular vector field

$$\begin{aligned} \dot{x} &= B(x,y), \\ \dot{y} &= \delta, \end{aligned}$$

multiplied by the function y. The phase portrait is represented in Fig. 2.7.

We do not wish to consider flat behavior in the C^∞ case. We now continue the study of semi-hyperbolic singularities with non-flat center behavior by

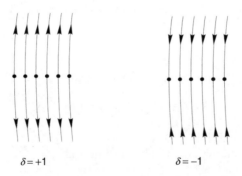

$\delta = +1$ $\delta = -1$

Fig. 2.7. Flows of system (2.28)

first applying the Formal Normal Form Theorem of Sect. 2.1. As usual we take $\lambda > 0$ and divide by λ in the Lie bracket calculations

$$\left[y\frac{\partial}{\partial y}, x^m y^n \frac{\partial}{\partial x}\right] = n x^m y^n \frac{\partial}{\partial x},$$

$$\left[y\frac{\partial}{\partial y}, x^m y^n \frac{\partial}{\partial y}\right] = (n-1) x^m y^n \frac{\partial}{\partial y}.$$

This shows that in the Taylor series of X at 0 all terms can be removed except for the terms $x^m(\partial/\partial x)$ and $x^m y(\partial/\partial y)$.

Using Borel's Theorem, which permits one to write every formal series as the Taylor development of a C^∞ function, we obtain the existence of C^∞ coordinates (x, y) in which we can write X as

$$\dot{x} = x^k f(x) + x\alpha(x, y),$$
$$\dot{y} = y(\lambda + g(x)) + x\beta(x, y), \tag{2.29}$$

where f, g, α and β are C^∞ functions, $j_\infty \alpha(0, 0) = j_\infty \beta(0, 0) = 0$, $f(0) \neq 0$ and $k \geq 2$. Besides the Formal Normal Form Theorem, we have also used the existence of a C^∞ unstable manifold, that we can hence change into $\{x = 0\}$, and we have linearized $X|_{\{x=0\}}$ in a C^∞ way, as made possible by Proposition 2.9.

We can take $f(0) \neq 0$ since we suppose that X has non-flat center behavior; by a linear change in x we can take $f(0) = 1$.

If we write

$$\alpha(x, y) = y\alpha_0(y) + x\overline{\alpha}_0(x, y),$$

and let $\gamma(y)$ be a C^∞ solution of

$$\lambda\gamma'(y) = -\alpha_0(y)\gamma(y),$$
$$\gamma(0) = 1,$$

then $j_\infty(\gamma(y) - 1)(0) = 0$ and $(x, y) = (X\gamma(y), y)$ is a C^∞ coordinate change transforming (2.29) into a similar expression with

$$\alpha(x, y) = x\alpha_1(x, y).$$

Further adaptation of the x-coordinates will lead to

$$\alpha(x, y) = x^k A(x, y),$$

where A is a C^∞ function.

This can be done by inductively showing that

$$\alpha(x, y) = x^l \alpha_l(x, y),$$

where α_l is a C^∞ function and $2 \leq l \leq k + 1$.

We start with an expression

$$\dot{x} = x^k f(x) + x^l (y\alpha_l(y) + x\overline{\alpha}_l(x,y)),$$
$$\dot{y} = y(\lambda + g(x)) + x\beta(x,y),$$

for some $2 \le l \le k$. By using the C^∞ flat function

$$\gamma(y) = -\frac{1}{\lambda} \int_0^y \alpha_l(u)du,$$

we find that $(x + x^l\gamma(y), y)$ are new C^∞ variables for which X gets a similar expression as (2.29) with

$$\alpha(x,y) = x^{l+1}\alpha_{l+1}(x,y)$$

for some C^∞ function α_{l+1}.

As such, the expression that we get after the previous steps is

$$\dot{x} = x^k(f(x) + xA(x,y)),$$
$$\dot{y} = y(\lambda + g(x)) + xB(x,y), \tag{2.30}$$

where f, g, A and B are C^∞ functions, $j_\infty A(0,0) = j_\infty B(0,0) = g(0) = 0$, $f(0) = 1$, $k \ge 2$ and $\lambda > 0$.

We are now ready to prove the existence of (one or more) invariant C^∞ manifolds that are tangent to the x-axis. Such an invariant manifold is called a *center manifold*. We can represent it as the graph $(x, y(x))$ of a C^∞ function, with $y(0) = y'(0) = 0$, and which is solution of the scalar differential equation related to (2.30) and having an expression

$$x^k \frac{dy}{dx} = y(\lambda + h(x) + xC(x,y)) + xD(x,y), \tag{2.31}$$

where h, C and D are C^∞ functions, $j_\infty C(0,0) = j_\infty D(0,0) = h(0) = 0$, $k \ge 2$ and $\lambda > 0$. We now study the required solutions of (2.31) in the next lemma.

Lemma 2.11 *Consider a scalar differential equation*

$$x^k \frac{dy}{dx} = y(\lambda + xF(x,y)) + G(x), \tag{2.32}$$

where F and G are C^∞ functions, $j_\infty G(0) = 0$ and $\lambda \ne 0$. Then there is at least one C^∞ solution $y = \alpha(x)$ with $\alpha : (-\varepsilon, \varepsilon) \to \mathbb{R}$ for some $\varepsilon > 0$ sufficiently small, such that $j_\infty\alpha(0) = 0$. Moreover, whenever $\beta : (0, \varepsilon) \to \mathbb{R}$ (respectively $\beta : (-\varepsilon, 0) \to \mathbb{R}$) is a solution, for some $\varepsilon > 0$, with $\lim_{x \to 0} \beta(x) = 0$, then $\overline{\beta} : [0, \varepsilon) \to \mathbb{R}$ (respectively $\overline{\beta} : (-\varepsilon, 0] \to \mathbb{R}$) defined by $\overline{\beta}(x) = \beta(x)$ for $x \ne 0$ and $\overline{\beta}(0) = 0$, is everywhere C^∞ on its domain of definition and $j_\infty\overline{\beta}(0) = 0$.

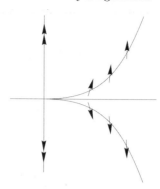

Fig. 2.8. Flow near a center manifold when $\lambda > 0$

Proof. All solutions are clearly C^∞ for $x \neq 0$. We need only deal with $x = 0$. As in Lemma 2.5, we now attach to the differential equation (2.32) the planar system

$$\dot{x} = x^k,$$
$$\dot{y} = y(\lambda + xF(x,y)) + G(x).$$

We first take $\lambda > 0$ and restrict to $\{x \geq 0\}$. If we choose $l \in \mathbb{N}$ with $l \geq 1$ and $c \in \mathbb{R} \setminus \{0\}$, we see that at any point of the curves (x, cx^l), and for $x > 0$ sufficiently small,

$$\left| \frac{dy}{dx}(x, cx^l) \right| > |c|l x^{l-1}. \tag{2.33}$$

See Fig. 2.8.

As such all solutions $y(x)$, for $x > 0$, have 0 as their limit as $x \to 0$. They are clearly C^∞ for $x > 0$ and they are flat at 0 in the sense that $y(x) = O(x^l)$, for any $l \in \mathbb{N}$ with $l \geq 1$.

Since they are solutions of equation (2.32), the flatness property also holds for the derivative $y'(x)$, and by differentiating (2.32) we can prove inductively that it holds for all derivatives $y^{(n)}(x)$. As such, on $x \geq 0$, $\overline{y}(x)$, defined by $y(x)$ for $x > 0$ and by $\overline{y}(0) = 0$, also has to be C^∞ at $x = 0$ and $j_\infty y(0) = 0$.

Second we consider the case $\lambda < 0$, still keeping $x \geq 0$. Property (2.33) still holds along the curves (x, cx^l) for any $l \in \mathbb{N}$ with $l \geq 1$ and any $c \in \mathbb{R} \setminus \{0\}$. It induces a picture as in Fig. 2.9.

It easily follows that there exists at least one solution $y(x)$, for $x > 0$ sufficiently small, such that $\lim_{x \to 0} y(x) = 0$. Based on the same reasoning as in the previous case it follows that $\overline{y}$, with $\overline{y}(x) = y(x)$ for $x > 0$ and $\overline{y}(0) = 0$, is also C^∞ at 0 and $j_\infty \overline{y}(0) = 0$.

The side $\{x \leq 0\}$ can be reduced to $\{x \geq 0\}$ by a reflection in x.   □

The lemma hence proves that there exists at least one C^∞ center manifold for (2.30). It also proves that every center manifold is necessarily C^∞. All center manifolds are also infinitely tangent to the x-axis, and hence mutually infinitely tangent.

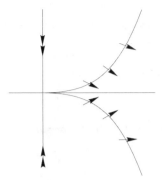

Fig. 2.9. Flow near a center manifold when $\lambda < 0$

We choose any of these center manifolds, given as a graph $\{y = \psi(x)\}$ for some C^∞ function ψ with $j_\infty\psi(0) = 0$. The change of variables $(x, y) \longmapsto (x, y - \psi(x))$ changes expression (2.30) into

$$\dot{x} = x^k(f(x) + xA(x, y)),$$
$$\dot{y} = y(\lambda + g(x) + xB(x, y)), \tag{2.34}$$

for some f, g, A and B that are C^∞ functions, with $\lambda \neq 0$, $f(0) = 1$, $g(0) = 0$ and $j_\infty A(0, 0) = j_\infty B(0, 0) = 0$.

By expression (2.34) it is now easy to see that in case $\lambda < 0$, there is a unique center manifold on the side $\{x \geq 0\}$. The same happens on the side $\{x \leq 0\}$ in case k is odd. If k is even, then in sufficiently small neighborhoods of the origin, every point (x, y) with $x < 0$ lies on a center manifold formed by the orbit through (x, y) together with the origin.

In Sect. 2.7 we will show that the flat terms A and B in expression (2.34) can be removed by a C^∞ coordinate change φ with the property that φ is infinitely tangent to the identity at $(0, 0)$. Since the proof of this fact is independent of what follows, we will take this for granted in our further investigation of the semi-hyperbolic singularities, in the sense that we suppose the semi-hyperbolic vector field X to have the following simple normal form near the origin:

$$\dot{x} = x^k f(x),$$
$$\dot{y} = y(\lambda + g(x)). \tag{2.35}$$

This is the general C^∞ normal form for semi-hyperbolic singularities with a non-flat behavior on the center manifold.

Contrary to the situation for a stable (respectively unstable) manifold, there is no reason to believe that there should always exists an analytic center manifold in the analytic case, not even if the center manifold is unique. We will see a counterexample in a moment. Before concentrating on the analytic case however, we describe a simplification of the center behavior comparable to the linearization on the stable (respectively unstable) manifold as given in Proposition 2.9.

Proposition 2.12 *Let*

$$\dot{u} = u^k(1 + g(u)), \tag{2.36}$$

be a differential equation on $\mathbb{R}$*, where* $k \in \mathbb{N}$ *with* $k \geq 2$*,* g *is a* C^∞ *function and* $g(0) = 0$*. Then there exists a change of coordinate*

$$u = x(1 + \alpha(x)), \tag{2.37}$$

with α *a* C^∞ *function and* $\alpha(0) = 0$*, changing (2.36) into*

$$\dot{x} = x^k(1 + ax^{k-1}), \tag{2.38}$$

for some $a \in \mathbb{R}$*. If* g *is analytic then* α *can be chosen to be analytic.*

Proof. In fact we will not prove the proposition for the indicated normal form (2.38) but for the related one

$$\dot{x} = \frac{x^k}{1 - ax^{k-1}}. \tag{2.39}$$

The latter result, of course, implies the required one. We will treat both the C^∞ and the C^ω results at once, each time writing C^∞ (respectively C^ω). So we start with an equation (2.36) where g is C^∞ (respectively C^ω). We write

$$g(u) = \sum_{i=1}^{N-1} a_i u^i + O(u^N),$$

for some arbitrarily chosen $N \geq k$. By inductively using a coordinate change

$$u = x(1 + \alpha x^l),$$

for a well chosen α, and $1 \leq l \leq k - 2$, we can change g into

$$g(u) = au^{k-1} + O(u^k).$$

The coefficient in front of u^{k-1} cannot be changed, but continuing the induction on l we can, for each $N > k$, get

$$g(u) = au^{k-1} + O(u^N).$$

We can also adapt the power series of g in such a way that for (2.36) we get:

$$\dot{u} = \frac{u^k}{1 - au^{k-1}}(1 + h(u)), \tag{2.40}$$

where h is C^∞ (respectively C^ω) and $h(u) = O(u^N)$. We leave this as an exercise, but take it now for granted.

We do not yet need to take a precise value for N, but we will see that $N \geq 2k - 2$ will permit us a proof of the proposition.

We now try a coordinate change

$$u = x(1 + y(x)), \tag{2.41}$$

where $y(x)$ is C^∞ (respectively C^ω), $y(0) = 0$, and check the necessary condition. Substituting (2.41) into (2.40) we immediately see that y needs to be solution of

$$(1 + y + x\frac{dy}{dx}) = \frac{1 - ax^{k-1}}{1 - ax^{k-1}(1 + y)^{k-1}}(1 + y)^k(1 + O(x^N)),$$

with $O(x^N)$ some function in (x, y) which is C^∞ (respectively C^ω).

A straightforward calculation reduces this to:

$$x\frac{dy}{dx} = y((k - 1) + A(x, y)) + B(x), \tag{2.42}$$

where A and B are C^∞ (respectively C^ω functions), $A(0,0) = 0$, $B(x) = O(x^{2k-2})$.

If we write $y = x^{2k-3}Y$, then (2.42) changes into

$$x\frac{dY}{dx} = Y((2 - k) + A(x, x^{2k-3}Y)) + C(x), \tag{2.43}$$

for a function C which is C^∞ (respectively C^ω) and $C(0) = 0$.

The differential equation (2.43) can be represented by the system

$$\begin{aligned}\dot{x} &= x, \\ \dot{Y} &= Y((2 - k) + A(x, x^{2k-3}Y)) + C(x).\end{aligned} \tag{2.44}$$

This system has a hyperbolic saddle at $(0,0)$ when $k \geq 3$, while it has a semi-hyperbolic singularity for $k = 2$. In any case there exists a C^∞ (respectively C^ω) unstable manifold given by a graph $(x, Y(x))$. The function $y(x) = x^{2k-3}Y(x)$ provides a solution of (2.42). $\qquad\square$

Applying Proposition 2.12 to expression (2.35) we can put a semi-hyperbolic singularity with non-flat center behavior in a very simple normal form for C^∞-equivalence. We first multiply (2.35) by $(\lambda + g(x))^{-1}$, leading to

$$\begin{aligned}\dot{x} &= x^k\overline{f}(x), \\ \dot{y} &= y,\end{aligned} \tag{2.45}$$

for some C^∞ function $\overline{f}$. We can require that $\overline{f}(0) = 1$ by a linear coordinate change. Applying Proposition 2.12 to $x^k\overline{f}(x)$ we can change (2.45) into:

$$\begin{aligned}\dot{x} &= x^k(1 + ax^{k-1}), \\ \dot{y} &= y.\end{aligned} \tag{2.46}$$

This is the final normal form for C^∞ equivalence, in the sense that the coefficient a is an invariant if we want to keep the expression as it is.

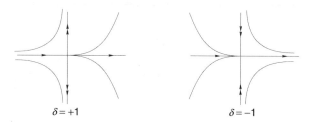

Fig. 2.10. Saddle–nodes

2.5.2 Topological Results

Topological models, also for C^0-conjugacy, are even simpler: there are three hyperbolic cases of, respectively, attracting, repelling and saddle type, besides the saddle-node type singularities

$$\dot{x} = x^2,$$
$$\dot{y} = \delta y,$$

with $\delta = \pm 1$. See Fig. 2.10.

We remark that it is a common practice, in representing local phase portraits of semi-hyperbolic singularities, to use double arrows for the flow on the stable/unstable invariant manifolds, and to use simple arrows on the center manifold. This rule of double arrows is not always used when representing global phase portraits.

To prove our claim about topological models we start with the C^∞ models (for C^∞-conjugacy):

$$\dot{x} = x^k(1 + h(x)),$$
$$\dot{y} = y(\lambda + g(x)), \tag{2.47}$$

where h and g are C^∞, $h(0) = g(0) = 0 \neq \lambda$. The method of proof is a combination of the proofs employed in the topological study of the hyperbolic singularities. The most delicate part deals with the monotonicity of the time function in the hyperbolic sectors. We study it on $\{x \geq 0, y \geq 0\}$ for an expression (2.47) with $\lambda < 0$. Without loss of generality we can take $\lambda = -1$.

We can also restrict to the calculation of the time that orbits spend between sections $C_1 = \{y = a\}$ and $C_2 = \{x = b\}$ (see Fig. 2.11) with $a > 0$ and $b > 0$ and using the x-coordinate as a regular parameter on C_1.

Along the orbits we have

$$dt = \frac{dx}{x^k(1 + h(x))}.$$

So starting at a point (v, a), the time needed to arrive at $\{x = b\}$ is given by

$$T(v) = \int_v^b \frac{dx}{x^k(1 + h(x))}.$$

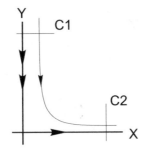

Fig. 2.11. Transition close to a semi-hyperbolic point

For the derivative we get

$$\frac{dT}{dv}(v) = -\frac{1}{v^k(1 + h(v))},$$

which clearly tends to $-\infty$ as $v \to 0$.

Other sectors of saddle-type (hyperbolic sectors) can be treated similarly, while parabolic sectors rely on a method similar to that used in the C^0 study of attracting/repelling hyperbolic singularities. We leave the details of the elaboration as an exercise.

2.5.3 More About Center Manifolds

To trace a center manifold in a specific example, it is often necessary to calculate its Taylor development. As we can expect from the previous observations, such a Taylor development is uniquely determined. To formally prove this statement we can use the simple normal forms (for C^∞–equivalence) that we have obtained in (2.46). They have a related scalar differential equation

$$(1 + ax^{k-1})x^k\frac{dy}{dx} = y$$

for which it is clear that the only formal solution $\sum\limits_{i=1}^{\infty} a_i x^i$ is given by $a_i = 0$ for all i.

To finish this section about semi-hyperbolic singularities, we add some more detailed information about analytic vector fields. We first provide an example (due to Euler) in which there is no analytic center manifold,

$$\begin{aligned} \dot{x} &= x^2, \\ \dot{y} &= y - x. \end{aligned} \tag{2.48}$$

In view of the previous observation, it suffices to prove that the unique formal center manifold is not convergent.

To that end, consider a formal sum

$$\hat{y} = \sum_{n=1}^{\infty} y_n x^n,$$

such that $\hat{y}$ is a formal center manifold of (2.48), i.e.,

$$x^2 \frac{d\hat{y}}{dx} = \hat{y} - x.$$

This equation leads to the easy recursion $\{ny_n - y_{n+1}, y_1 = 1\}$. Hence $y_n = n!$ is the unique solution.

The formal expansion

$$\hat{y} = \sum_{n=1}^{\infty} n! \, x^n,$$

is clearly divergent at all non-zero values of x.

Although not convergent, the series in the example is not too divergent. It is Gevrey-1 in x of type 1.

Definition 2.13 *A formal power series*

$$\hat{f}(x) = \sum_{n=0}^{\infty} f_n x^n,$$

is Gevrey-$1/\sigma$ in x of type A if there exist positive constants A, C and α such that

$$|f_n| \leq CA^{n/\sigma} \, \Gamma\left(\alpha + \frac{n}{\sigma}\right),$$

for all $n \geq 0$; Γ is the traditional gamma function.

For semi-hyperbolic singularities of analytic vector fields there is the following result, which we state without proof.

Proposition 2.14 *Let X be an analytic vector field having an expression (2.27) whose center behavior starts with the leading term x^m, or more precisely, such that*

$$G(x, f(x)) = x^m + o(x^m).$$

Then there exists a unique formal center manifold $y = \hat{f}(x)$, and this unique formal power series is Gevrey-$1/m$ in x of type $m/|\lambda|$.

The n th coefficient of a Gevrey-$1/\sigma$ series is roughly of the order given by $|Ax^\sigma|^{n/\sigma} \, \Gamma(\alpha + n/\sigma)$. It should be clear that for small $|x|$, the factor $|A^{n/\sigma} x^n|$ competes against the factor $\Gamma(\alpha + n/\sigma)$. This explains the behavior of the upper bound: it is decreasing for small n, up to a point $N(x)$, from which the upper bound rapidly tends to infinity. The order $N(x)$ at which the upper bound is the smallest can be approximated, using Stirling's formula, by

$$N(x) = \left[\frac{\sigma}{A|x|^\sigma}\right],$$

$$\left(|Ax^\sigma|^{n/\sigma}\,\Gamma(\alpha + n/\sigma)\right)_{n=N(x)} = O\left(\exp\left(-\frac{1}{A|x|^\sigma}\right)\right).$$

One can introduce now the notion of a *Gevrey function*: we say that a function is Gevrey-$1/\sigma$ asymptotic to a power series $\hat{f}$ if its N th order Taylor development coincides with the N th order truncation $\hat{f}_N$ of $\hat{f}$, and if $|f - \hat{f}_N|$ is of the order $|Ax^\sigma|^{N/\sigma}\,\Gamma(\alpha + N/\sigma)$, for all N.

One can now approximate a function f by its N th order Taylor development, and by choosing the cut-off point equal to $N(x)$, one finds an approximation for which the error is exponentially small:

$$\left|f(x) - \hat{f}_{N(x)}\right| = O\left(\exp\left(-\frac{1}{A|x|^\sigma}\right)\right).$$

The finite sum $\hat{f}_{N(x)}$ is called the *summation to the least term*.

Let f be a function whose graph represents a center manifold of a vector field of the earlier type. It is known that such a function is Gevrey-asymptotic to its formal counterpart. The approximation earlier is the best one can hope for, and this is reflected in the fact that these center manifolds are in general non-unique and can differ by a function of the order $O(\exp(-1/A|x|^\sigma))$.

2.6 Summary on Elementary Singularities

Here we state two specific and practical theorems which summarize the previous results and which are very helpful in determining the behavior of elementary singular points.

We characterize the local phase portrait at a hyperbolic singular point in the following theorem. Since we are mainly interested in polynomial differential systems, we present the theorem for analytic systems.

Theorem 2.15 (Hyperbolic Singular Points Theorem) *Let $(0,0)$ be an isolated singular point of the vector field X, given by*

$$\begin{aligned}
\dot{x} &= ax + by + A(x,y),\\
\dot{y} &= cx + dy + B(x,y),
\end{aligned} \tag{2.49}$$

where A and B are analytic in a neighborhood of the origin with $A(0,0) = B(0,0) = DA(0,0) = DB(0,0) = 0$. Let λ_1 and λ_2 be the eigenvalues of the linear part $DX(0)$ of the system at the origin. Then the following statements hold.

(i) If λ_1 and λ_2 are real and $\lambda_1\lambda_2 < 0$, then $(0,0)$ is a saddle (see Fig. 2.12(a)). If we denote by E_1 and E_2 the eigenspaces of respectively λ_1 and λ_2, then

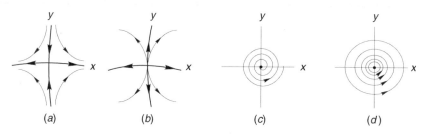

Fig. 2.12. Phase portraits of non–degenerate singular points

one can find two invariant analytic curves, tangent respectively to E_1 and E_2 at 0, on one of which points are attracted towards the origin, and on one of which points are repelled away from the origin. On these invariant curves X is C^ω-linearizable. There exists a C^∞ coordinate change transforming (2.49) into one of the following normal forms:

$$\dot{x} = \lambda_1 x,$$
$$\dot{y} = \lambda_2 y,$$

in the case $\lambda_2/\lambda_1 \in \mathbb{R} \setminus \mathbb{Q}$, and

$$\dot{x} = x(\lambda_1 + f(x^k y^l)),$$
$$\dot{y} = y(\lambda_2 + g(x^k y^l)),$$

in the case $\lambda_2/\lambda_1 = -k/l \in \mathbb{Q}$ with $k, l \in \mathbb{N}$ and where f and g are C^∞ functions. All systems (2.49) are C^0-conjugate to

$$\dot{x} = x,$$
$$\dot{y} = -y.$$

(ii) *If λ_1 and λ_2 are real with $|\lambda_2| \geq |\lambda_1|$ and $\lambda_1 \lambda_2 > 0$, then $(0,0)$ is a node (see Fig. 2.12(b)). If $\lambda_1 > 0$ (respectively < 0) then it is repelling or unstable (respectively attracting or stable). There exists a C^∞ coordinate change transforming (2.49) into*

$$\dot{x} = \lambda_1 x,$$
$$\dot{y} = \lambda_2 y,$$

in case $\lambda_2/\lambda_1 \notin \mathbb{N}$, and into

$$\dot{x} = \lambda_1 x,$$
$$\dot{y} = \lambda_2 y + \delta x^m,$$

for some $\delta = 0$ or 1, in case $\lambda_2 = m\lambda_1$ with $m \in \mathbb{N}$ and $m \geq 1$. All systems (2.49) are C^0-conjugate to

$$\dot{x} = \delta x,$$
$$\dot{y} = \delta y,$$

with $\delta = \pm 1$ and $\lambda_1 \delta > 0$.

(iii) If $\lambda_1 = \alpha + i\beta$ and $\lambda_2 = \alpha - i\beta$ with $\alpha, \beta \neq 0$, then $(0,0)$ is a "strong" focus (see Fig. 2.12(c)). If $\alpha > 0$ (respectively < 0), it is repelling or unstable (respectively attracting or stable). There exists a C^∞ coordinate change transforming (2.49) into

$$\dot{x} = \alpha x + \beta y,$$
$$\dot{y} = - \beta x + \alpha y.$$

All systems (2.49) are C^0-conjugate to

$$\dot{x} = \delta x,$$
$$\dot{y} = \delta y,$$

with $\delta = \pm 1$ and $\alpha \delta > 0$.

(iv) If $\lambda_1 = i\beta$ and $\lambda_2 = -i\beta$ with $\beta \neq 0$, then $(0,0)$ is a linear center, topologically, a "weak" focus or a center (see Figs. 2.12(c) and (d)).

Remark 2.16 The proofs concerning the C^∞- and C^0-normal forms rely on the results presented in Sect. 2.7 about the "removal of flat terms." In fact we could have given proofs of the C^0 results not depending on it, but we decided not to do so, since they can be found in many other texts.

Remark 2.17 The denomination *strong focus* in (iii) is used in describing singularities of which the linear part $DX(0)$ is already a focus, while the denomination *weak focus* is used in the case $DX(0)$ is a center. All results presented in this theorem are proven in this chapter (including Sect. 2.7), except for the last statement of (iv), namely that X is topologically a focus or a center. This proof will be given in Chap. 3. We added it here only for sake of completeness.

Remark 2.18 The origin of system (2.49) is a hyperbolic singular point in cases (i), (ii) and (iii) of Theorem 2.15.

Theorem 2.15 can also be stated in terms of the determinant (det), the trace (tr) and the discriminant (dis) of the linear part at the singular point. Thus, statement (i) corresponds to det < 0; statement (ii) corresponds to det > 0, tr $\neq 0$ and dis ≥ 0; statement (iii) corresponds to det > 0, tr $\neq 0$ and dis < 0; and statement (iv) corresponds to det > 0, tr $= 0$ and dis < 0.

We characterize the local phase portrait at a semi-hyperbolic singular point in the following theorem, for whose proof we again rely on Sect. 2.7.

Theorem 2.19 (Semi-Hyperbolic Singular Points Theorem) *Let* $(0,0)$ *be an isolated singular point of the vector field* X *given by*

$$\dot{x} = A(x,y),$$
$$\dot{y} = \lambda y + B(x,y), \tag{2.50}$$

where A *and* B *are analytic in a neighborhood of the origin with* $A(0,0) = B(0,0) = DA(0,0) = DB(0,0) = 0$ *and* $\lambda > 0$. *Let* $y = f(x)$ *be the solution of the equation* $\lambda y + B(x,y) = 0$ *in a neighborhood of the point* $(0,0)$, *and suppose that the function* $g(x) = A(x, f(x))$ *has the expression* $g(x) = a_m x^m + o(x^m)$, *where* $m \geq 2$ *and* $a_m \neq 0$. *Then there always exists an invariant analytic curve, called the strong unstable manifold, tangent at* 0 *to the* y–*axis, on which* X *is analytically conjugate to*

$$\dot{y} = \lambda y;$$

it represents repelling behavior since $\lambda > 0$. *Moreover the following statements hold.*

(i) If m *is odd and* $a_m < 0$, *then* $(0,0)$ *is a topological saddle (see Fig. 2.13(a)). Tangent to the* x-*axis there is a unique invariant* C^∞ *curve, called the center manifold, on which* X *is* C^∞-*conjugate to*

$$\dot{x} = -x^m(1 + ax^{m-1}), \tag{2.51}$$

for some $a \in \mathbb{R}$. *If this invariant curve is analytic, then on it* X *is* C^ω-*conjugate to* (2.51).
System X *is* C^∞-*conjugate to*

$$\dot{x} = -x^m(1 + ax^{m-1}),$$
$$\dot{y} = \lambda y,$$

and is C^0-*conjugate to*

$$\dot{x} = -x,$$
$$\dot{y} = y.$$

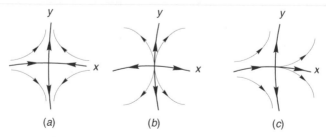

Fig. 2.13. Phase portraits of semi-hyperbolic singular points

(ii) *If m is odd and $a_m > 0$, then $(0,0)$ is a unstable topological node (see Fig. 2.13(b)). Every point not belonging to the strong unstable manifold lies on an invariant C^∞ curve, called a center manifold, tangent to the x-axis at the origin, and on which X is C^∞-conjugate to*

$$\dot{x} = x^m(1 + ax^{m-1}), \tag{2.52}$$

for some $a \in \mathbb{R}$. All these center manifolds are mutually infinitely tangent to each other, and hence at most one of them can be analytic, in which case X is C^ω-conjugate on it to (2.52).
System X is C^∞-conjugate to

$$\dot{x} = x^m(1 + ax^{m-1}),$$
$$\dot{y} = \lambda y,$$

and is C^0-conjugate to

$$\dot{x} = x,$$
$$\dot{y} = y.$$

(iii) *If m is even, then $(0,0)$ is a saddle-node, that is, a singular point whose neighborhood is the union of one parabolic and two hyperbolic sectors (see Fig. 2.13(c)). Modulo changing x into $-x$, we suppose that $a_m > 0$. Every point to the right of the strong unstable manifold (side $x > 0$) lies on an invariant C^∞ curve, called a center manifold, tangent to the x-axis at the origin, and on which X is C^∞-conjugate to*

$$\dot{x} = x^m(1 + ax^{m-1}), \tag{2.53}$$

for some $a \in \mathbb{R}$. All these center manifolds coincide on the side $x \leq 0$ and are hence infinitely tangent at the origin. At most one of these center manifolds can be analytic, in which case X is C^ω-conjugate on it to (2.53).
System X is C^∞-conjugate to

$$\dot{x} = x^m(1 + ax^{m-1}),$$
$$\dot{y} = \lambda y,$$

and is C^0-conjugate to

$$\dot{x} = x^2,$$
$$\dot{y} = y.$$

Remark 2.20 The case $\lambda < 0$ can be reduced to $\lambda > 0$ by changing X into $-X$. Besides the results on semi-hyperbolic singularities with non-flat center behavior, as stated in Theorem 2.19, we also recall that in case $g(x) = A(x, f(x))$, as defined in Theorem 2.19, is identically zero, then there exists an analytic curve consisting of singularities.

We leave it as an exercise to prove that, in that case, X is C^ω-conjugate to

$$\dot{x} = 0,$$
$$\dot{y} = yh(x),$$

for some C^ω function h with $h(0) \neq 0$. It is hence C^ω-equivalent to

$$\dot{x} = 0,$$
$$\dot{y} = \delta y,$$

with $\delta = \pm 1$.

2.7 Removal of Flat Terms

2.7.1 Generalities

In the previous sections we have left three problems concerning removal of flat terms in order to get the required C^∞ normal forms. In each case we have two C^∞ vector fields X and $X + Y$, with the property that $j_\infty Y(0) = 0$; both vector fields are defined in a neighborhood of $(0,0)$. We want to find a local C^∞ diffeomorphism $\varphi : \mathbb{R}^2 \to \mathbb{R}^2$ with $\varphi(0) = 0$, $\varphi_* X = Y$ and $j_\infty(\varphi - I)(0,0) = 0$, for I the identity isomorphism; we call it a *near-identity transformation*.

To that end we will use the so called *homotopic method* (see, e.g., [91]). We introduce a parameter $\tau \in [0, 1]$, consider $X_\tau = X + \tau Y$, and look for a family φ_τ of near-identity transformations with the property that

$$(\varphi_\tau)_* X = X + \tau Y. \tag{2.54}$$

It now suffices to look for a τ-dependent vector field Z_τ with $j_\infty(Z_\tau)(0,0) = 0$ and satisfying

$$[X + \tau Y, Z_\tau] = Y, \tag{2.55}$$

where $[\cdot, \cdot]$ denotes the usual Lie-bracket operation that we already used in Sect. 2.1. We prove that this is sufficient in the next lemma, in which we use $z = (x, y)$, $Z(z, \tau) = Z_\tau(z)$ and $\varphi(z, \tau) = \varphi_\tau(z)$.

Lemma 2.21 *Let Z_τ be a solution to equation (2.55). Then φ_τ defined by*

$$\frac{\partial \varphi}{\partial \tau}(z, \tau) = Z(\varphi(z, \tau), \tau), \tag{2.56}$$

is a solution of equation (2.54).

Proof. It clearly follows from $j_\infty(Z_\tau)(0,0) = 0$ that $j_\infty(\varphi - I)(0,0) = 0$. We furthermore have to check condition (2.54), which we can write as

$$\begin{pmatrix} \dfrac{\partial \varphi_1}{\partial x}(z,\tau) & \dfrac{\partial \varphi_1}{\partial y}(z,\tau) \\ \dfrac{\partial \varphi_2}{\partial x}(z,\tau) & \dfrac{\partial \varphi_2}{\partial y}(z,\tau) \end{pmatrix} \begin{pmatrix} X_1(z) \\ X_2(z) \end{pmatrix} = \begin{pmatrix} (X_1 + \tau Y_1)(\varphi(z,\tau)) \\ (X_2 + \tau Y_2)(\varphi(z,\tau)) \end{pmatrix}. \qquad (2.57)$$

To find the relation with equation (2.55), we differentiate (2.57) with respect to τ

$$\begin{pmatrix} Y_1(\varphi(z,\tau)) \\ Y_2(\varphi(z,\tau)) \end{pmatrix} = \begin{pmatrix} \dfrac{\partial}{\partial x}(\dfrac{\partial \varphi_1}{\partial \tau})(z,\tau) & \dfrac{\partial}{\partial y}(\dfrac{\partial \varphi_1}{\partial \tau})(z,\tau) \\ \dfrac{\partial}{\partial x}(\dfrac{\partial \varphi_2}{\partial \tau})(z,\tau) & \dfrac{\partial}{\partial y}(\dfrac{\partial \varphi_2}{\partial \tau})(z,\tau) \end{pmatrix} \begin{pmatrix} X_1(z) \\ X_2(z) \end{pmatrix} -$$

$$\begin{pmatrix} \dfrac{\partial(X_1 + \tau Y_1)}{\partial x}(\varphi(z,\tau)) & \dfrac{\partial(X_1 + \tau Y_1)}{\partial y}(\varphi(z,\tau)) \\ \dfrac{\partial(X_2 + \tau Y_2)}{\partial x}(\varphi(z,\tau)) & \dfrac{\partial(X_2 + \tau Y_2)}{\partial y}(\varphi(z,\tau)) \end{pmatrix} \begin{pmatrix} (\dfrac{\partial \varphi_1}{\partial \tau})(z,\tau) \\ (\dfrac{\partial \varphi_2}{\partial \tau})(z,\tau) \end{pmatrix}. \qquad (2.58)$$

In this expression we can write

$$\frac{\partial \varphi_i}{\partial \tau}(z,\tau) = Z_{\tau,i}(\varphi(z,\tau)), \qquad (2.59)$$

and by differentiating (2.59) we get

$$\begin{pmatrix} \dfrac{\partial}{\partial x}(\dfrac{\partial \varphi_1}{\partial \tau})(z,\tau) & \dfrac{\partial}{\partial y}(\dfrac{\partial \varphi_1}{\partial \tau})(z,\tau) \\ \dfrac{\partial}{\partial x}(\dfrac{\partial \varphi_2}{\partial \tau})(z,\tau) & \dfrac{\partial}{\partial y}(\dfrac{\partial \varphi_2}{\partial \tau})(z,\tau) \end{pmatrix} =$$

$$\begin{pmatrix} \dfrac{\partial Z_{\tau,1}}{\partial x}(\varphi(z,\tau)) & \dfrac{\partial Z_{\tau,1}}{\partial y}(\varphi(z,\tau)) \\ \dfrac{\partial Z_{\tau,2}}{\partial x}(\varphi(z,\tau)) & \dfrac{\partial Z_{\tau,2}}{\partial y}(\varphi(z,\tau)) \end{pmatrix} \begin{pmatrix} \dfrac{\partial \varphi_1}{\partial x}(z,\tau) & \dfrac{\partial \varphi_1}{\partial y}(z,\tau) \\ \dfrac{\partial \varphi_2}{\partial x}(z,\tau) & \dfrac{\partial \varphi_2}{\partial y}(z,\tau) \end{pmatrix}.$$

This relation, together with (2.57) and (2.59), shows that (2.58) is the same as (2.55). As such (2.57) follows from (2.55) by integration, i.e., solving (2.56). $\qquad \square$

Remark 2.22 If instead of having $j_\infty Y(0,0) = 0$, we had only $j_\infty Y(x,0) = 0$ (respectively $j_\infty Y(0,y) = 0$), then clearly the "homotopic method" would imply that $j_\infty(\varphi - I)(x,0) = 0$ (respectively $j_\infty(\varphi - I)(0,y) = 0$). It suffices in equation (2.55) to look for a solution which has the property that $j_\infty Z_\tau(x,0) = 0$ (respectively $j_\infty Z_\tau(0,y) = 0$).

Taking into account the previous remark, and knowing that we will need it in the sequel, we will from now on suppose that $j_\infty Y(z) = 0$ for all $z \in M$, where M is either $\{(0,0)\}$, $\{x = 0\}$ or $\{y = 0\}$.

For solving (2.55) in the traditional way, it is important for $X + \tau Y$ to be a τ-family of complete vector fields, or at least, vector fields whose orbits exist for $t \to \infty$. For the moment we take this for granted and denote the orbits by

$\gamma(z, \tau, t)$. We also suppose that, uniformly in $\tau \in [0, 1]$ and for initial values z in some ε-neighborhood of M, all orbits tend to M in an exponentially fast way

$$\|\gamma(z, \tau, t)\|_M \le Ce^{-\lambda t}\|z\|_M, \tag{2.60}$$

for some $\lambda > 0$, $C > 0$, and where $\|\cdot\|_M$ represents the distance to M. Under some extra (rather mild) conditions, we will see that (2.55) can be solved by defining

$$Z_\tau(z) = -\int_0^\infty (F(\gamma(z, \tau, t)))^{-1}(Y(\gamma(z, \tau, t)))dt \quad \text{for} \quad z \notin M,$$
$$Z_\tau(z) = 0 \quad \text{for} \quad z \in M, \tag{2.61}$$

where $F(\gamma(z, \tau, t))$ represents, along an orbit $\gamma(z, \tau, t)$, the matrix solution of the related variational equation

$$\frac{d}{dt}F(\gamma(z, \tau, t)) = [D_z(X + \tau Y)(\gamma(z, \tau, t))](F(\gamma(z, \tau, t))),$$
$$F(\gamma(z, \tau, 0)) = I. \tag{2.62}$$

From now on we suppose that Z_τ is defined as in (2.61). In order to prove that Z_τ is C^∞, with $j_\infty Z_\tau(z) = 0$, for all $z \in M$, we will have to impose some extra conditions on $X + \tau Y$. Once these conditions are proven, it is easy to check that Z_τ satisfies the Lie-bracket relation required in (2.55), since (2.55) can clearly be written as

$$\frac{d}{dt}(Z_\tau(\gamma(z, \tau, t))) = (D_z(X + \tau Y)(\gamma(z, \tau, t)))(Z_\tau(\gamma(z, \tau, t))) + Y(\gamma(z, \tau, t)),$$

from which it is clear that Z_τ, as defined in (2.62), is a solution.

To prove the smoothness of Z_τ and the flatness at points $z \in M$, it will turn out to be necessary to find good bounds on the norm of $(F(\gamma(z, \tau, t)))^{-1}$ as well as on the norms of successive derivatives of $(F(\gamma(z, \tau, t)))^{-1}$ with respect to (z, τ). Since we know that

$$\|(F(\gamma(z, \tau, t)))^{-1}\| = \frac{\|F(\gamma(z, \tau, t))\|}{|\det(F(\gamma(z, \tau, t)))|}, \tag{2.63}$$

it clearly shows that we essentially need to control the value of $\|F(\gamma(z, \tau, t))\|$ and $|\det(F(\gamma(z, \tau, t)))|$. Upper bounds on $\|F(\gamma(z, \tau, t))\|$ will be obtained from (2.62) by considering appropriate upper bounds on the expression $D_z(X + \tau Y)(\gamma(z, \tau, t))$. To find strictly positive lower bounds on the determinant $|\det(F(\gamma(z, \tau, t)))|$ we will use Liouville's formula stating that

$$|\det(F(\gamma(z, \tau, t)))| = \exp\left(\int_0^\infty \text{div}(D_z(X + \tau Y)(\gamma(z, \tau, t)))dt\right). \tag{2.64}$$

This shows the importance that $\text{div}(D_z(X + \tau Y)(\gamma(z, \tau, t)))$ has in this calculation. We now treat the specific cases encountered in the previous sections of this chapter.

2.7.2 Hyperbolic Case

The simplest case is the attracting one, so we start with it.

The divergence at $(0,0)$ is a strictly negative number, so on some neighborhood V of the origin we can suppose that the divergence stays bounded, implying, in view of (2.64), the existence of some $\nu > 0$ such that

$$|\det(F(\gamma(z,\tau,t)))| \geq e^{-\nu t}. \tag{2.65}$$

Together with (2.62) this proves the existence of some $\mu > 0$ with the property that

$$\|(F(\gamma(z,\tau,t)))^{-1}\| \leq e^{\mu t}. \tag{2.66}$$

Taking into account (2.60) with $M = \{(0,0)\}$, and the fact that $j_\infty Y(0,0) = 0$, we see at once that the integral in (2.61) converges and that

$$\lim_{z \to (0,0)} Z_\tau(z) = 0.$$

Concerning the different derivatives $(\partial^{j+k+l} Z_{\tau,i})/(\partial x^j \partial y^k \partial \tau^l)$, we can either rely on uniform expressions of these derivatives (or their related tensors), or we can work inductively. In any case, from (2.62)–(2.64), and using the fact that Y is C^∞ with $j_\infty Y(0,0)$, similar estimates as earlier will permit us to prove that Z_τ is C^∞ on $V \setminus \{(0,0)\}$ and that all derivatives tend to zero for z tending to the origin. We do not work it out, but leave it as an exercise. In any case, the flatness at the origin of all derivatives implies that Z_τ is also C^∞ at the origin, with of course $j_\infty Z_\tau(0,0) = 0$.

The repelling case can be reduced to the attracting one by reversing time, hence looking at $-(X + \tau Y)$.

The treatment of the saddle case is more subtle, since we do not have property (2.60) with respect to the origin. The idea now is to proceed in two steps, relying on the following lemma that we state without proof. The proof can be obtained by using a partition of unity in combination with a polar blow-up at the origin. We do not work this out but refer to [91].

Lemma 2.23 *Let $f : \mathbb{R}^2 \to \mathbb{R}$ be a C^∞ function defined on a neighborhood V of $(0,0)$ with $j_\infty f(0) = 0$. Then there is a decomposition $f = g + h$ such that both g and h are C^∞ functions on V with $j_\infty g(p) = 0$ and $j_\infty h(q) = 0$ for all $p = (x,0)$ or $q = (0,y)$ in V.*

We consider a saddle point with expression (2.4) of Sect. 2.3, implying that $\{y = 0\}$ represents the stable manifold while $\{x = 0\}$ represents the unstable one.

Based on Lemma 2.23 we decompose the flat terms Y as $Y = Y_s + Y_u$ with respectively $j_\infty Y_u(x,0) = 0$ and $j_\infty Y_s(0,y) = 0$ and we apply the homotopic method in order to subsequently remove Y_s and Y_u.

We start by removing Y_s.

We recall that the family $X + \tau Y$ with $Y = Y_s$ has an expression

$$\dot{x} = x(-\lambda_1 + A(x,y) + \tau Y_1),$$
$$\dot{y} = y(\lambda_2 + B(x,y) + \tau Y_2),$$

with $\lambda_1 > 0$, $\lambda_2 > 0$, $A(0,0) = B(0,0) = 0$ and $j_\infty Y_1(0,y) = j_\infty Y_2(0,y) = 0$. We can chose a neighborhood $V = V_1 \times V_2$ of the origin in which we have

$$-\lambda_1 + A(x,y) + \tau Y_1 < -\frac{\lambda_1}{2}.$$

We now consider a C^∞ function $\alpha(y)$ with compact support inside V_2 and equal to 1 in a neighborhood of 0. We change our focus to the study of

$$\dot{x} = x(-\lambda_1 + \alpha(y)(A(x,y) + \tau Y_1)),$$
$$\dot{y} = y\alpha(y)(\lambda_2 + B(x,y) + \tau Y_2).$$

For this vector field, it is clear that there exists an ε-neighborhood of $M = \{x = 0\}$ on which not only all orbits exist for $t \to +\infty$, but also condition (2.60) holds with respect to M.

From now on it will be rather unimportant to know the exact behavior of $X + \tau Y$ on M. So in view of an application of the same method to the semi-hyperbolic case, as we will do in the next subsection, we prove the following proposition, which is of independent interest.

Proposition 2.24 *Consider the differential equation*

$$\dot{x} = x(f(x,y) + \tau Y_1),$$
$$\dot{y} = g(x,y) + \tau Y_2,$$
(2.67)

on some ε-neighborhood V of $M = \{x = 0\}$, where f, g, Y_1 and Y_2 are C^∞ functions and $j_\infty Y_1(0,y) = j_\infty Y_2(0,y) = 0$. Suppose moreover that g is supported in $C \times \mathbb{R}$ for a compact set $C \subset M$, and that $f(x,y) + \tau Y_1 < -\lambda < 0$, for some $\lambda > 0$, $\tau \in [0,1]$ and $(x,y) \in V$. Then the vector field Z_τ, as defined in (2.61), with respect to $X + \tau Y$ as in (2.65), is a C^∞ vector field satisfying $j_\infty Z_\tau(0,y) = 0$ for all y.

Proof. The proof is essentially similar to the one we have given in studying the attracting hyperbolic case, i.e., $M = \{(0,0)\}$. Of course, all estimates need to be made using $\|\cdot\|_M$ instead of $\|\cdot\|$, which is possible because of the fact that $j_\infty Y_1$ and $j_\infty Y_2$ are infinitely flat along M. Again the conditions on (2.67) imply that the divergence of $X + \tau Y$ stays bounded, implying condition (2.65) on $|\det(F(\gamma(z,\tau,t)))|$ and condition (2.66) on $\|(F(\gamma(z,\tau,t)))^{-1}\|$. The proof is then completely analogous to the one in the case $M = \{(0,0)\}$ and we leave it as an exercise.   □

After removing Y_s we can remove Y_u in exactly the same way, by reversing time and working with respect to $M = \{y = 0\}$.

2.7.3 Semi-Hyperbolic Case

To remove flat terms in the semi-hyperbolic case, we start with the following expression for X

$$\dot{x} = - x^k(1 + \overline{f}(x) + yw_1(x, y)),$$
$$\dot{y} = y(\lambda + \overline{g}(x) + yw_2(x, y)),$$

$$(2.68)$$

where $\lambda \neq 0$, $\overline{f}$ and $\overline{g}$ are C^∞ functions and $\overline{f}(0) = \overline{g}(0) = j_\infty w_1(0, 0) = j_\infty w_2(0, 0) = 0$. We claim that there exists a C^∞ near-identity change of coordinates at $(0, 0)$ adapting (2.68) in a way that $j_\infty w_i(0, x) = 0$ for $i = 1, 2$. In this way, the flat perturbation $w = (w_1, w_2)$ can be considered to be flat along $\{y = 0\}$. Since (2.68) is exponentially contracting towards $\{y = 0\}$ (up to changing X into $-X$), Proposition 2.24 will permit us to completely remove it.

To prove the claim we work with Taylor expansions in y, having as coefficients C^∞ functions in x. The proof relies first on an induction procedure on the powers of y^n followed by an application of Borel's Theorem for representing such semi-formal developments by genuine C^∞ functions in two variables (see for example [19] and [115]).

We now work out the induction procedure on n. Suppose that we can write (2.68) as

$$\dot{x} = - x^k(1 + \overline{f}(x) + y^n w_{1,n}(x, y)),$$
$$\dot{y} = y(\lambda + \overline{g}(x) + y^n w_{2,n}(x, y)),$$

$$(2.69)$$

for some $n \geq 1$. We will now show how transformations of the form $\mathrm{Id} + O(y^n)$ can change (2.69) into a similar expression with n replaced by $n + 1$.

We first try

$$(x, y) = (X(1 + X^{k-1}y^n\alpha(X)), y),$$

$$(2.70)$$

proving the existence of a flat C^∞ function $\alpha(X)$ that permits us to change $y^n w_{1,n}(x, y)$ into $y^{n+1}\overline{w}_{1,n}(X, y)$ for some C^∞ function $\overline{w}_{1,n}(X, y)$ such that $j_\infty \overline{w}_{1,n}(0, 0) = 0$. We write $f_n(x) = w_{1,n}(x, 0)$.

Substituting (2.70) into the first equation of (2.69), and writing x instead of X, gives

$$\dot{x}(1 + kx^{k-1}y^n\alpha(x) + x^k y^n\alpha'(x)) + nx^k y^{n-1}\alpha(x)\dot{y} =$$
$$- x^k(1 + x^{k-1}y^n\alpha(x))^k(1 + \overline{f}(x(1 + x^{k-1}y^n\alpha(x)))) +$$
$$y^n f_n(x(1 + x^{k-1}y^n\alpha(x))) + O(y^{n+1})).$$

Working modulo $O(y^{n+1})$, this gives the following equation:

$$\dot{x}(1 + kx^{k-1}y^n\alpha(x) + x^k y^n\alpha'(x)) =$$
$$- x^k \left[(1 + kx^{k-1}y^n\alpha(x))(1 + \overline{f}(x) + x^k y^n\alpha(x)\overline{f}'(x) + y^n f_n(x)) + \right.$$
$$\left. ny^n\alpha(x)(\lambda + \overline{g}(x)) \right].$$

A straightforward calculation of the terms in y^n shows that we have to solve the following equation:

$$(1 + \overline{f}(x))x^k \alpha'(x) = (n(\lambda + \overline{g}(x)) + x^k \overline{f}'(x))\alpha(x) + f_n(x).$$

Seeing that $j_\infty f_n(0) = 0$, Lemma 2.11 guarantees the existence of a C^∞ solution $\alpha(x)$ with $j_\infty \alpha(0) = 0$.

Substituting (2.70) into the second equation of (2.69) does not change the form of it.

Coordinate change (2.70) hence reduces (2.69) to some expression

$$\begin{aligned}
\dot{x} &= -x^k(1 + \overline{f}(x) + y^{n+1}w_{1,n+1}(x,y)), \\
\dot{y} &= y(\lambda + \overline{g}(x) + y^{n+1}w_{2,n+1}(x,y)),
\end{aligned} \tag{2.71}$$

of which we do not specify the relation with the previous expression, except for the fact that the ∞-jet is unchanged.

Next we try

$$(x, y) = (x, Y(1 + Y^n \beta(x))), \tag{2.72}$$

proving the existence of a flat C^∞ function $\beta(x)$ that permits us to change $y^n w_{2,n}(x, y)$ into $Y^{n+1}\overline{w}_{2,n}(x, Y)$ for a convenient C^∞ function $\overline{w}_{2,n}$ such that $j_\infty \overline{w}_{2,n}(0,0) = 0$. We write $g_n(x) = w_{2,n}(x, 0)$

Substituting (2.72) into the second equation of (2.71), and writing y instead of Y, gives:

$$\begin{aligned}
&\dot{y}(1 + (n+1)y^n \beta(x)) + y^{n+1}\beta'(x)\dot{x} = \\
&y(1 + y^n \beta(x))(\lambda + \overline{g}(x) + y^n g(x) + O(y^{n+1})).
\end{aligned}$$

Working modulo $O(y^{n+2})$ this gives the following equation:

$$\begin{aligned}
&\dot{y}(1 + (n+1)y^n \beta(x)) = \\
&y(\lambda + \overline{g}(x) + y^n g_n(x) + (\lambda + \overline{g}(x))y^n \beta(x) + y^n \beta'(x)x^k(1 + \overline{f}(x))).
\end{aligned}$$

A straightforward calculation of the terms in y^{n+1} shows that we have to solve the following equation:

$$\beta'(x)x^k(1 + \overline{f}(x)) = n(\lambda + \overline{g}(x))\beta(x) - g_n(x).$$

Seeing that $j_\infty g_n(0) = 0$, Lemma 2.11 guarantees the existence of a C^∞ solution $\beta(x)$ with $j_\infty \beta(0) = 0$.

Substituting (2.72) into the first equation of (2.71) does not change its form.

The successive coordinate changes (2.70) and (2.72) prove that the induction works.

We hence know that a semi-hyperbolic singularitiy is C^∞-conjugate to some normal form

$$\dot{x} = -x^k(1 + \overline{f}(x)),$$
$$\dot{y} = y(\lambda + \overline{g}(x)),$$
(2.73)

We are now going to prove that two such normal forms are C^∞-conjugate if they have the same formal development at $(0,0)$. We therefore consider

$$\dot{x} = -x^k(1 + \overline{f}(x) + v_1(x)),$$
$$\dot{y} = y(\lambda + \overline{g}(x) + v_2(x)),$$
(2.74)

with $j_\infty v_1(0) = j_\infty v_2(0) = 0$, and will show that (2.73) and (2.74) are mutually C^∞-conjugate.

We therefore first eliminate $v_1(x)$ by means of a coordinate change

$$(x, y) = (X(1 + \alpha(X)), y).$$
(2.75)

Substituting (2.75) into (2.74) and writing x instead of X gives

$$\dot{x}(1 + \alpha(x) + x\alpha'(x)) = -x^k(1 + \alpha(x))^k(1 + \overline{f}(x(1 + \alpha(x))))$$

leading to the equation

$$(1 + \overline{f}(x) + v_1(x))(1 + \alpha(x) + x\alpha'(x)) = (1 + \alpha(x))^k(1 + \overline{f}(x(1 + \alpha(x)))).$$

A straightforward calculation leads to

$$(1 + \overline{f}(x) + v_1(x))x\alpha'(x) = ((k-1) + O(x) + O(\alpha(x)))\alpha(x) - v_1(x); \quad (2.76)$$

by Lemma 2.6 and since $j_\infty v_1(0) = 0$, we know that (2.76) has a C^∞ solution with $j_\infty \alpha(0) = 0$.

As such, the inverse transformation of (2.75) has changed (2.74) into

$$\dot{x} = -x^k(1 + \overline{f}(x)),$$
$$\dot{y} = y(\lambda + \overline{g}(x) + w(x)),$$
(2.77)

for some C^∞ function $w(x)$ with $j_\infty w(0) = 0$.

We now consider the coordinate change

$$(x, y) = (x, Y(1 + \beta(x))).$$
(2.78)

Substituting (2.78) into (2.77) and writing y instead of Y gives

$$\dot{y}(1 + \beta(x)) + y\beta'(x)\dot{x} = y(1 + \beta(x))(\lambda + \overline{g}(x) + w(x)).$$

This leads to

$$\dot{y}(1 + \beta(x)) = y((1 + \beta(x))(\lambda + \overline{g}(x) + w(x)) + (1 + \overline{f}(x))x^k\beta'(x)).$$

A straightforward calculation gives

$$(1 + \overline{f}(x))x^k\beta'(x) = -(1 + \beta(x))w(x),$$

which can be written as

$$(1 + \overline{f}(x))\beta'(x) = -\frac{w(x)}{x^k}(1 + \beta(x)).$$

This equation clearly has a C^∞ solution with $j_\infty \beta(0) = 0$.

2.8 Exercises

Exercise 2.1 Find stable, unstable and center subspaces E^s, E^u and E^c for the linear systems given by the following matrices:

$$\begin{pmatrix} 2 & -1 \\ 0 & 3 \end{pmatrix} \quad \begin{pmatrix} 3 & 6 \\ 2 & 4 \end{pmatrix} \quad \begin{pmatrix} -3 & 6 \\ 2 & -4 \end{pmatrix}.$$

Exercise 2.2 Given the system $\dot{x} = x + 2y$, $\dot{y} = y$, show that there is a unique invariant regular C^∞ curve through the origin.

Exercise 2.3 Suppose that X is a vector field on $\mathbb{R}^2$ with $R_*(X) = X$ and $X(0) = 0$, where $R(x, y) = (-x, y)$.

(i) Check all possible non-zero linear parts (1-jets) $DX(0)$ of X at the origin.
*(ii)*Give the formal normal form Y of $j_\infty X(0)$, show that it also satisfies
$R_*(Y) = Y$, and that it can be obtained by a coordinate change $\varphi =$
Id $+ O(\|(x, y)\|^2)$ such that $\varphi \circ R = R \circ \varphi$.

Exercise 2.4 For nilpotent singularities (i.e., $X(0) = 0$ and $j_1 X(0) \sim y(\partial/\partial x)$), we have found two interesting formal normal forms, namely

$$(y + \sum_{i=2}^{\infty} a_i x^i) \frac{\partial}{\partial x} + (\sum_{j=2}^{\infty} b_j x^j) \frac{\partial}{\partial y}$$

and

$$y \frac{\partial}{\partial x} + (\sum_{j=2}^{\infty} \bar{b}_j x^j + y \sum_{i=1}^{\infty} \bar{a}_i x^i) \frac{\partial}{\partial y}.$$

Calculate the relation between (a_i, b_j) and $(\bar{a}_i, \bar{b}_j)$.

Exercise 2.5 Given a formal power series $\sum_{i=0}^{\infty} a_i x^i$, we know by Borel's Theorem that there exists a C^∞ function f such that $j_\infty f(0) = \sum_{i=0}^{\infty} a_i x^i$. Show that such a C^∞ function f is never uniquely determined.

Exercise 2.6 Check that for the inner product in expression (2.1) then $(ad_m A^T) = (ad_m A)^T$ indeed holds as claimed.

Exercise 2.7 Given a smooth two-dimensional vector field X such that $X(0) = 0$ and $DX(0) = A$ with A semi-simple (i.e., A is $\mathbb{C}$-diagonalizable), show that there exists a formal normal form Y for X at 0 such that $(e^{tA})_* Y = Y$ for all $t \in \mathbb{R}$.

Exercise 2.8 Let X and Y be two-dimensional smooth vector fields with $X(0) = Y(0) = 0$ and let h be a local C^0-conjugacy between X and Y at 0, which is differentiable at the origin ($Dh(0)$ exists). Show that $DX(0)$ and $DY(0)$ are linearly conjugate.

  Hint: Use Exercise 1.1.

Exercise 2.9 Let X and Y be two-dimensional smooth vector fields with $X(0) = Y(0) = 0$. Suppose that X and Y are both hyperbolic attractors at 0 with diagonal linear part and respective eigenvalues $\lambda_1 \leq \lambda_2 < 0$ and $\mu_1 \leq \mu_2 < 0$. If X and Y are C^0-equivalent at 0, by means of a homeomorphism h that it is differentiable at the origin ($Dh(0)$ exists), then it can be proven that $\lambda_1/\lambda_2 = \mu_1/\mu_2$. Show this in case $\lambda_1 = \lambda_2$.

Exercise 2.10 Show that the mapping h, defined in (2.2) and related to Fig. 2.1, indeed defines a C^0-conjugacy between X and L.

Exercise 2.11 Calculate a Taylor approximation of order 4 for the stable and unstable manifold of

$$\dot{x} = 2x + 3x^2 + 4xy + 6y^3,$$
$$\dot{y} = -y + x + 2x^2 + 3xy^2.$$

Exercise 2.12 Show that all vector fields $(\lambda + \varepsilon y)(\partial/\partial x) + \lambda y(\partial/\partial y)$, for a fixed $\lambda \in \mathbb{R}$ and variable $\varepsilon \neq 0$, are mutually linearly conjugate.

Exercise 2.13 Prove the claim, made in the proof of Lemma 2.6 (see Fig. 2.2), that there exists at least one point on $\{x_0\} \times [-y_0, y_0]$ whose ω-limit set is $(0, 0)$.

Exercise 2.14 Solve next items:

(i) Prove the claim, made at the end of the proof of Lemma 2.6, that the property $f(x) = O(x^n)$, for every $n \in \mathbb{N}$ with $n \geq 1$, also holds for the derivative f' of f.
(ii) Prove the same claim for all successive derivatives of f.
(iii) Give an example of a C^∞ function f on $(0, 1)$ with $f(x) = O(x^n)$ for all $n \in \mathbb{N}$ with $n \geq 1$, but such that this property does not hold for f'.

Exercise 2.15 Check in Sect. 2.4 (see Fig. 2.6) the details of the construction of a C^0-conjugacy between two hyperbolic saddles.

Exercise 2.16 Consider a C^∞ vector field $X = (X_1, X_2)$ defined on $\mathbb{R}^2$ and having both coordinate axes as invariant curves. Suppose that for $X|_{\{x \geq 0, y \geq 0\}}$ the following properties hold: $X_1(x, y) > 0$ for $x > 0$, $X_2(x, y) < 0$ for $y > 0$, $(\partial X_2/\partial y)(0,0) < 0$ and $(\partial X_1/\partial x)(0,0) > 0$. Let $\varphi(p, t)$ denote the flow of X, with $\varphi = (\varphi_1, \varphi_2)$. Show that for $T > 0$ sufficiently large there exists a unique $p \in \{x \geq 0, y \geq 0\}$ such that $\varphi_2(p, 0) = 1$ and $\varphi_1(p, T) = 1$.

Exercise 2.17 Calculate a Taylor approximation of order 4 of the center manifold of the following vector fields X at the origin:

$$(y + 2x + 3xy)\frac{\partial}{\partial y} + (x^2 + 2y^2)\frac{\partial}{\partial x},$$

$$(3x^3 + y^3)\frac{\partial}{\partial y} + (x + 4y^3)\frac{\partial}{\partial x}.$$

Also calculate in each of the foregoing cases the 3-jet of the center behavior at the origin, i.e., the 3-jet at 0 of the vector field $X|_{W^c}$, where W^c represents a C^∞ center manifold.

Exercise 2.18 Calculate the precise expression of all possible center manifolds at 0 for the following vector fields:

$$-y\frac{\partial}{\partial y} + x^2\frac{\partial}{\partial x},$$

$$-y\frac{\partial}{\partial y} - x^3\frac{\partial}{\partial x}.$$

Exercise 2.19 Solve the exercise given in Remark 2.10.

Exercise 2.20 Let X be an analytic vector field on some open neighborhood of $(0, 0)$ in $\mathbb{R}^2$, having the origin as a non-isolated singularity and with the property that the divergence $\operatorname{div} X(0) \neq 0$.

(i) Show that X near the origin is C^ω-conjugate to

$$\dot{x} = 0$$
$$\dot{y} = yh(x)$$

for some C^ω function h with $h(0) = 0$.

  Hint: Starting from (2.23), divide the vector field by y and use the flow box theorem.

(ii) Show that X near the origin is C^ω-equivalent to

$$\dot{x} = 0$$
$$\dot{y} = \delta y$$

with $\delta = \pm 1$.

Exercise 2.21 Let X be a C^∞ vector field on some open set $U \subset \mathbb{R}^2$, having at some point $p \in U$ a hyperbolic saddle with eigenvalues λ_1 and λ_2 satisfying $\lambda_2/\lambda_1 \in \mathbb{R} \setminus \mathbb{Q}$. Show that if $f : U \to \mathbb{R}$ is C^∞ with $Xf = 0$ and $f(p) = 0$, then $j_\infty f(p) = 0$.

Exercise 2.22 Let X be a C^∞ vector field on an open set $U \subset \mathbb{R}^2$, having at p a hyperbolic saddle with eigenvalues λ_1 and λ_2 and such that $\lambda_2/\lambda_1 \in \mathbb{R} \setminus \mathbb{Q}$. Show that there exists a neighborhood V of p in U and C^∞ mappings $\pi_s : V \to W^s$, $\pi_u : V \to W^u$, with $\pi_s|W^s = \mathrm{Id}$ and $\pi_u|W^u = \mathrm{Id}$ and with the property that inside V both the foliations $\{\pi_s^{-1}(y)\}$ and $\{\pi_u^{-1}(x)\}$ are preserved under the flow of X, where x and y denote a regular parameter on respectively the unstable and stable manifolds W^u and W^s.

REMARK: The foliations $\{\pi_s^{-1}(y)\}$ and $\{\pi_u^{-1}(x)\}$ are called *invariant C^∞ foliations*.

Exercise 2.23 Let X be a C^∞ vector field on an open set $U \subset \mathbb{R}^2$, having at p a hyperbolic saddle with respectively W^s and W^u as stable and unstable manifolds. Suppose that there exist C^∞ mappings $\pi_s : V \to W^s$ and $\pi_u : V \to W^u$ with the properties described in the previous exercise. Show that X near p is (locally) C^∞-linearizable.

REMARK: A similar technique can be used in any dimension to prove C^∞-linearization of a hyperbolic saddle.

Exercise 2.24 Let U be an open subset of $\mathbb{R}^2$, let $f : U \to \mathbb{R}$ be a C^r with $r \geq 2$ function and let $X(p) = \nabla f(p)$ be the associated gradient vector field given by

$$\dot{x} = \frac{\partial f}{\partial x}(x, y)$$
$$\dot{y} = \frac{\partial f}{\partial y}(x, y).$$

Suppose that p is a singularity of X, then

(i) the eigenvalues of $DX(p)$ are real.
(ii) X has a hyperbolic singularity at p if and only if $Df(p) = 0$ and $D^2 f_p(\cdot, \cdot)$ is a non–degenerate bilinear form.

Exercise 2.25 Determine the Lyapunov stability and the asymptotic stability (see Sect. 1.8) of the following list of singularities: hyperbolic saddle, hyperbolic focus, center, semi-hyperbolic singularity with non-flat center behavior.

Exercise 2.26 Let X_1 and X_2 be two C^∞ vector fields having a hyperbolic saddle at the origin. Then it is possible to prove that they are C^0-equivalent, without using the monotonicity of the time function, as we did. One such proof goes as follows:

(i) Use the existence of C^∞ invariant stable and unstable manifolds to prove the existence of C^∞ coordinates in which $W^s = \{x = 0\}$ and $W^u = \{y = 0\}$.

(ii) Consider both $X_i^\perp$, where $X_i^\perp = g_i(x,y)(\partial/\partial x) - f_i(x,y)(\partial/\partial y)$ if $X_i = f_i(x,y)\frac{\partial}{\partial x} + g_i(x,y)\frac{\partial}{\partial y}$.

Show that a mapping h is a C^∞-equivalence between X_1 and X_2 on a neighborhood of $(0,0)$ if h is defined as follows:

(i) $h|_{\{x=0\}\cup\{y=0\}}$ is the identity,
(ii) h sends X_1-orbits to X_2-orbits,
(iii) h sends $X_1^\perp$-orbits to $X_2^\perp$-orbits,
(iv) for some y_0 sufficiently small, let $y = \gamma_i^\pm(x)$ denote the $X_i^\perp$-orbit with $\gamma_i^\pm(x) = \pm y_0$. Then we require that $h(x, \gamma_1^\pm(x)) = (x, \gamma_2^\pm(x))$ for x in a sufficiently small neighborhood of 0.

2.9 Bibliographical Comments

Although the idea of simplifying ordinary differential equations through changes of variables can already be found in earlier works, we can call Poincaré the founding father of normal form theory as we use it nowadays. After Poincaré the theory was developed by many people, among others Lyapunov, Dulac, Birkhoff, Siegel, Sternberg, Chen, Moser, Arnol'd, Pliss, Belitskii, Bruno, Takens, Cushman, Sanders, Elphick, van der Meer, Yoccoz, Pérez Marco and Stolovitch. In February 2005, MathSciNet provides 1618 hits for "Normal forms" as element in the title. Maybe not all deal with differential equations, but many do. Some deal with formal aspects or with computational aspects. Others impose extra conditions on the normal forms or the normalizing transformations (like symmetries, preservation of extra structure). Some deal with C^r normal forms, some with C^∞ or analytic normal forms, with individual vector fields or with families of vector fields. To cite a few books, we mention in alphabetic order the books by Arnol'd [5], Arnol'd and Il'Yashenko [6], Bronstein and Kopanskii [21], Bruno [23], Chow and Hale [35], Chow, Li and Wang [36], Golubitsky and Schaeffer [74, 75], Guckenheimer and Holmes [77], Murdock [114] and Wiggins [164]. We also would like to mention the paper of Stowe [154] for many results concerning C^s-linearization of resonant saddles, focusing on the problem of finding the best s possible.

The theory of invariant manifolds also has a long history, although as with normal form theory, it is still a very active branch of research. Introducing "invariant manifold" on MathSciNet as element in the title gave 468 hits in February 2005, besides another 144 with "invariant curve" in the title, 91 with "stable manifold," 140 with "center manifold" and 42 with "centre manifold." A number of these papers merely deal with the use of these objects but many deal with more theoretical aspects.

Besides mentioning such names as Hadamard and Perron concerning the stable (and unstable) manifold, we want to mention the book of Hartman [80], as well as refer to some books that we have listed before concerning normal forms. We also cite the paper by Meyer [110] for the analyticity of these objects.

Concerning center manifolds we would like to refer to Carr [26], Fenichel [64], Kelley [96], Hirsch, Pugh and Shub [83], Sijbrand [149] and especially to Vanderbauwhede [162], besides some of the books that we already mentioned earlier. We also want to refer to van Strien [161] for an example of a polynomial vector field not having a C^∞ center manifold.

For a general introduction to the theory of dynamical systems, paying significant attention to normal forms and invariant manifolds, we can also refer to Palis and de Melo [118], Robinson [136] and Katok and Hasselblatt [95].

The usual methods of proving results about normal forms and invariant manifolds are somewhat different than the ones that we have used here. Our choice of proof is more elementary but is perhaps less appropriate for generalization to higher dimensions. For the two-dimensional case we believe that it is quite complete in comparison to other texts since we not only prove the existence of C^∞ center manifolds, but also include good C^∞ normal forms for the elementary singularities. We also proved the existence of analytic stable/unstable manifolds for elementary analytic singularities as well as good analytic normal forms on one dimensional analytic invariant manifolds. The theory of analytic normal forms for hyperbolic two-dimensional saddles is however beyond the scope of this book.

3

Desingularization of Nonelementary Singularities

In this chapter we provide the basic tool for studying nonelementary singularities of a differential system in the plane. This tool is based on changes of variables called blow-ups. We use this technique for classifying the nilpotent singularities; i.e., the singularities having both eigenvalues zero but whose linear part is not identically zero. Blow-up is also used to show that at isolated singularities an analytic system has a finite sectorial decomposition.

3.1 Homogeneous Blow-Up

Before describing the effective algorithm that we use in the program P4 [9], and which is based on the use of quasihomogeneous blow-up, we will first explain the basic ideas only introducing *homogeneous blow-up*, which essentially means using polar coordinates. We position the singularity that we want to study at the origin.

Let 0 be a singularity of a C^∞ vector field X on $\mathbb{R}^2$. Consider the map

$$\phi: \mathbb{S}^1 \times \mathbb{R} \to \qquad \mathbb{R}^2$$
$$(\theta, r) \mapsto (r\cos\theta, r\sin\theta) .$$

We can define a C^∞ vector field $\hat{X}$ on the cylinder $\mathbb{S}^1 \times \mathbb{R}$ such that $\phi_*(\hat{X}) = X$, in the sense that $D\phi_v(\hat{X}(v)) = X(\phi(v))$. It is called the pull back of X by ϕ. It is nothing else but X written down in polar coordinates. The map ϕ is a C^∞ diffeomorphism, hence a genuine C^∞ coordinate change on $\mathbb{S}^1 \times (0, \infty)$, but not on $\{r = 0\}$; ϕ sends $\{r = 0\}$ to $(0, 0)$, and as such, the inverse mapping ϕ^{-1} blows up the origin to a circle. In order to study the phase portrait of X in a neighborhood V of the origin, it clearly suffices to study the phase portrait of $\hat{X}$ on the neighborhood $\phi^{-1}(V)$ of the circle $\mathbb{S}^1 \times \{0\}$, and we can even restrict to $\{r \geq 0\}$. A priori this might seem a more difficult problem than the original one, but as we will see in this chapter, the construction is very helpful. If the k-jet $j_k(X)(0)$ is zero, then $j_k(\hat{X})(u) = 0$ for all $u \in \mathbb{S}^1 \times \{0\}$.

Although the cylinder is a good surface for getting a global view of $\hat{X}$ and its phase portrait, it is often less appropriate for making calculations, since we constantly have to deal with trigonometric expressions. For that reason it is often preferable to make the calculations in different charts.

On the parts of the cylinder given, respectively, by $\theta \in (-\pi/2, \pi/2)$ and $\theta \in (\pi/2, 3\pi/2)$ use a chart given by

$$K^x : (\theta, r) \mapsto (r \cos \theta, \tan \theta) = (\overline{x}, \overline{y}).$$

In this chart the expression of the blow-up map ϕ is given by

$$\phi^x : (\overline{x}, \overline{y}) \mapsto (\overline{x}, \overline{x}\overline{y}). \tag{3.1}$$

Indeed we see that

$$\phi = \phi^x \circ K^x : (\theta, r) \overset{K^x}{\mapsto} (r \cos \theta, \tan \theta)$$
$$\overset{\phi^x}{\mapsto} (r \cos \theta, r \cos \theta \tan \theta) = (r \cos \theta, r \sin \theta). \tag{3.2}$$

We call (3.1) a "blow-up in the x-direction"; the pull-back of X by means of ϕ^x is denoted by $\hat{X}^x$, i.e., $(\phi^x)_*(\hat{X}^x) = X$.

On the parts of the cylinder given, respectively, by $\theta \in (0, \pi)$ and $\theta \in (\pi, 2\pi)$, we use a chart given by

$$K^y : (\theta, r) \mapsto (\cot \theta, r \sin \theta) = (\overline{x}, \overline{y}).$$

In this chart the expression of the blow-up map ϕ is given by

$$\phi^y : (\overline{x}, \overline{y}) \mapsto (\overline{x}\overline{y}, \overline{y}), \tag{3.3}$$

in the sense that $\phi = \phi^y \circ K^y$. We call (3.3) a "blow-up in the y–direction"; the pullback of X by means of ϕ^y is denoted by $\hat{X}^y$, i.e., $(\phi^y)_*(\hat{X}^y) = X$.

Both ϕ^x and ϕ^y are called "directional blow-ups."

If $j_k(X)(0) = 0$ and $j_{k+1}(X)(0) \neq 0$, then again $j_k(\hat{X}^x)(z) = 0$ and $j_k(\hat{X}^y)(z) = 0$ for, respectively, $z \in \{\overline{x} = 0\}$ or $z \in \{\overline{y} = 0\}$.

In case $j_k(X)(0) = 0$ and $j_{k+1}(X)(0) \neq 0$ the pullback $\hat{X}$ and likewise $\hat{X}^x$ and $\hat{X}^y$, are quite degenerate, and to make the situation less degenerate we consider $\overline{X}$ with

$$\overline{X} = \frac{1}{r^k} \hat{X}.$$

Then $\overline{X}$ also is a C^∞ vector field on $\mathbb{S}^1 \times \mathbb{R}$. On $\{r > 0\}$ this division does not change the orbits of $\hat{X}$ or their sense of direction, but only the parametrization by t. From the formulas it is clear that singularities of $\overline{X}|_{\{r=0\}}$ come in pairs of opposite points.

For the related directional blow-up we use $(1/\overline{x}^k)\hat{X}^x$ in case (3.1) and $(1/\overline{y}^k)\hat{X}^y$ in case (3.3). On $\{\overline{x} \neq 0\}$ (respectively $\{\overline{y} \neq 0\}$) the vector fields

$(1/r^k)\hat{X}$ and $(1/\bar{x}^k)\hat{X}^x$ (respectively $(1/\bar{y}^k)\hat{X}^y$) are no longer equal up to analytic coordinate change, as were $\hat{X}$ and $\hat{X}^x$ (respectively, $\hat{X}^y$), but they are the same up to analytic coordinate change and multiplication by a nonzero analytic function.

We work this out for the blow-up in the x-direction: since $\phi = \phi^x \circ K^x$, we see that $(K^x)_*(\hat{X}) = \hat{X}^x$.

As such

$$(K^x)_*(\overline{X}) = (K^x)_*(\hat{X}/r^k) = \frac{1}{r^k}(K^x)_*(\hat{X}) = \frac{1}{r^k}\hat{X}^x = \overline{X}^x\left(\frac{\bar{x}}{r}\right)^k.$$

Seen in (θ, r)-coordinates we have $\bar{x}/r = \cos\theta$, which is strictly positive on the part of the cylinder given by $\theta \in (-\pi/2, \pi/2)$.

Similarly in the y-direction, we have $(K^y)_*(\hat{X}) = \hat{X}^y$ and $(K^y)_*(\overline{X}) = \overline{X}^y(\sin\theta)^k$, with $\sin\theta > 0$ on the part of the cylinder given by $\theta \in (0, \pi)$.

The directional blow-up ϕ^x can also be used for making a study on $\{(\theta, r) : \theta \in (\pi/2, 3\pi/2), r \geq 0\}$, but in that case we have $\cos\theta < 0$.

For odd k, this means that in the phase portraits that we find for $\overline{X}^x|_{\{\bar{x} \leq 0\}}$ we have to reverse time. A similar observation has to be made in using $\overline{X}^y$ for studying $\overline{X}$ on $\{(\theta, r) : \theta \in (\pi, 2\pi), r > 0\}$.

Such a time reversal could be avoided in using ϕ^x (respectively, ϕ^y) only for $\bar{x} \geq 0$ (respectively, $\bar{y} \geq 0$), and adding two extra directional blow-ups

$$\phi^{-x} : (\bar{x}, \bar{y}) \mapsto (-\bar{x}, -\bar{x}\bar{y}),$$
$$\phi^{-y} : (\bar{x}, \bar{y}) \mapsto (-\bar{x}\bar{y}, -\bar{y}),$$

that we limit to, respectively, $\bar{x} \geq 0$ and $\bar{y} \geq 0$. Of course the number of calculations can be limited by using both ϕ^x and ϕ^y on a full neighborhood of, respectively, $\{\bar{x} = 0\}$ and $\{\bar{y} = 0\}$, avoiding having to work with ϕ^{-x} and ϕ^{-y}.

We now treat a few examples.

Example 3.1 First we present an example in which we use one blow-up to obtain quite easily the topological picture of the orbit structure of the singularity:

$$X = (x^2 - 2xy)\frac{\partial}{\partial x} + (y^2 - xy)\frac{\partial}{\partial y} + O(\|(x, y)\|^3).$$

The formulas for (polar) blow-up are

$$\bar{X} = \eta_1\frac{\partial}{\partial\theta} + \eta_2 r\frac{\partial}{\partial r},$$

with

$$\eta_1(\theta, r) = \frac{1}{r^{k+2}} \left\langle X, x\frac{\partial}{\partial y} - y\frac{\partial}{\partial x} \right\rangle (\phi(r, \theta))$$

$$= \frac{1}{r^{k+2}} (-r\sin\theta X_1(r\cos\theta, r\sin\theta) + r\cos\theta X_2(r\cos\theta, r\sin\theta)),$$

$$\eta_2(\theta, r) = \frac{1}{r^{k+2}} \left\langle X, x\frac{\partial}{\partial x} + y\frac{\partial}{\partial y} \right\rangle (\phi(r, \theta))$$

$$= \frac{1}{r^{k+2}} (r\cos\theta X_1(r\cos\theta, r\sin\theta) + r\sin\theta X_2(r\cos\theta, r\sin\theta)),$$

In our example $k = 1$ and the result is

$$\bar{X}(\theta, r) = (\cos\theta\sin\theta(3\sin\theta - 2\cos\theta) + \mathrm{O}(r))\frac{\partial}{\partial\theta}$$

$$+ r(\cos^3\theta - 2\cos^2\theta\sin\theta - \cos\theta\sin^2\theta + \sin^3\theta + \mathrm{O}(r))\frac{\partial}{\partial r}.$$

Zeros on $\{r = 0\}$ are located at

$$\theta = 0, \pi; \ \theta = \pi/2, 3\pi/2; \ \tan\theta = 2/3.$$

At these singularities, the radial eigenvalue is given by the coefficient of $r\partial/\partial r$ while the tangential eigenvalue can be found by differentiating the $\partial/\partial\theta$-component with respect to θ. One thus obtains Fig. 3.1. In this figure we represent the half cylinder $\mathbb{S}^1 \times [0, \infty)$ as $E = \{(x, y) : x^2 + y^2 \geq 1\}$. This visualization will also be used in the sequel. The phase portrait which we see on E near the circle $C = \{x^2 + y^2 = 1\}$ gives a very good idea of the phase portrait of X near the origin. It suffices to shrink the circle to a point (see Fig. 3.2).

All the singularities on $S^1 \times \{0\}$ are hyperbolic. We say that we have desingularized X at 0 since all singularities of $\bar{X}|_{\{r=0\}}$ are elementary. The

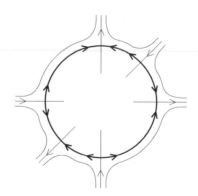

Fig. 3.1. Blow-up of Example 3.1

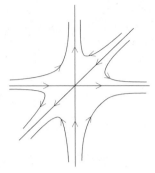

Fig. 3.2. Local phase portrait of Example 3.1

exact value of the eigenvalues at the different singularities depends only on the 2-jet of X. Using techniques similar to the ones that served to study the C^0-conjugacy classes of elementary singularities, one can now prove that X near the origin is C^0-conjugate to the vector field $Y = j_2 X(0)$, which has a similar blow-up. We will not work this out. In this chapter we will only take care of determining the sectorial decomposition of the singularity. Although in the line of what we did explicitly for the elementary singularities, it is rather tedious to show that for each kind of sectorial decomposition of an analytic vector field there only exists one model for C^0-conjugacy. We refer to [53] or [33] for a detailed elaboration. The proof on the unicity of the model for C^0-equivalence is simpler, but nevertheless we do not wish to pay attention to such constructions in general. We will treat some specific situations in the exercises.

Concerning the sectorial decomposition, we remark that in the case of the example, it is not hard to show, using the blow-up, that the vector field X indeed has a finite sectorial decomposition near 0 as defined in Sect. 1.5. We will come back to the proof of this in Sect. 3.5.

Concerning Fig. 3.1, we remark that the exact position of the invariant manifolds, transverse to C, of the six hyperbolic singularities in the blow-up can be approximated by Taylor approximation. After blowing down it leads to an accurate presentation of the six "separatrices" in the local phase portrait; see Fig. 3.2. □

Example 3.2 Second we present an example for which blowing up once is not sufficient to desingularize the singularity. There remain nonelementary singularities of $\overline{X}|_{\{r=0\}}$ at which we need to repeat the blow-up construction, leading to successive blowing up. The starting vector field is

$$y\frac{\partial}{\partial x} + (x^2 + xy)\frac{\partial}{\partial y} + O(\|(x,y)\|^3).$$

Blowing up in the y-direction gives no singularities on $\{\bar{y} = 0\}$. Direct calculations show that the singularities of $\overline{X}$ (and equally for $\overline{X}^x$ and $\overline{X}^y$), as well as their eigenvalues, depend only on the first nonzero jet, hence on $y\partial/\partial x$ in this example. We now perform a blow-up in the x-direction, working out the calculations explicitly. Writing

$$x = \bar{x}, \quad y = \bar{x}\bar{y},$$

or

$$\bar{x} = x, \quad \bar{y} = y/x,$$

we get

$$
\begin{aligned}
\dot{\bar{x}} &= \dot{x} \\
&= y + \mathrm{O}(\|(x,y)\|^3) \\
&= \bar{y}\bar{x} + \mathrm{O}(|\bar{x}|^3), \\
\dot{\bar{y}} &= \frac{\dot{y}}{x} - y\frac{\dot{x}}{x^2} \\
&= (x+y) + \frac{1}{x}\mathrm{O}(\|(x,y)\|^3) - \frac{y^2}{x^2} - \frac{y}{x^2}\mathrm{O}(\|(x,y)\|^3) \\
&= \bar{x} + \bar{y}\,\bar{x} - \bar{y}^2 + \mathrm{O}(|\bar{x}|^2).
\end{aligned}
$$

(3.4)

The only singularity on $\bar{x} = 0$ occurs for $\bar{y} = 0$, where the 1-jet of the vector field $\bar{X}^x$ at this singularity is $\bar{x}\partial/\partial\bar{y}$.

As the singularity is nonelementary, we are going to perform an extra blow-up in order to study it. Blowing up in the $\bar{x}$-direction gives no singularities. Blowing up (3.4) in the $\bar{y}$-direction ($\bar{x} = \bar{\bar{y}}\,\bar{\bar{x}}, \bar{y} = \bar{\bar{y}}$) gives

$$
\begin{aligned}
\dot{\bar{\bar{y}}} &= \dot{\bar{y}} \\
&= (\bar{x} + \bar{y}\,\bar{x} - \bar{y}^2 + \mathrm{O}(|\bar{x}|^2)) \\
&= \bar{\bar{x}}\,\bar{\bar{y}} - \bar{\bar{y}}^2 + \mathrm{O}(\|(\bar{\bar{x}},\bar{\bar{y}})\|^3), \\
\dot{\bar{\bar{x}}} &= \frac{\dot{\bar{x}}}{\bar{y}} - \bar{x}\frac{\dot{\bar{y}}}{\bar{y}^2} \\
&= \bar{x} + \frac{1}{\bar{y}}\mathrm{O}(|\bar{x}|^3) - \frac{\bar{x}}{\bar{y}^2}(\bar{x} + \bar{y}\bar{x} - \bar{y}^2 + \mathrm{O}(|\bar{x}|^2)) \\
&= \bar{\bar{y}}\bar{\bar{x}} - \bar{\bar{x}}^2 + \bar{\bar{y}}\bar{\bar{x}} + \mathrm{O}(\|(\bar{\bar{x}},\bar{\bar{y}})\|^2).
\end{aligned}
$$

The 2-jet is now $(\bar{\bar{x}}\bar{\bar{y}} - \bar{\bar{y}}^2)\partial/\partial\bar{\bar{y}} + (2\bar{\bar{x}}\bar{\bar{y}} - \bar{\bar{x}}^2)\partial/\partial\bar{\bar{x}}$. This singularity is not elementary, but as we have seen in Example 3.1, it can be studied by blowing up once. This succession of blowing up is schematized in Fig. 3.3. At each step we blow-up a point to a circle, not forgetting that singularities of $\overline{X}$ on $\{r = 0\}$ always come in pairs of opposite points. If we need to blow-up one, we also apply the same blow-up procedure to the second. As we already observed in discussing the directional blow-up $\overline{X}^x$ and $\overline{X}^y$ in relation to $\overline{X}$, the study of both singularities at a pair of opposite points can be done on the same expressions by treating $\bar{x} \leq 0$ as well as $\bar{x} \geq 0$, or, respectively, $\bar{y} \leq 0$ as well

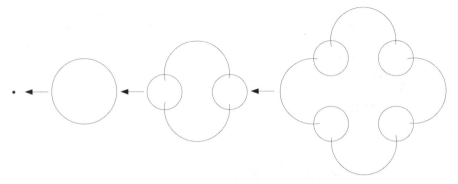

Fig. 3.3. Successive blowing up

as $\overline{y} \geq 0$. In terms of $\overline{X}$ it also means that we only have to consider one of the singularities, but considering $r \leq 0$ as well as $r \geq 0$. It is possible that we have to use a time-reversal, when k is odd (with $\overline{X} = \hat{X}/r^k$), when transporting information to the other singularity. At each step of the succession of blow-ups we only need to keep part of the information, sufficient to cover a full neighborhood of the origin after blowing down.

This procedure of successive blowing up can be formalized as follows, providing an overall geometric view like in Fig. 3.3. Instead of using the polar blow-up ϕ and dividing by some power of r, we use the map

$$\tilde{\phi} : \left\{ z \in \mathbb{R}^2 \ : \ \|z\| > \frac{1}{2} \right\} \subset \mathbb{R}^2 \to \mathbb{R}^2, z \mapsto z - \frac{z}{\|z\|},$$

and divide by the same power of $(\|z\| - 1)$.

The vector field we so obtain is analytically equivalent to $\overline{X}$, but the second is now defined on an open domain in $\mathbb{R}^2$ and therefore it becomes easier to visualize how we can blow-up again at some point $z_0 \in \{z \in \mathbb{R}^2 \ : \ \|z\| = 1\}$: we just use the mapping $T_{z_0} \circ \phi$ where T_{z_0} denotes the translation $z \mapsto z + z_0$.

As we again end up on an open domain of $\mathbb{R}^2$ we can repeat the construction if necessary. For simplicity in notation we denote the first blow-up by ϕ_1, the second by ϕ_2 and so on.

After a sequence of n blow-ups we find some C^∞-vector field $\bar{X}^n$ defined on a domain $U_n \subset \mathbb{R}^2$. $\bar{X}^n$ is even analytic if we start with an analytic X. We write $\Gamma_n = (\phi_1 \circ \ldots \circ \phi_n)^{-1}(0) \subset U_n$. Only one of the connected components of $\mathbb{R}^2 \setminus \Gamma_n$, call it A_n, has a noncompact closure. Furthermore $\partial A_n \subset \Gamma_n$ and ∂A_n, which is homeomorphic to $\mathbb{S},^1$ consists of a finite number of analytic regular closed arcs meeting transversely. The mapping $(\phi_1 \circ \ldots \circ \phi_n)|_{A_n}$ is an analytic diffeomorphism sending A_n onto $\mathbb{R}^2 \setminus \{0\}$. There exists a strictly positive function F_n on A_n such that $\hat{X}^n = F_n \bar{X}^n$ and $\hat{X}^n|_{A_n}$ is analytically diffeomorphic to $X|_{\mathbb{R}^2 \setminus \{0\}}$ by means of the diffeomorphism $(\phi_1 \circ \ldots \circ \phi_n)|_{A_n}$. The function F_n extends in a C^ω way to ∂A_n where in general it is 0.

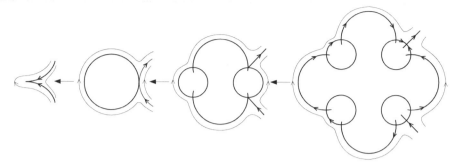

Fig. 3.4. Blowing up Example 3.2

Fig. 3.5. Local phase portrait of Example 3.2

The reconstruction of the local phase portrait of Example 3.2 is represented in Fig. 3.4. To understand the figure one has to start from the right, representing the phase portrait of a vector field $\overline{X}^3$ obtained after three blow-ups. One must not forget that at the second step one simultaneously blows up two (opposite) singularities and at the third step their 4 counter images. The calculations show that all the singularities of $\overline{X}^3$ on ∂A_3 are hyperbolic, and hence, that $\overline{X}^3$ is a desingularization of X. We say that X has been desingularized after three successive blow-ups. The structure of the desingularization of $\overline{X}^3$ is as represented in Fig. 3.4. In following the arrows to the left, we successively represent the phase portraits of $\overline{X}^2$ near ∂A_2, $\overline{X}^1$ near ∂A_1, and finally X near the origin. The sectorial decomposition of X near the origin is clear from its desingularization $\overline{X}^3$.

Again the method permits us to show that the vector fields of Example 3.2 are topologically determined by the 2-jet in the sense that such X near 0 is C^0-conjugate to $Y = j_2X(0)$. A precise drawing of the two separatrices of the *cusp* can be obtained by using Taylor approximations of the invariant manifolds in the desingularization followed by a blowing down, as shown in Fig. 3.5.   □

3.2 Desingularization and the Łojasiewicz Property

To control whether a sequence of blow-ups finally leads to a desingularization we use the notion of a *Łojasiewicz inequality*. We say that a vector field X on

$\mathbb{R}^2$ satisfies a Lojasiewicz inequality at 0 if there is a $k \in \mathbb{N}$ with $k \geq 1$, and a $c > 0$ such that $\|X(x)\| \geq c\|x\|^k$ on some neighborhood of 0.

For analytic vector fields at isolated singularities, a Lojasiewicz inequality always holds (see [18]).

Theorem 3.3 ([52]) *If X at 0 satisfies a Lojasiewicz inequality, then there exists a finite sequence of blowing ups $\phi_1 \circ \ldots \phi_n$ leading to a vector field $\bar{X}^n$ defined in the neighborhood of ∂A_n of which the singularities on ∂A_n are elementary.*

These elementary singularities can be as follows:

(i) Isolated singularities p which are hyperbolic or semi-hyperbolic with non-flat behavior on the center manifold;

(ii) Regular analytic closed curves (or possibly the whole of ∂A_n when $n = 1$) along which $\bar{X}^n$ is normally hyperbolic.

The position on ∂A_n as well as the determinating properties of the singularities as used in the classification presented in the Theorems 2.15 and 2.19 depend only on a finite jet of X.

We do not give a proof of this theorem. We merely consider blow-up as a technique to desingularize singularities. The technique turns out to be successful, at least if we apply it to a singularity of Lojasiewicz type, such as an isolated singularity of an analytic system.

Taking a close look at the singularities of $\overline{X}^n$ on ∂A_n, we see that some lie on regular arcs of ∂A_n, while others lie in corners. At the former we see, because of the Theorems 2.15 and 2.19, that there always exists an invariant C^∞ curve, transversely cutting ∂A_n, unless the singularity is a resonant hyperbolic node. The most degenerate one is such that the linear part of the singularity consists of a single Jordan block. We represent the attracting case in Fig. 3.6a. In any case, near all singularities on the regular part of ∂A_n we find at least one orbit that blows down to a characteristic orbit of X. We repeat, from Sect. 1.6, that a characteristic orbit is an orbit which tends to the singularity, either in positive or negative time, with a well-defined slope. We see therefore that X necessarily has a characteristic orbit at 0 if $\overline{X}^n$ has at least one singularity on the regular parts of ∂A_n.

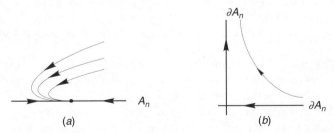

Fig. 3.6. Some singularities of $\overline{X}$ on ∂A_n

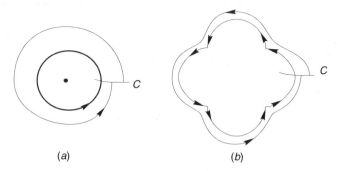

(a)   (b)

Fig. 3.7. Samples of desingularizations of monodromic orbits

Singularities at corners of ∂A_n also lead to the existence of characteristic orbits, except when the singularity has a hyperbolic sector with both separatrices in A_n, as represented in Fig. 3.6b.

Because of Theorem 3.3, and the observation just made, we thus see that a Łojasiewicz singularity either has a characteristic orbit or, if it does not have a characteristic orbit, is a center or a focus. The latter situation is also called *monodromic*. This can only happen if either $\overline{X} = \overline{X}^1$ has no singularities on ∂A_1 (see Fig. 3.7a) or if all singularities are corners of saddle type (see Fig. 3.7b). In that case there is a segment C lying in $A_n \cup \partial A_n$ that is transverse to the flow of $\overline{X}$ and cuts ∂A_n transversely at a regular point p. A first return map can be defined for values $q \in C$ for q sufficiently close to p.

We now treat the two cases separately, starting with the monodromic one. We consider only analytic systems, and we choose C to be an analytic curve with an analytic choice of a regular parameter s on it; we let $s = 0$ coincide with ∂A_n and $s > 0$ with A_n. In case there are no singularities (see Fig. 3.7a) the return map f is analytic in s, and as such, either $f(s) = s$ or $j_n(f(s) - s) \neq 0$ for some $n \in \mathbb{N}$ with $n \geq 1$. In the former case, X represents a center, and in the latter case a focus. The focus need not to be a hyperbolic one, but is at least C^0-conjugate to a hyperbolic focus.

In the case there are singularities (see Fig. 3.7b), then we enter into a really difficult subject. Although X as well as $\overline{X}^n$ are analytic, the first return map f does not need to be analytic. Nevertheless it is possible to prove that in this case as well the system is either a center or a focus, excluding the possibility of having accumulation of limit cycles at the singularities (which can occur in the C^∞ case).

An important paper dealing with the proof is [51]. The paper contains valuable results on which subsequent work still relies. It does not however provide a complete proof, leaving a gap that was detected only in the mid-seventies. For a while this gap was called Dulac's problem (see e.g., [113]). In the meantime the proof has been completed independently by Ecalle [59] and Ilyashenko [88].

In the case that characteristic orbits occur, we show how to prove that such singularities have a "finite sectorial decomposition" as defined in Sect. 1.6. We do not have to restrict to analytic systems, but can consider C^∞ singularities of Lojasiewicz type. The proof relies completely on Theorem 3.3, together with Theorems 2.15 and 2.19; we provide only a rough sketch, referring to the exercises for working out some of the details.

Choosing an orientation for ∂A_n we get a cyclic order on the singularities of $\overline{X}^n|_{\partial A_n}$. To fix the ideas, we suppose that ∂A_n is oriented in a clockwise way. We denote the cyclic order by "$<$." The only way to get a hyperbolic sector is by having two singularities p and q, neither lying in a corner of ∂A_n, such that:

(i) Every singularity r with $p < r < q$ is a corner of saddle type;
(ii) There is an invariant C^∞ curve C_1 (respectively, C_2), transversely cutting ∂A_n at p (respectively, q), which, together with $\partial A_n \cap [p,q]$ borders a hyperbolic sector.

For a general picture we refer to Fig. 3.8.

Based on the normal form given in the Theorems 2.15 and 2.19, it is an easy exercise to prove the existence of a C^∞ curve that transversely cuts both C_1 and C_2 and that meets, inside the hyperbolic sector in between C_1 and C_2, the requirements expressed in the definition of "finite sectorial decomposition."

The way to encounter an elliptic sector is by having two singularities p and q such that:

(i) Every singularity r with $p < r < q$ is a corner of saddle type;
(ii) Both at p and q there is a parabolic sector adherent to $[p,q] \subset \partial A_n$,

of which one is attracting and the other is repelling. We refer to Fig. 3.9 for an example. In this picture we cannot however guarantee that we see the full elliptic sector, and surely not the maximal one as defined in Sect. 1.5.

In fact the curve C_1 in Fig. 3.9 could be transverse to ∂A_n, but it could also belong to ∂A_n. It is also possible that C_2 (or its blow down) is not a good

Fig. 3.8. Blowing up a hyperbolic sector

Fig. 3.9. Blowing up an elliptic sector

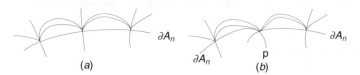

Fig. 3.10. Blowing up of part of adjacent elliptic sectors

choice for bordering a maximal elliptic sector, since it is possible that the orbits to the right of C_2 also tend in negative time to ∂A_n, and as such, belong to the elliptic sector if we want it to be maximal (in the sense that it is an elliptic sector of a minimal sectorial decomposition). From the desingularized vector field $\overline{X}^n$ it is easy to find a maximal elliptic sector containing the part near $[p, q]$ as given in Fig. 3.9. We will treat some examples in the exercises.

We find two kind of bordering curves for a maximal elliptic sector. On the one hand there are the bordering curves which also border a hyperbolic sector. On the other hand there are bordering curves separating two adjacent elliptic sectors; their choice is not unique, as we see in the examples given in Fig. 3.10.

In any case, for an elliptic sector, the normal forms from the Theorems 2.15 and 2.19 permit us an easy proof of the existence of a C^∞ curve, transversely cutting the bordering curves and having the exact properties as described in the definition of "finite sectorial decomposition" (see Sect. 1.5).

In between two hyperbolic sectors one can encounter a unique maximal parabolic sector, whose desingularization can be quite complicated; however, based on the normal forms in the Theorems 2.15 and 2.19 one can easily find, inside any a priori chosen neighborhood of 0, a C^∞ curve, the portion of which that lies inside the parabolic sector is everywhere transverse to the orbits, including the bordering orbits. We again refer to the exercises for the details of the construction.

For a minimal sectorial decomposition it is always possible to find bordering curves or separatrices (see Sect. 1.5) which are images by a blow down mapping of a C^∞ curve. The ones bordering a hyperbolic sector are of *finite type* in the sense that they possess a C^∞ parametrization $\gamma : [0, \varepsilon] \mapsto \mathbb{R}^2$ with $j_r\gamma(0) \neq 0$ for some $r \in \mathbb{N}$. They can also be seen as graphs of a C^∞ function in the variable $x^{1/n}$ for some $n \in \mathbb{N}$ with $n \geq 1$ in suitable C^∞ coordinates (x, y); see [57]. The separatrices between two elliptic sectors do not need to have this property, which is the case for example in Fig. 3.10b if the corner point p is a semi-hyperbolic point. For more information see [57].

3.3 Quasihomogeneous blow-up

Although the method of successive homogeneous blow-ups is sufficient for studying isolated singularities of an analytic vector field, it turns out to be

much more efficient to include *quasihomogeneous blow-ups*. In fact the algorithm that we have implemented in the program P4 [9] relies on the systematic
approach presented in [124], and which is based on the use of quasihomogeneous blow-ups; see also [23] and [22]. We first present the technique before
describing the algorithm.

Let 0 be a singularity of a C^∞ vector field X on $\mathbb{R}^2$. Consider the map

$$\phi: \mathbb{S}^1 \times \mathbb{R} \to \mathbb{R}^2$$
$$(\theta, r) \mapsto (r^\alpha \cos\theta, r^\beta \sin\theta) ,$$

for some well chosen $(\alpha, \beta) \in \mathbb{N} \times \mathbb{N}$ with $\alpha, \beta \geq 1$. Exactly as in the "homogeneous case," where $(\alpha, \beta) = (1, 1)$, we can define a C^∞ vector field $\hat{X}$ on
$\mathbb{S}^1 \times \mathbb{R}$ with $\phi_*(\hat{X}) = X$. We will divide it by r^k, for some $k \in \mathbb{N}$ with $k \geq 1$,
in order to get a C^∞ vector field $\bar{X} = \frac{1}{r^k}\hat{X}$, which is as non–degenerate as
possible along the invariant circle $\mathbb{S}^1 \times \{0\}$.

In practice one again uses directional blow-ups:

positive x-direction: $(\bar{x}, \bar{y}) \mapsto (\bar{x}^\alpha, \bar{x}^\beta \bar{y})$,   leading to $\hat{X}^x_+$,
negative x-direction: $(\bar{x}, \bar{y}) \mapsto (-\bar{x}^\alpha, \bar{x}^\beta \bar{y})$, leading to $\hat{X}^x_-$,
positive y-direction: $(\bar{x}, \bar{y}) \mapsto (\bar{x}\bar{y}^\alpha, \bar{y}^\beta)$,   leading to $\hat{X}^y_+$,
negative y-direction: $(\bar{x}, \bar{y}) \mapsto (\bar{x}\bar{y}^\alpha, -\bar{y}^\beta)$, leading to $\hat{X}^y_-$,

inducing also the systems $\bar{X}^x_-$, $\bar{X}^x_+$, $\bar{X}^y_-$ and $\bar{X}^y_+$ that we obtain dividing,
respectively, by $\bar{x}^k$ and $\bar{y}^k$.

If α is odd (respectively, β is odd), the information found in the positive x-direction (respectively, y-direction) also covers the one in the negative
x-direction (respectively, y-direction).

To show by an example that this technique can be quite efficient, we again
study the cusp-singularity

$$y\frac{\partial}{\partial x} + (x^2 + xy)\frac{\partial}{\partial y} + \mathrm{O}(\|(x,y)\|^3), \tag{3.5}$$

this time using a quasihomogeneous blowing up with $(\alpha, \beta) = (2, 3)$.

In the positive x-direction we consider the transformation $(x, y) = (\bar{x}^2, \bar{x}^3\bar{y})$.
In this case we have $\dot{x} = 2\bar{x}\dot{\bar{x}}$, hence $\dot{\bar{x}} = \frac{1}{2}\bar{x}^2\bar{y} + \mathrm{O}(\bar{x}^3)$ and $\dot{y} = 3\bar{x}^2\bar{y}\dot{\bar{x}} + \bar{x}^3\dot{\bar{y}}$,
hence $\dot{\bar{y}} = (1 - \frac{3}{2}\bar{y}^2)\bar{x} + \mathrm{O}(\bar{x}^2)$. We divide by $\bar{x}$ and find

$$\dot{\bar{x}} = \frac{\bar{x}\bar{y}}{2} + \mathrm{O}(\bar{x}^2),$$

$$\dot{\bar{y}} = 1 - \frac{3}{2}\bar{y}^2 + \mathrm{O}(\bar{x}).$$

We find two hyperbolic singularities of saddle type, situated at the points
$(\bar{x}, \bar{y}) = (0, \pm\sqrt{2/3})$.

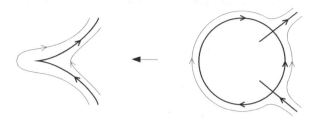

Fig. 3.11. Quasihomogeneous blow-up of the cusp singularity

Similar calculations in the negative $\bar{x}$-direction, as well as in the $\bar{y}$-direction show that no other singularities are present.

As such blowing up once suffices to desingularize the singularity leading to the picture in Fig. 3.11.

Again an accurate positioning of the invariant separatrices can be obtained by Taylor approximation of the stable and unstable manifolds.

A question one might ask is how to effectively find the coefficient (α, β) to use in a quasihomogeneous blow-up. This can be obtained by using the so called Newton diagram. We first define the Newton diagram.

Let $X = P(x,y)\frac{\partial}{\partial x} + Q(x,y)\frac{\partial}{\partial y}$ be a polynomial vector field with an isolated singularity at the origin.

Let $P(x,y) = \sum\limits_{i+j\geq 1} a_{ij}x^iy^j$ and $Q(x,y) = \sum\limits_{i+j\geq 1} b_{ij}x^iy^j$. The *support* of X is defined to be

$$S = \{(i-1, j) : a_{ij} \neq 0\} \cup \{(i, j-1) : b_{ij} \neq 0\} \subset \mathbb{R}^2,$$

and the *Newton polygon* of X is the convex hull Γ of the set

$$P = \bigcup_{(r,s)\in S} \{(r', s') : r' \geq r, s' \geq s\}.$$

The *Newton diagram* of X is the union γ of the compact faces γ_k of the Newton polygon Γ, which we enumerate from the left to the right. If there exists a face γ_k which lies completely on the half-plane $\{r \leq 0\}$, then we start the enumeration with $k = 0$, otherwise we start with $k = 1$. Since the origin is an isolated singularity we have that at least one of the points $(-1, s)$ or $(0, s)$ is an element of S for some s, and also at least one of the points $(r, 0)$ or $(r, -1)$ is an element of S for some r. Hence there always exists a face γ_1 in the Newton diagram.

Suppose that γ_1 has equation $\alpha r + \beta s = d$, with $\gcd(\alpha, \beta) = 1$. As a first step in the desingularization process we use a quasihomogeneous blow-up of degree (α, β). As an example we calculate the Newton diagram of the vector field (3.5), providing the best choice of coefficients (α, β).

The support of (3.5) surely contains $(-1, 1)$, $(2, -1)$, and $(1, 0)$, coming, respectively, from $y\frac{\partial}{\partial x}$, $x^2\frac{\partial}{\partial y}$, and $xy\frac{\partial}{\partial y}$. Besides these three points it can contain many other points, which are in fact not essential since they all lie

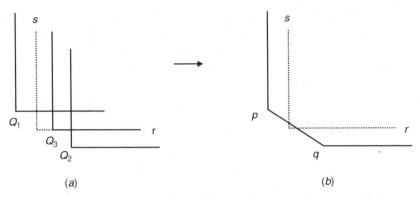

Fig. 3.12. Calculating the Newton polygon

in the convex hull Q of $Q_1 \cup Q_2 \cup Q_3$ with $Q_1 = \{(r,s) : r \geq -1, s \geq 1\}$, $Q_2 = \{(r,s) : r \geq 2, s \geq -1\}$, and $Q_3 = \{(r,s) : r \geq 1, s \geq 0\}$. In Fig. 3.12 we represent Q_i for $i = 1, 2, 3$ in (a) as well as $P = Q$ in (b).

We see that the Newton diagram consists of one compact face, that we denote by γ_1 and which is the line segment joining the points $p = (-1, 1)$ to $q = (2, -1)$. The line segment lies on the line of equation $2r + 3s = 1$ inducing the choice $(\alpha, \beta) = (2, 3)$.

In view of an efficient use of a quasihomogeneous blow-up with coefficients (α, β) we will no longer decompose a vector field in homogeneous components, but in adapted quasihomogeneous components. Based on this decomposition we will now describe an algorithm for blowing up.

We write

$$X = \sum_{j \geq d} X_j, \text{ where } X_j = P_j(x,y)\frac{\partial}{\partial x} + Q_j(x,y)\frac{\partial}{\partial y}$$

is the *quasihomogeneous component* of X of type (α, β) and quasihomogeneous degree j, that is to say $P_j(r^\alpha x, r^\beta y) = r^{j+\alpha}P_j(x,y)$ and $Q_j(r^\alpha x, r^\beta y) = r^{j+\beta}Q_j(x,y)$. After blowing up we will divide by r^d. In practice we first blow-up the vector field in the positive x-direction, yielding, after multiplying the result with $\alpha \bar{x}^{-d}$:

$$\bar{X}^x_+ : \quad \begin{aligned} \dot{\bar{x}} &= \sum_{\delta \geq d} \bar{x}^{\delta + 1 - d} P_\delta(1, \bar{y}), \\ \dot{\bar{y}} &= \sum_{\delta \geq d} \bar{x}^{\delta - d}(\alpha Q_\delta(1, \bar{y}) - \beta \bar{y} P_\delta(1, \bar{y})). \end{aligned}$$

We determine the singularities on the line $\{\bar{x} = 0\}$.

(1) If $\alpha Q_d(1, \bar{y}) - \beta \bar{y} P_d(1, \bar{y}) \not\equiv 0$, the points $(0, \bar{y}_0)$ satisfying the equation $\alpha Q_d(1, \bar{y}) - \beta \bar{y} P_d(1, \bar{y}) = 0$ are isolated singularities of $\bar{X}$ on the line $\{\bar{x} = 0\}$, at which

$$d(\bar{X}^x_+)_{(0,\bar{y}_0)} = \begin{pmatrix} P_d(1, \bar{y}_0) & 0 \\ \star & \alpha \frac{\partial Q_d}{\partial \bar{y}}(1, \bar{y}_0) - \beta(P_d(1, \bar{y}_0) + \bar{y}_0 \frac{\partial P_d}{\partial \bar{y}}(1, \bar{y}_0)) \end{pmatrix},$$

which immediately gives the eigenvalues as the diagonal entries. If the singularity is hyperbolic, we are done. If the singularity is semi-hyperbolic, we have to determine the behavior on the center manifold. If the singularity is nonelementary, we introduce $\tilde{y} = \bar{y} - \bar{y}_0$, and blow-up this vector field again in the positive $\bar{x}$-direction as well as in the positive and negative $\tilde{y}$-direction with a certain degree (α', β'), which we determine from the Newton diagram associated to the vector field.

(2) If $\alpha Q_d(1, \bar{y}) - \beta \bar{y} P_d(1, \bar{y}) \equiv 0$, we have a line of singularities. Since

$$D(\bar{X}_+^x)_{(0, \bar{y}_0)} = \begin{pmatrix} P_d(1, \bar{y}_0) & 0 \\ \star & 0 \end{pmatrix},$$

all the singularities are semi-hyperbolic, except those singularities $(0, \bar{y}_0)$ for which $P_d(1, \bar{y}_0) = 0$. The latter will require further blow-up.

Next we blow-up the vector field in the negative x-direction and study this vector field in the same way as in the previous case.

Finally we have to blow-up the vector field in the positive and the negative y-direction, and determine whether or not $(0, 0)$ is a singular point, since the others have been studied in the previous charts.

It is easy to see that $(0, 0)$ is a singularity if and only if γ_1 lies completely in the half-plane $\{r \geq 0\}$. If this is the case then $(0, 0)$ is elementary. Indeed, blowing up the vector field in the positive y-direction yields, after multiplying the result by $\beta \bar{y}^{-d}$:

$$\bar{X}_+^y : \quad \begin{aligned} \dot{\bar{x}} &= \sum_{\delta \geq d} \bar{y}^{\delta - d} (\beta P_\delta(\bar{x}, 1) - \alpha \bar{x} Q_\delta(\bar{x}, 1)), \\ \dot{\bar{y}} &= \sum_{\delta \geq d} \bar{y}^{\delta + 1 - d} Q_\delta(\bar{x}, 1). \end{aligned}$$

Hence $(0, 0)$ is a singular point if $P_d(0, 1) = 0$, i.e., if $P_d(x, y) = xF(x, y)$, implying that γ_1 lies completely in the half-plane $\{r \geq 0\}$. Suppose now that $(0, 0)$ is a singular point of $\bar{X}_+^y$; then we have

$$D(\bar{X}_+^y)_{(0,0)} = \begin{pmatrix} \beta \frac{\partial P_d}{\partial \bar{x}}(0, 1) - \alpha Q_d(0, 1) & \star \\ 0 & Q_d(0, 1) \end{pmatrix}.$$

Let $(0, s)$ be the intersection of the line γ_1 and the line $r = 0$. Then $P_d(x, y) = axy^s + G(x, y)$ and $Q_d(x, y) = by^{s+1} + H(x, y)$, with $a^2 + b^2 \neq 0$, $\deg_x G(x, y) \geq 2$ and $\deg_x H(x, y) \geq 1$. Hence $\beta \frac{\partial P_d}{\partial \bar{x}}(0, 1) - \alpha Q_d(0, 1) = a\beta - b\alpha$. So if $a\beta - b\alpha \neq 0$ then $(0, 0)$ is nonelementary. If $a\beta - b\alpha = 0$, then $Q_d(0, 1) = b \neq 0$, and $(0, 0)$ is elementary, too.

In [124] it has been proven that the algorithm, as presented here, leads to a desingularization. It is also more efficient than the usual one.

In the program P4 [9] we not only perform a detailed study near the singular points in $\mathbb{R}^2$, but also near singular points at infinity. In Chap. 5 we will describe how polynomial vector fields on $\mathbb{R}^2$ can be extended to infinity. We now apply the blow-up technique to study nilpotent singularities.

3.4 Nilpotent Singularities

In this section we study singularities, positioned at the origin, at which the linear approximation $DX(0)$ of the vector field X is linearly conjugate to $y\dfrac{\partial}{\partial x}$.

As usual we take X to be at least of class C^∞; we recall that such a singularity is called a nilpotent singularity or nilpotent singular point.

Using the Formal Normal Form Theorem presented in Sect. 2.1 and, more specifically, the example treated there, we find the following normal form for C^∞-conjugacy:

$$\dot{x} = y + A(x,y),$$
$$\dot{y} = f(x) + yg(x) + y^2 B(x,y),$$

where f, g, A, and B are C^∞ functions, $j_1 f(0) = g(0) = j_\infty A(0,0) = j_\infty B(0,0) = 0$. By introducing the new variable $Y = y + A(x,y)$, we change the former expression into

$$\dot{x} = y,$$
$$\dot{y} = f(x) + yg(x) + y^2 B(x,y), \tag{3.6}$$

for appropriately adapted f, g, and B with similar properties as before. If $B \equiv 0$, then the system comes from the Liénard equation $\ddot{x} = f(x) + \dot{x}g(x)$. We now make a complete local topological study of all cases in which $j_\infty f(0)$ is not zero. This includes the local study of the related Liénard equations. Either $j_\infty g(0) \neq 0$ and

$$f(x) = ax^m + o(x^m),$$
$$g(x) = bx^n + o(x^n), \tag{3.7}$$

with $ab \neq 0$, or $j_\infty g(0) = 0$. The dual 1-form of (3.6) is given by

$$-ydy + (f(x) + yg(x) + y^2 B(x,y))dx,$$

which is equal to

$$- ydy + d\overline{f}(x) + yd\overline{g}(x) + y^2 B(x,y)dx, \tag{3.8}$$

for some C^∞ functions $\overline{f}$ and $\overline{g}$ such that

$$\overline{f}(x) = \frac{ax^{m+1}}{m+1} + o(x^{m+1}) \quad \text{and}$$
$$\overline{g}(x) = \frac{bx^{n+1}}{n+1} + o(x^{n+1}),$$

provided $j_\infty g(0) \neq 0$.

A linear change in x permits $\overline{f}$ to be changed into

$$\overline{f}(x) = \frac{\delta x^{m+1}}{m+1} + o(x^{m+1}),$$

with $\delta = 1$ in case m is even and $\delta = \pm 1$ in case m is odd. Changing y by $-y$ if necessary we may suppose that $b > 0$. Instead of reducing a to δ, we could also perform on $\overline{g}$ an operation similar to the one performed on $\overline{f}$ to reduce b to $+1$, hence obtaining

$$\overline{g}(x) = \frac{x^{n+1}}{n+1} + o(x^{n+1}),$$

provided $j_\infty g(0) \neq 0$.

So up to linear (not necessarily orientation preserving) equivalence, we can suppose that in expression (3.7)

$$\text{either } a = \delta \text{ and } b > 0, \text{ or } b = 1, \tag{3.9}$$

with $\delta = 1$ when m is even and $\delta = \pm 1$ when m is odd.

If $j_\infty g(0) \neq 0$, we observe that a coordinate change

$$y = Y + \int_0^x g(u)du = Y + G(x), \tag{3.10}$$

permits changing an expression (3.6) with $B \equiv 0$ into

$$\dot{x} = Y + G(x),$$
$$\dot{y} = f(x).$$

If $B \not\equiv 0$, then (3.10) changes expression (3.6) into

$$\dot{x} = Y + G(x),$$
$$\dot{Y} = F(x) + YH(x) + Y^2 D(x,Y), \tag{3.11}$$

where G, F, H, and D are C^∞ functions, $j_\infty F(0) = j_\infty f(0)$, $j_\infty H(0) = j_\infty D(0,0) = 0$. By a well chosen C^∞ coordinate change $Y = y + \alpha(x)$ with $j_\infty \alpha(0) = 0$ we can even change (3.11) into

$$\dot{x} = y + \overline{G}(x),$$
$$\dot{y} = \overline{F}(x) + y^2 C(x,y), \tag{3.12}$$

with $j_\infty \overline{G}(0) = j_\infty G(0)$, $j_\infty \overline{F}(0) = j_\infty F(0)$ and $j_\infty C(0,0) = 0$. Then expression (3.12) is also valid if $j_\infty g(0) = 0$.

We write (3.12) as

$$\dot{x} = y + H(x),$$
$$\dot{y} = F(x) + y^2 C(x,y), \tag{3.13}$$

with $H(0) = 0$. The relation between (3.12) and (3.13) is given by

$$j_\infty H'(0) = j_\infty g(0),$$
$$j_\infty F(0) = j_\infty f(0).$$

The study of these singularities now relies on (quasihomogeneous) blow-up. We systematically work it out, depending on the values of m and n, where $m \geq 2$ and $n \geq 1$ including $n = \infty$, the latter of which means that we accept $j_\infty g(0) = 0$ in expression (3.6).

We distinguish three cases.

Hamiltonian like case: $m < 2n + 1$. If m is odd we use the blow-up

$$x = u, \quad y = u^k \overline{y}, \tag{3.14}$$

with $k = (m + 1)/2$ and we divide by u^{k-1}.

If m is even we use

$$x = u^2, \quad y = u^{m+1} \overline{y}, \tag{3.15}$$

and we divide by u^{m-1}.

Singular like case: $m > 2n + 1$. We use the blow-up

$$x = u, \quad y = u^{n+1} \overline{y}, \tag{3.16}$$

and we divide by u^n.

Mixed case: $m = 2n + 1$. We use again the blow-up (3.16)

In all cases there is no need to check the directional charts $\{\overline{y} = \pm 1\}$ because of the specific expression of the linear part. We may also restrict to $\{\overline{x} = 1\}$, since the n-exponent in front of $\overline{x}$ is odd, except for blow-up (3.15).

3.4.1 Hamiltonian Like Case ($m < 2n + 1$)

We start with expression (3.8) for the dual 1-form and change $\overline{f}(x)$ to $\delta x^{m+1}/(m+1)$ by a coordinate change in x. This changes expression (3.6) by C^∞ equivalence into

$$\dot{x} = y,$$
$$\dot{y} = \delta x^m + y(bx^n + o(x^n)) + O(y^2).$$

Case m odd: We know that $\delta = \pm 1$ and we use blow-up (3.14) in the $\overline{x}$-direction $(x, y) \longmapsto (u, u^k \overline{y})$, with $k = (m + 1)/2$. After division by u^{k-1} we get

$$\dot{u} = u \overline{y},$$
$$\dot{\overline{y}} = (\delta - k \overline{y}^2) + O(u).$$

The blow-up and related phase portraits for this system that can be seen in Fig. 3.13.

For $\delta = 1$ we get a singularity of saddle-type. One can prove that it is C^0-conjugate to a hyperbolic saddle, but we will not work it out, as announced before. The contact between the different separatrices is described by the blowing up mapping.

For $\delta = -1$ we get a singularity of center or focus type. It is not a simple problem to determine whether it is a center or a focus.

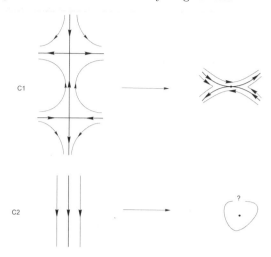

Fig. 3.13. Desingularization of Hamiltonian like case when m is odd

Case m even: In this case we can take $\delta = 1$ and we use blow-up (3.15) in the $\overline{x}$-direction as well as in the $-\overline{x}$-direction

$$(x, y) = (u^2, u^{m+1}\overline{y}), \quad \text{or}$$
$$(x, y) = (-u^2, u^{m+1}\overline{y}).$$

After division by u^{m-1} we get, respectively,

$$\dot{u} = \frac{u\overline{y}}{2},$$
$$\dot{\overline{y}} = \left(1 - \frac{m+1}{2}\overline{y}^2\right) + O(u), \qquad \text{and} \qquad \begin{aligned} \dot{u} &= -\frac{u\overline{y}}{2}, \\ \dot{\overline{y}} &= \left(1 + \frac{m+1}{2}\overline{y}^2\right) + O(u). \end{aligned}$$

The blow-up and related phase portraits for this system can be seen in Fig. 3.14.

We get a singularity consisting of two hyperbolic sectors. It is called a *cusp point*. The contact between the separatrices is described by the blowing up mapping.

We note that in the Hamiltonian like case we can also take $n = +\infty$, meaning that $j_\infty g(0) = 0$ in expression (3.6).

3.4.2 Singular Like Case ($m > 2n + 1$)

We start with expression (3.8) for the dual 1-form and, making analogous changes as in the Hamiltonian like case but on $\overline{g}$ instead of $\overline{f}$, we change expression (3.6) by C^∞ equivalence, and possible time reversal, into

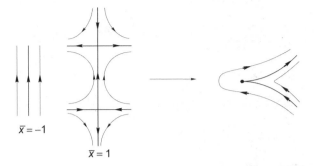

Fig. 3.14. Desingularization of Hamiltonian like case when m even

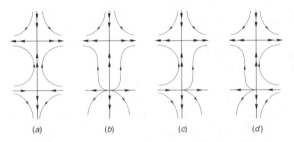

Fig. 3.15. Blow-ups of the singular like case

$$\dot{x} = y,$$
$$\dot{y} = ax^m + y(x^n + o(x^n)) + O(y^2),$$

with $a \neq 0$.

We use blow-up (3.16) in the $\bar{x}$-direction

$$(x, y) = (u, u^{n+1}\bar{y}).$$

After division by u^n we get

$$\dot{u} = u\bar{y},$$
$$\dot{\bar{y}} = \bar{y}(1 - (n+1)\bar{y} + O(u)) + au^{m-2n-1}.$$

On $\{u = 0\}$ we find two singularities, situated, respectively, at $\bar{y} = 1/(n+1)$ and at $\bar{y} = 0$. The former is clearly a hyperbolic saddle. The latter is a semi-hyperbolic point with $\{u = 0\}$ as unstable manifold and having a center manifold transverse to it. We leave it as an exercise to prove that the center behavior is not flat, because of the presence of the term au^{m-2n-1}. This leads to the possible portraits for the blow-up given in Fig. 3.15.

In getting the $\{\bar{x} = -1\}$–chart out of these pictures, we must take care concerning the parity of n. Depending on the parity of n, cases (b) and (d) in Fig. 3.15 will induce two different phase portraits (see, respectively, (a), (c) and (e), (f) in Fig. 3.16).

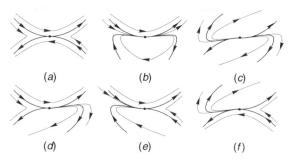

Fig. 3.16. Phase portraits of the singular like case

The totality of phase portraits obtained is represented in Fig. 3.16. Again the contact in between the separatrices is clear from the blow-up mapping.

The cases (d) and (f) are topologically equivalent, so we have five different phase portraits.

Figure 3.16 may give the impression that a global attractor, similar to the global repellor in (c), might not be possible in the singular like case, but this is merely because of the choice of the coefficients in the normal form. Multiplying the vector field by -1, corresponding to a time reversal, creates the possibility of getting such an attractor. Thus there are six topologically distinct phase portraits in all.

3.4.3 Mixed Case ($m = 2n + 1$)

We start with expression (3.8) for the dual 1-form and, making analogous changes as in the previous cases, we change expression (3.6) by C^∞ equivalence, and possible time reversion, into

$$\dot{x} = y,$$
$$\dot{y} = ax^{2n+1} + y(x^n + o(x^n)) + O(y^2),$$

with $a \neq 0$.

We use blow-up (3.16) in the $\bar{x}$-direction $(x, y) = (u, u^{n+1}\bar{y})$. After division by u^n we get

$$\dot{u} = u\bar{y},$$
$$\dot{\bar{y}} = a + \bar{y} - (n+1)\bar{y}^2 + \bar{y}O(u),$$

with $a \neq 0$.

On $\{u = 0\}$ the singularities are given by the solution of

$$a + \bar{y} - (n+1)\bar{y}^2 = 0.$$

We either have zero, one, or two solutions. No solution can be situated at the origin, since $a \neq 0$. The blow-ups are shown in Fig. 3.17.

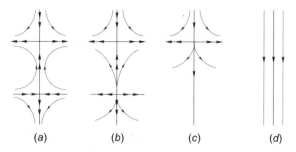

Fig. 3.17. Blow-ups of the mixed case

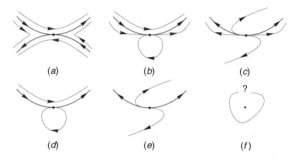

Fig. 3.18. Phase portraits of the mixed case

Again, depending on the parity of n, cases (b) and (c) give rise to two different phase portraits.

The blow-ups in Fig. 3.17 hence induce the phase portraits of Fig. 3.18.

Although contacts of separatrices might differ, these phase portraits are not topologically different from the ones we already found before. In case (f) we are again left with a center or focus problem.

Remark 3.4 One can prove that for nilpotent singularities with $j_\infty f(0) \neq 0$, there are eight different classes for topological conjugacy. As we have seen, the multiplicity is clearly given by the number m.

With respect to (3.6) it remains to consider the cases in which $j_\infty f(0) = 0$.

We perform this study only for analytic vector fields. We start with an analytic vector field X given by

$$\dot{x} = y + \alpha(x,y),$$
$$\dot{y} = \beta(x,y),$$

with α and β analytic, $j_1\alpha(0,0) = j_1\beta(0,0) = 0$.

If we introduce the new variable $Y = y + \alpha(x,y)$, and write y instead of Y, then we get

$$\dot{x} = y,$$
$$\dot{y} = \delta(x) + y\gamma(x,y). \tag{3.17}$$

We now prove that

$$j_\infty f(0) = 0 \quad \text{if and only if} \quad \delta(x) \equiv 0. \tag{3.18}$$

The Normal Form Theorem relies on transformations of the form

$$(u, v) = (x + A(x, y), y + B(x, y)), \tag{3.19}$$

with A and B polynomials of a given degree n; the procedure uses an induction on n. To prove (3.18) we need only make a formal calculation, that is simply look at ∞-jets at $(0, 0)$. The operation to transform (3.17) to a normal form by means of (3.19) implies that

$$\left(1 + \frac{\partial A}{\partial x}\right) y + \frac{\partial A}{\partial y}(\delta(x) + y\gamma(x, y)) = y + B(x, y),$$

$$\frac{\partial B}{\partial x}y + \left(1 + \frac{\partial B}{\partial y}\right)(\delta(x) + y\gamma(x, y)) = f(x + A(x, y)) \tag{3.20}$$

$$+ (y + B(x, y))\Gamma(x, y),$$

for some function Γ.

If we consider the second equation at $y = 0$, this gives

$$\left(1 + \frac{\partial B}{\partial y}(x, 0)\right)\delta(x) = f(x + A(x, 0)) + B(x, 0)\Gamma(x, 0).$$

Taking into account the degree of the terms, we see that $j_n\delta(0) = j_nf(0)$ when $j_{n-1}\delta(0) = j_{n-1}f(0) = 0$.

An induction argument hence shows that both f and δ have the same term of lowest degree, with the same coefficient.

Therefore the condition $j_\infty f(0) = 0$ on the normal form implies that $\delta(x) \equiv 0$ in expression (3.17), and we get the vector field X given by

$$\dot{x} = y,$$
$$\dot{y} = y\gamma(x, y), \tag{3.21}$$

for which $\{y = 0\}$ is a line of singularities.

We now prove that

$$j_\infty g(0) = 0 \quad \text{if and only if} \quad \gamma(x, 0) = 0. \tag{3.22}$$

This means that, under the condition $j_\infty f(0) = 0$, $j_\infty g(0) = 0$ as well if and only if the divergence of X is identically zero along the curve of singularities.

The proof of (3.22) is similar to the proof of (3.18). Instead of (3.20) we now find the equations (at the ∞-jet level)

$$\left(1 + \frac{\partial A}{\partial x}\right) y + \frac{\partial A}{\partial y}y\gamma(x, y) = y + B(x, y),$$

$$\frac{\partial B}{\partial x}y + \left(1 + \frac{\partial B}{\partial y}\right)y\gamma(x, y) = (y + B(x, y))g(x + A(x, y)).$$

In the second equation we see that

$$B(x,0)g(x + A(x,0))$$

has to be zero. So either $B(x,0) = 0$ or $j_\infty g(0) = 0$.

We first take $B(x,0) = 0$. If we now divide the second equation by y, and put $y = 0$ we then get

$$\left(1 + \frac{\partial B}{\partial y}(x,0)\right) \gamma(x,0) = g(x + A(x,0)),$$

implying that the n-jet of $\gamma(x,0)$ is given by $j_n g(0)$ when their $(n-1)$-jets are both zero. If however $j_\infty g(0) = 0$, then we get

$$\frac{\partial B}{\partial x}(x,0) + \left(1 + \frac{\partial B}{\partial y}(x,0)\right) \gamma(x,0) = 0,$$

implying inductively that the ∞-jet of $\gamma(x,0)$ also has to be zero.

We hence find two different situations for expressions (3.21). On the one hand we have

$$\begin{aligned}
\dot{x} &= y, \\
\dot{y} &= y^2 \varphi(x,y),
\end{aligned} \tag{3.23}$$

describing the cases whose formal normal form is zero. The phase portrait is obtained by drawing the flow box

$$\begin{aligned}
\dot{x} &= 1, \\
\dot{y} &= y\varphi(x,y),
\end{aligned}$$

and multiplying it by the function y. The result is presented in Fig. 3.19.

On the other hand we have the analytic expressions

$$\begin{aligned}
\dot{x} &= y, \\
\dot{y} &= yx^n(1 + r(x)) + y^2\psi(x,y),
\end{aligned} \tag{3.24}$$

for some $n \geq 1$.

It is again a flow box multiplied by the function y. The flow box is

$$\begin{aligned}
\dot{x} &= 1, \\
\dot{y} &= x^n(1 + r(x)) + y\psi(x,y).
\end{aligned} \tag{3.25}$$

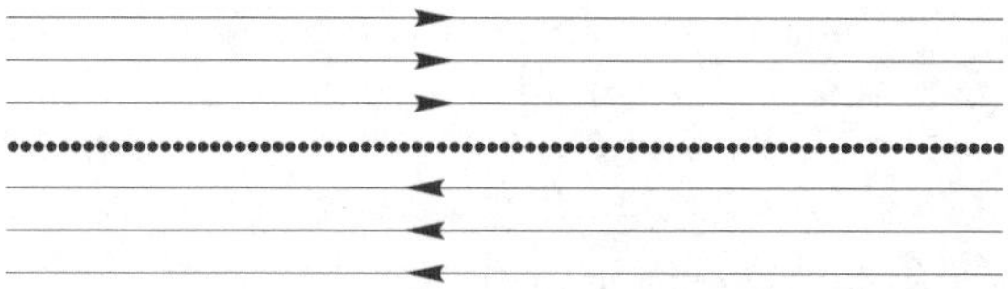

Fig. 3.19. Phase portrait of (3.23)

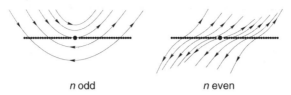

n odd n even

Fig. 3.20. Phase portrait of (3.24)

Along $\{y = 0\}$ the contact of (3.25) with the x-axis is described by

$$\dot{y} = x^n(1 + r(x)) \sim x^n.$$

This leads to the phase portraits described in Fig. 3.20.

3.5 Summary on Nilpotent Singularities

Here we give a specific and practical theorem which summarizes the previous results and which is very helpful for determining the local behavior at a nilpotent singular point; see [2] for more details.

Theorem 3.5 (Nilpotent Singular Points Theorem) *Let $(0,0)$ be an isolated singular point of the vector field X given by*

$$\begin{aligned}
\dot{x} &= y + A(x,y), \\
\dot{y} &= B(x,y),
\end{aligned} \qquad (3.26)$$

where A and B are analytic in a neighborhood of the point $(0,0)$ and also $j_1A(0,0) = j_1B(0,0) = 0$. Let $y = f(x)$ be the solution of the equation $y + A(x,y) = 0$ in a neighborhood of the point $(0,0)$, and consider $F(x) = B(x,f(x))$ and $G(x) = (\partial A/\partial x + \partial B/\partial y)(x,f(x))$. Then the following holds:

(1) If $F(x) \equiv G(x) \equiv 0$, then the phase portrait of X is given by Fig. 3.21a.

(2) If $F(x) \equiv 0$ and $G(x) = bx^n + o(x^n)$ for $n \in \mathbb{N}$ with $n \geq 1$ and $b \neq 0$, then the phase portrait of X is given by Fig. 3.21b or c.

(3) If $G(x) \equiv 0$ and $F(x) = ax^m + o(x^m)$ for $m \in \mathbb{N}$ with $m \geq 1$ and $a \neq 0$, then

  (i) If m is odd and $a > 0$, then the origin of X is a saddle (see Fig. 3.21d) and if $a < 0$, then it is a center or a focus (see Fig. 3.21e–g);

  (ii) If m is even then the origin of X is a cusp as in Fig. 3.21h.

(4) If $F(x) = ax^m + o(x^m)$ and $G(x) = bx^n + o(x^n)$ with $m \in \mathbb{N}$, $m \geq 2$, $n \in \mathbb{N}$, $n \geq 1$, $a \neq 0$ and $b \neq 0$, then we have

  (i) If m is even, and

    (i1) $m < 2n + 1$, then the origin of X is a cusp as in Fig. 3.21h;

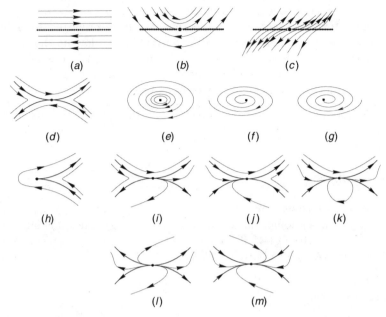

Fig. 3.21. Phase portraits of nilpotent singular points

(i2) $m > 2n + 1$, *then the origin of X is a saddle-node as in Fig. 3.21i or j;*

(ii) *If m is odd and $a > 0$ then the origin of X is a saddle as in Fig. 3.21d;*

(iii) *If m is odd, $a < 0$ and*

(iii1) *Either $m < 2n + 1$, or $m = 2n + 1$ and $b^2 + 4a(n + 1) < 0$, then the origin of X is a center or a focus (see Fig. 3.21e–g);*

(iii2) *n is odd and either $m > 2n + 1$, or $m = 2n + 1$ and $b^2 + 4a(n + 1) \geq 0$, then the phase portrait of the origin of X consists of one hyperbolic and one elliptic sector as in Fig. 3.21k;*

(iii3) *n is even and either $m > 2n+1$, or $m = 2n+1$ and $b^2 + 4a(n+1) \geq 0$, then the origin of X is a node as in Fig. 3.21l, m. The node is attracting if $b < 0$ and repelling if $b > 0$.*

Remark 3.6 In Fig. 3.21 we have represented all possible phase portraits of nilpotent singularities. In the pictures we have paid attention to the fact that the separatrices have certain contacts, but we have of course not stressed the exact order of contact they have. This easily follows from the blow-up procedures. We have also ignored to the exact position of the different separatrices in relation to $\{y = 0\}$ in case that we change the expression of (3.26), by means of an analytic coordinate change, to a new one in which $x = y$. Such information can also be easily been obtained by the blow-up procedure.

Remark 3.7 Section 3.4 has not really been arranged in such a way as to contain a systematic proof of this theorem. We have preferred to treat the different nilpotent singularities according to the kind of blow-up that is needed to study them. It is however clear that in order to get a precise proof of Theorem 3.5 it is sufficient to go through the different cases and to apply the blow-ups as indicated in Sect. 3.4. We can leave this as an exercise.

The remaining center-focus problem, on the other hand, is an open problem.

3.6 Exercises

Exercise 3.1 Consider the vector field $(y + 2x^3)\frac{\partial}{\partial x} + (x^2 + xy + y^3)\frac{\partial}{\partial y}$. Make an appropriate quasihomogeneous blow-up at the origin and calculate a parametrization $\gamma : (\mathbb{R}, 0) \to (\mathbb{R}^2, 0)$ of the two separatrices, up to terms of order 5, i.e., with a remainder of order $O(t^5)$.

Exercise 3.2 Study the following vector fields by means of an appropriate quasihomogeneous blow-up

(i) $-y\frac{\partial}{\partial x} + x^3\frac{\partial}{\partial y}$,
(ii) $(y^3 + xy^2)\frac{\partial}{\partial x} + x^2\frac{\partial}{\partial y}$.

Exercise 3.3 Let X be a C^∞ vector field satisfying a Łojasiewicz inequality at p and suppose that the C^∞ vector field Y at q is C^∞-conjugate to X at p. Show that Y at q also satisfies a Łojasiewicz inequality.

Exercise 3.4 Check that the following vector fields satisfy a Łojasiewicz inequality at the origin:

(i) $(y + x^3)\frac{\partial}{\partial x} - x^3\frac{\partial}{\partial y}$.
(ii) $(x + y)\frac{\partial}{\partial x} + y^4\frac{\partial}{\partial y}$.

Exercise 3.5 Prove that a Łojasiewicz inequality holds for

(i) all hyperbolic singularities.
(ii) Singularities whose 1-jet is a center.
(iii) Semi-hyperbolic singularities with nonflat center behavior.
(iv) Nilpotent singularities for which in (3.6) $j_\infty f(0) \neq 0$ holds.

Exercise 3.6 Prove that a C^∞ vector field X satisfies a Łojasiewicz inequality at a singularity p if and only if there exists a finite jet $j_k X(p)$ with the property that $||j_k X(p)|| \geq c||x||^k$ for some $c > 0$.

Exercise 3.7 Prove that no Łojasiewicz inequality holds for the vector field

$$X = (y + x^2)^2 \frac{\partial}{\partial x} + (y + x^2)\frac{\partial}{\partial y},$$

but its 2-jet $j_2X(0)$ does satisfy a Lojasiewicz inequality $||j_2X(0)|| \geq c||x||^4$ for some $c > 0$.

Exercise 3.8 Prove the existence of a finite sectorial decomposition for the following singularities at $(0,0)$.

(i) $y\frac{\partial}{\partial x} + x^2\frac{\partial}{\partial y}$,

(ii) $(3x^2 - 2xy)\frac{\partial}{\partial x} + (y^2 - 3xy)\frac{\partial}{\partial y}$,

(iii) $(x^2 - y^2)\frac{\partial}{\partial x} + 2xy\frac{\partial}{\partial y}$,

(iv) $x^2\frac{\partial}{\partial x} + y(2x - y)\frac{\partial}{\partial y}$

Exercise 3.9 Consider the vector field in Example 3.1, of which the local phase portrait is represented in Fig. 3.2,

$$X = (x^2 - 2xy)\frac{\partial}{\partial x} + (y^2 - xy)\frac{\partial}{\partial y} + O(||(x,y)||^3).$$

(i) Show that for ε sufficiently small, the six separatrices of this system cut $S_\varepsilon = \{(x,y)|x^2 + y^2 = \varepsilon\}$ transversely.

(ii) Show that for each of these systems the "finite sectorial decomposition" property holds on some neighborhood V of the origin.

(iii) Prove that any two of the above systems are locally C^0-equivalent.

(iv) Show that inside any hyperbolic sector, the time to go from the boundary ∂V to itself tends monotonically to infinity when the orbit approaches the separatrices.

(v) Use (iv) to prove that any two of the above systems are mutually C^0-conjugate.

Exercise 3.10 Show that for any C^∞ singularity of Lojasiewicz type (satisfying a Lojasiewicz inequality) with a characteristic orbit, there is a finite sectorial decomposition whose boundary is C^∞.

Remark: The proof of the existence of a C^∞ boundary relies on the use of a "C^∞-partition of the unity." As a first step in the proof we suggest proving the existence of a C^0 boundary.

Exercise 3.11 Check that every C^∞ singularity that satisfies a Lojasiewicz inequality but does not have a characteristic orbit is necessarily monodromic.

Hint: Provide a proof based on the theorems that are cited in the book, even those whose proof is not incorporated.

Exercise 3.12 Use Theorem 3.5 to check the nature of the singularity at the origin of

$$(-x^2 + ay^2 - xy - 2xy^2)\frac{\partial}{\partial x} + (y^2 + xy + y^3)\frac{\partial}{\partial y},$$

with $a \neq 0$ and $1 + 2a > 0$.

3.7 Bibliographical Comments

The desingularization theorem for planar vector fields has a long history. It was first stated by Bendixson in 1901, however without proof. The paper also included the topological classification of the elementary singular points.

In 1968, Seidenberg gave the first rigorous proof of the theorem for the analytic case. The desingularization procedure was extended to C^∞ vector fields of Lojasiewicz type in [52]. This paper is based on the Ph.D.–thesis of Dumortier from 1973. In the mid seventies, van den Essen found a transformed proof of the desingularization theorem for analytic vector fields; see [159].

In all previous papers, the desingularization was based on quadratic transformations, or in the real case, on polar blow-up, hence in the terminology of this chapter, on the homogeneous blow-up.

Quasihomogeneous blow-up was already used in the book of Lyapunov [106] but was essentially put forward as a systematic and a more powerful technique in the paper by Brunella and Miari [22] and especially in the book of Bruno [23]. A proof of the desingularization theorem for C^∞ vector fields of Lojasiewicz type, based on quasihomogeneous blow-ups, was given by Pelletier in her thesis [124]; see also [125]. The desingularization procedure used in the program P4 [9] is based on the algorithm presented in that thesis. The classification of nilpotent singularities can be found in the papers of Andreev [2] and of Arrowsmith [8]. The elaboration that we provide in this chapter is based on quasihomogeneous blow-up and is by far the simplest that seems possible.

4

Centers and Lyapunov Constants

One of the more classical problems in the qualitative theory of planar differential equations is the problem of distinguishing between a focus and a center. This problem is unsolved in general, but in the case that the singular point is a linear center there are algorithms for solving it. The present chapter provides one of the best of these algorithms.

4.1 Introduction

Suppose that P and Q are analytic functions defined in a neighborhood of the origin, and that $(x, y) = (0, 0)$ is a singular point, linearly a focus or a center, for the system

$$\dot{x} = P(x, y), \qquad \dot{y} = Q(x, y).$$

Doing a suitable linear change of variables, the system can written in the form

$$\dot{x} = \lambda x + y + p(x, y), \qquad \dot{y} = -x + \lambda y + q(x, y), \qquad (4.1)$$

where p and q are analytic functions without constant or linear terms. If $\lambda \neq 0$ the origin is a strong focus, stable if $\lambda < 0$, and unstable if $\lambda > 0$. If $\lambda = 0$ the origin is either a weak focus or a center. The search for necessary and sufficient conditions on p and q for the origin to be a center has a long history and many methods have been developed. When the origin is a center there exists an analytic first integral; this result is due to Poincaré and for a proof see [90].

Recently, the problem of distinguishing between a weak focus and a center has been stimulated considerably by the use of computer algebra systems. In particular, it has been made possible to find necessary and sufficient conditions for a singularity to be a center for many classes of systems of the form (4.1) which had previously been intractable.

In this chapter we will provide an algorithm for distinguishing between a focus and a center when the system is linearly a center, that is, when the eigenvalues at the origin are purely imaginary. In the first section of this chapter we present the normal form for singularities having a linear center. The rest of the chapter is dedicated to the algorithm and to its applications.

4.2 Normal Form for Linear Centers

In this section we will follow the notations and definitions introduced in Sect. 2.1.

We calculate the normal forms for singularities having a linear center. Although it is not necessary, we use complex coordinates, since it simplifies the calculations.

Take $z = x + iy$ and $\bar{z} = x - iy$; as we know

$$\frac{\partial}{\partial z} = \frac{1}{2}\left(\frac{\partial}{\partial x} - i\frac{\partial}{\partial y}\right), \quad \frac{\partial}{\partial \bar{z}} = \frac{1}{2}\left(\frac{\partial}{\partial x} + i\frac{\partial}{\partial y}\right)$$

and

$$\alpha\left(x\frac{\partial}{\partial y} - y\frac{\partial}{\partial x}\right) = i\alpha\left(z\frac{\partial}{\partial z} - \bar{z}\frac{\partial}{\partial \bar{z}}\right),$$

$$\left[z\frac{\partial}{\partial z} - \bar{z}\frac{\partial}{\partial \bar{z}}, z^i\bar{z}^j\frac{\partial}{\partial z}\right] = (i - 1 - j)z^i\bar{z}^j\frac{\partial}{\partial z},$$

$$\left[z\frac{\partial}{\partial z} - \bar{z}\frac{\partial}{\partial \bar{z}}, z^i\bar{z}^j\frac{\partial}{\partial \bar{z}}\right] = (i + 1 - j)z^i\bar{z}^j\frac{\partial}{\partial \bar{z}}.$$

We have that $\operatorname{Ker} ad_{2m}A = \{0\}$, while the kernel of $ad_{2m+1}A$ is spanned by
$\left\{(z\bar{z})^m z\frac{\partial}{\partial z}, (z\bar{z})^m\bar{z}\frac{\partial}{\partial \bar{z}}\right\}$, or by

$$\left\{(z\bar{z})^m i\left(z\frac{\partial}{\partial z} - \bar{z}\frac{\partial}{\partial \bar{z}}\right), (z\bar{z})^m\left(z\frac{\partial}{\partial z} + \bar{z}\frac{\partial}{\partial \bar{z}}\right)\right\} =$$
$$\left\{(x^2 + y^2)^m\left(x\frac{\partial}{\partial y} - y\frac{\partial}{\partial x}\right), (x^2 + y^2)^m\left(x\frac{\partial}{\partial x} + y\frac{\partial}{\partial y}\right)\right\}.$$

This leads to the ∞-jet

$$(\alpha + f(x^2 + y^2))\left(x\frac{\partial}{\partial y} - y\frac{\partial}{\partial x}\right) + g(x^2 + y^2)\left(x\frac{\partial}{\partial x} + y\frac{\partial}{\partial y}\right),$$

with $f(0) = g(0) = 0$, which is the normal form for C^∞ conjugacy of a linear center. If we divide it by $(\alpha + f(x^2 + y^2))$ we get the following normal form for C^∞ equivalence:

$$\left(x\frac{\partial}{\partial y} - y\frac{\partial}{\partial x}\right) + h(x^2 + y^2)\left(x\frac{\partial}{\partial x} + y\frac{\partial}{\partial y}\right). \tag{4.2}$$

Both normal forms are only formal and not C^∞; it is in general not possible to remove extra flat terms, as we did in Sect. 2.7 for the elementary singularities. The formal series f, g, and h are also not convergent in general, not even for analytic (4.1). Nevertheless we now show how to prove that an analytic (4.1) represents a center if and only if h is identically zero.

We consider a planar analytic differential equation

$$
\begin{aligned}
\dot{x} &= -y + P(x,y) = -y + \sum_{k=2}^{\infty} P_k(x,y), \\
\dot{y} &= \quad x + Q(x,y) = \quad x + \sum_{k=2}^{\infty} Q_k(x,y),
\end{aligned}
\tag{4.3}
$$

where P_k and Q_k are homogeneous polynomials of degree k. We can also write it in polar coordinates (r, θ):

$$
\frac{dr}{d\theta} = \sum_{k=2}^{\infty} S_k(\theta) r^k,
\tag{4.4}
$$

where $S_k(\theta)$ are trigonometric polynomials in the variables $\sin(\theta)$ and $\cos(\theta)$; or in complex coordinates $(z, \bar{z})$:

$$
\dot{z} = iz + \sum_{k=2}^{\infty} R_k(z, \bar{z}),
\tag{4.5}
$$

where R_k are homogeneous polynomials of degree k in $z, \bar{z}$.

If we denote by $r(\theta, r_0)$ the solution of (4.4) such that $r(0, r_0) = r_0$ then close to $r = 0$ we have

$$
r(\theta, r_0) = r_0 + \sum_{k=2}^{\infty} u_k(\theta) r_0^k,
$$

with $u_k(0) = 0$ for $k \geq 2$. The Poincaré return map near $r = 0$ is expressed by

$$
\Pi(r_0) = r(2\pi, r_0) = r_0 + \sum_{k=2}^{\infty} u_k(2\pi) r_0^k.
$$

Since Π is analytic it is clear that $\Pi(r_0) \equiv r_0$, and hence that (4.3) represents a center, if and only if $u_n(2\pi) = 0$ for all n. Suppose for a moment that this is not the case and let K be the smallest natural number such that $u_K(2\pi) \neq 0$. This K clearly does not depend on the analytic coordinates chosen, nor does it change when we multiply the vector field by a positive analytic function. We can hence use the fact that (4.3) is analytically equivalent to an analytic equation whose $(2k+1)$-jet is given by

$$
\left(x\frac{\partial}{\partial y} - y\frac{\partial}{\partial x} \right) + \sum_{l=1}^{k} a_l (x^2 + y^2)^l \left(x\frac{\partial}{\partial x} + y\frac{\partial}{\partial y} \right),
\tag{4.6}
$$

with $k = [K/2]$.

The analytic equation can be written in polar coordinates as

$$\frac{dr}{d\theta} = \sum_{l=1}^{k} a_l r^{2l} + O(r^{2k+2}). \tag{4.7}$$

It is a straightforward calculation to show that necessarily $a_1 = \ldots = a_{k-1} = 0$, that $K = 2k + 1$, hence is odd, and that

$$u_K(2\pi) = 2\pi a_k. \tag{4.8}$$

Our claim follows from relation (4.8), which also gives a nice relation between the first nonzero term of h and the first nonzero term of $\Pi-\mathrm{Id}$.

When P and Q are polynomials, by the Hilbert Basis Theorem, there exists a $m \in \mathbb{N}$ with $m \geq 1$ such that $h = 0$ if and only if $j_m h(0) = 0$.

Although (4.2) is quite simple, it turns out that in many applications the current status of computer technology does not permit its calculation, to the required order, within a reasonable amount of time. The formulas also become rather intractable if many parameters are involved. We therefore will now pay attention to different techniques.

4.3 The Main Result

In several papers starting with Poincaré and Lyapunov it has been shown how necessary conditions are obtained by calculating the "focal values." It is well known (see e.g., [106]) that there exists a formal series $V(x, y) = (x^2 + y^2)/2 + O(\|(x, y)\|^3)$ such that $\dot{V}$ (its rate of change along orbits) is of the form

$$\dot{V} = \eta_2 r^2 + \eta_4 r^4 + \ldots + \eta_{2k} r^{2k} + \ldots,$$

where $r^2 = x^2 + y^2$. The η_{2k} are called the *focal values* and are polynomials in the coefficients of p and q. We will not prove this, since we will merely work with the first nonzero term. So we prove the existence and uniqueness of some η_{2l} for which there exists $V(x, y) = (x^2 + y^2)/2 + O(\|(x, y)\|^3)$ with

$$\dot{V} = \eta_{2l} r^{2l} + O(r^{2l+1}).$$

Such an expression, including the value of η_{2l}, is invariant under near-identity transformations as well as under multiplication of the vector field with a function f, that has the property $f(0) = 1$. We can hence suppose that (4.3) has the form

$$\left(x\frac{\partial}{\partial y} - y\frac{\partial}{\partial x} \right) + a_k(x^2 + y^2)^k \left(x\frac{\partial}{\partial x} + y\frac{\partial}{\partial y} \right) + Y, \tag{4.9}$$

where $Y = O(\|(x, y)\|^{2k+3})$. We also consider

$$V(x, y) = (x^2 + y^2)/2 + \sum_{i=3}^{2l} P_i(x, y) + W(x, y),$$

where P_i are homogeneous polynomials of degree i and $W = O(\|(x, y)\|^{2l+1})$.
    The cited problem then reduces to:

$$\sum_{i=3}^{2l} \left(x\frac{\partial P_i}{\partial y} - y\frac{\partial P_i}{\partial x} \right) + a_k(x^2 + y^2)^{k+1} - \eta_{2l}(x^2 + y^2)^l = O(\|(x, y)\|^p), \quad (4.10)$$

where $p = \min(2l+1, 2k+3)$. We now suppose that $k = l - 1$ such that (4.10)
can be written as

$$\sum_{i=3}^{2l} \left(x\frac{\partial P_i}{\partial y} - y\frac{\partial P_i}{\partial x} \right) + (a_{l-1} - \eta_{2l})(x^2 + y^2)^l = O(\|(x, y)\|^{2l+1}). \quad (4.11)$$

A treatment similar to what we will do now would show that (4.10) has no
solution if $k \neq l - 1$.
    If we write

$$P_i(x, y) = \sum_{j=0}^{i} a_j x^j y^{i-j},$$

we see that

$$x\frac{\partial P_i}{\partial y} - y\frac{\partial P_i}{\partial x} = -a_1 y^i + \sum_{k=1}^{i-1}((i - k + 1)a_{k-1} - (k + 1)a_{k+1})x^k y^{i-k} + a_{i-1}x^i.$$

$$(4.12)$$

For i odd, expression (4.12) needs to be zero, which is not possible if we take
$P_i = 0$.
    For i even and $i < 2l$ we also need (4.12) to be zero. This can be achieved
if we take $i = 2p$ and

$$P_i(x, y) = a_0(y^{2p} + \sum_{k=1}^{p} \frac{p - (k-1)}{k} \frac{p - (k-2)}{k - 1} \cdots \frac{p - 1}{2} px^{2k} y^{2(p-k)}). \quad (4.13)$$

For $i = 2l$ we may of course take a solution like in (4.13) that makes (4.12)
equal to zero, but we can also check that (4.12) is never of the form $c(x^2 + y^2)^l$
for some $c \neq 0$. We leave this as an exercise for the reader.
    In any case we see for the first nonzero focal value that

$$\eta_{2l} = a_{l-1}. \quad (4.14)$$

Together with (4.8) this gives:

$$u_{2l-1}(2\pi) = 2\pi\eta_{2l}. \tag{4.15}$$

It is hence clear that the origin is a center if and only if $\eta_{2k} = 0$ for all k. Moreover the stability of the origin is determined by the sign of the first nonzero focal value. As η_{2k} is relevant only when $\eta_{2l} = 0$ for $l < k$, we put $\eta_2 = \eta_4 = \ldots = \eta_{2k-2} = 0$ in the expression for η_{2k}. The quantities obtained in this way are called the *Lyapunov constants* and are denoted by $L(k)$, for $k = 1, 2, \ldots$. By convention, nonzero multiplicative factors can be omitted from the expressions given for the Lyapunov constants if we are interested only in knowing whether they are zero or nonzero. It turns out that in many applications it is easier to calculate Lyapunov constants than the normal form (4.2), although it still requires a great deal from the computer technology.

If P and Q are polynomials, by the Hilbert basis theorem there is a constant m such that $L(k) = 0$ for all k if and only if $L(k) = 0$ if $k \leq m$. Therefore it is necessary to compute only a finite number of the Lyapunov constants, though with few exceptions for any given case it is unknown a priori how many are required. Thus sets of necessary conditions are obtained, and the sufficiency of each set is then considered separately.

In general, the calculation of Lyapunov constants by hand is impossible except in the simplest cases, and several computational methods have been developed. In what follows we present one of the best methods for computing the Lyapunov constants.

Instead of working with the η_{2l} we prefer to work with

$$V_{2n+1} = u_{2n+1}(2\pi) = 2\pi\eta_{2n+2},$$

when $u_2(2\pi) = \ldots = u_{2n}(2\pi) = 0$ and call it, from now on, the n-th Lyapunov constant. Instead of using the name of Lyapunov constant other authors call them focal values. Moreover their definition is slightly different, but in the end the definitions only differ by a constant.

The main result of this chapter is the following.

Theorem 4.1 *We consider the analytic differential system (4.3) or the associated 1-form*

$$dH(x, y) + \omega_1(x, y) + \omega_2(x, y) + \ldots = 0,$$

where $H(x, y) = \frac{1}{2}(x^2 + y^2)$ and $\omega_l(x, y) = -Q_{l+1}(x, y)dx + P_{l+1}(x, y)dy$, for every $l \in \mathbb{N}$. Then

$$V_n = -\frac{1}{2^{\frac{n+1}{2}}} \frac{1}{\rho^{\frac{n+1}{2}}} \int_{H=\rho} \sum_{l=1}^{n-1} \omega_l h_{n-1-l},$$

where $h_0 \equiv 1$ and h_m for $m = 1, \ldots, n-2$ are polynomials defined recursively by the expression

$$d \left(\sum_{l=1}^{m} \omega_l h_{m-l} \right) = -d \left(h_m dH \right).$$

We remark that the integral that appears in the expression for V_{2n+1} and in other similar formulas in this section, must be over a level curve of H which is oriented in the direction of the Hamiltonian vector field $-y\partial/\partial x + x\partial/\partial y$.

We must note that, in Theorem 4.1, the key point in obtaining the Lyapunov constants lies in the computation of the polynomials h_m. In Lemma 4.5 it will be explained how they can be computed if the system is written either in polar coordinates, or if it is written in complex coordinates.

From Theorem 4.8 presented in Sect. 4.5, we shall deduce an algorithm which allows the computation of the expressions for V_3, V_5, V_7, V_9, and V_{11} for a general system (4.3) in an acceptable period of time; see Table 4.1.

In Sect. 4.6, Theorems 4.1 and 4.8 are used to compute Lyapunov constants for some families of differential equations in the plane. More explicitly, the speed of the method is shown in the case of systems with nonlinear homogeneous perturbation; see Tables 4.2–4.5. Finally, the algorithm is also applied to study the focus-center problem for systems

$$\begin{aligned} \dot{x} &= -y, \\ \dot{y} &= x + Q_n(x,y), \end{aligned} \qquad (4.16)$$

where $Q_n(x, y)$ is a homogeneous polynomial of degree n, for $n = 2, \ldots, 5$.

4.4 Basic Results

This section deals with the proof of the main result, Theorem 4.1. A first step consists in evaluating the first nonzero derivative of the return map associated to the perturbation of the Hamiltonian system $dH = 0$, when $H(x,y) = (x^2 + y^2)/2$

Theorem 4.2 *Consider the solution of the analytic differential equation in the plane*

$$\omega_\varepsilon = dH + \sum_{i=1}^{\infty} \varepsilon^i \omega_i = 0,$$

where ω_i are polynomial 1-forms, not necessarily homogeneous, which vanish at the origin, and $H(x,y) = \frac{1}{2}(x^2 + y^2)$.

Let $L(\rho, \varepsilon) = \rho + \varepsilon L_1(\rho) + \ldots + \varepsilon^k L_k(\rho) + O(\varepsilon^{k+1})$ be the map that defines the first return associated to the transverse section Σ (we choose $H = \rho > 0$

to parametrize the transverse section Σ). We assume that $L_1(\rho) \equiv \ldots \equiv L_{m-1}(\rho) \equiv 0$. Then there exist polynomials $h_0 \equiv 1, h_1, \ldots, h_{m-1}$ such that

$$d\left(\sum_{i=1}^{m} \omega_i h_{m-i}\right) = -d(h_m dH),$$

and the mth derivative of L with respect to ε is $m!L_m(\rho)$, with

$$L_m(\rho) = -\int_{H=\rho} \sum_{i=1}^{m} \omega_i h_{m-i}.$$

We must observe that the usefulness of these theorems in evaluating the Lyapunov constants primarily depends on the way that one can calculate the polynomials h_j. In [65], the author chooses complex coordinates. In [157] the authors use both complex and polar coordinates. The results which follow show how the polynomials h_j are found in these cases. The first is just a reformulation of the results of [65], while the second will allow us to extend Theorem 4.1 to other Hamiltonian different from $H = \frac{1}{2}(x^2 + y^2)$, like for instance $H = \frac{1}{4}(x^4 + 2y^2)$.

We first introduce a bit of notation, and we also prove a technical lemma.

Lemma 4.3 *Consider the 1-differential form $\omega = A(z, \bar{z})dz + B(z, \bar{z})d\bar{z}$, with A and B polynomials in z and $\bar{z}$. Then*

$$\int_{H=\rho} \omega = -2\pi i \sum_k coef\left(-\frac{\partial}{\partial \bar{z}}A + \frac{\partial}{\partial z}B, z^k \bar{z}^k\right) \frac{(2\rho)^{k+1}}{k+1},$$

where the function $coef(f, z^k\bar{z}^l)$ gives the coefficient of the monomial $z^k\bar{z}^l$, for every k and l, of f.

Proof. From Stokes' Theorem we have that

$$\int_{H=\rho} \omega = \int_{H=\rho} (A(z, \bar{z})dz + B(z, \bar{z})d\bar{z})$$

$$= \int_{H\leq\rho} d(A(z, \bar{z})dz + B(z, \bar{z})d\bar{z})$$

$$= \int_{H\leq\rho} \left(\frac{\partial A}{\partial \bar{z}}d\bar{z} \wedge dz + \frac{\partial B}{\partial z}dz \wedge d\bar{z}\right)$$

$$= \int_{H\leq\rho} \left(-\frac{\partial A}{\partial \bar{z}} + \frac{\partial B}{\partial z}\right) dz \wedge d\bar{z}.$$

If the functions A and B are polynomials, then we only need to know the value of $\int_{H \leq \rho} z^k \bar{z}^l dz \wedge d\bar{z}$, for every k, l. With the change $z = Re^{i\theta}$ we get

$$\int_{H \leq \rho} z^k \bar{z}^l dz \wedge d\bar{z} =$$

$$= \int_0^{\sqrt{2\rho}} \int_0^{2\pi} R^{k+l} e^{i(k-l)\theta} \left(e^{i\theta}dR + Rie^{i\theta}d\theta\right) \wedge \left(e^{-i\theta}dR - iRe^{-i\theta}d\theta\right)$$

$$= \int_0^{\sqrt{2\rho}} \int_0^{2\pi} R^{k+l} e^{i(k-l)\theta}(-2iR)dR \wedge d\theta$$

$$= -2i \int_0^{\sqrt{2\rho}} R^{k+l+1} dR \int_0^{2\pi} e^{i(k-l)\theta}d\theta$$

$$= \begin{cases} 0 & \text{if } k \neq l, \\ -2\pi i \frac{(2\rho)^{k+1}}{k+1} & \text{if } k = l. \end{cases} \qquad \square$$

This lemma suggests the next definition.

Definition 4.4 *Let $\mathcal{P}$ be the set of all polynomials in the variables $z, \bar{z}$ which vanish at the origin. We denote by $\mathcal{P}_0$ the subset of $\mathcal{P}$ formed by the polynomials which have no monomials of the form $z^k \bar{z}^k$ for every $k > 0$, and by $\mathcal{P}_1$ the subset of $\mathcal{P}$ formed by the polynomials whose derivative with respect to z lies in $\mathcal{P}_0$. On these subsets of $\mathcal{P}$ we may consider the following operators:*

(i)
$$\begin{aligned} \mathcal{G}: \quad & \mathcal{P}_0 \quad \longrightarrow \quad \mathcal{P}_0 \\ & R = \sum_{k \neq l} r_{kl} z^k \bar{z}^l \longmapsto \sum_{k \neq l} \tfrac{2}{k-l} r_{kl} z^k \bar{z}^l, \end{aligned}$$

(ii)
$$\begin{aligned} \mathcal{F}: \quad & \mathcal{P}_1 \longrightarrow \quad \mathcal{P}_0 \\ & R \longmapsto -\operatorname{Im}\left(\mathcal{G}\left(\frac{\partial R(z, \bar{z})}{\partial z}\right)\right). \end{aligned}$$

Lemma 4.5 *Consider the function $H(x, y) = \frac{1}{2}(x^2 + y^2)$. Let ω be a polynomial 1-form such that $\int_{H=\rho} \omega \equiv 0$. Then there exists a polynomial g such that $d\omega = d(gdH)$.*

(i) *If ω is written in complex coordinates as*

$$\omega = A(z, \bar{z})dz + B(z, \bar{z})d\bar{z},$$

then

$$g(z, \bar{z}) = \mathcal{G}\left(-\frac{\partial A(z, \bar{z})}{\partial \bar{z}} + \frac{\partial B(z, \bar{z})}{\partial z}\right),$$

where $\mathcal{G}$ is the function given in Definition 4.4(i).

(ii) If ω is written in polar coordinates as

$$\omega = \alpha(r,\theta)dr + \beta(r,\theta)d\theta,$$

then

$$g(r,\theta) = \frac{1}{r}\int_0^\theta \left(\frac{\partial\alpha(r,\psi)}{\partial\psi} - \frac{\partial\beta(r,\psi)}{\partial r}\right)d\psi.$$

We must note that in the proof of item *(i)* of Lemma 4.5 we strongly use the hypothesis $\int_{H=\rho}\omega \equiv 0$, in order to define the polynomial g. On the other hand, the construction of g in item *(ii)* does not depend on the hypothesis $\int_{H=\rho}\omega \equiv 0$. This hypothesis is needed only to assure that the g obtained is a trigonometric polynomial, using that g is 2π-periodic in θ.

From Lemma 4.5 we get that g satisfies $d\omega = d(gdH)$. Thus we can say that there exists a function S such that $\omega = gdH + dS$.

Proof of Lemma 4.5. *(i)* It is clear from Lemma 4.3 that $\int_{H=\rho}\omega \equiv 0$ if and only if the polynomial $-\frac{\partial}{\partial\bar{z}}A + \frac{\partial}{\partial z}B$ has no monomial of the form $(z\bar{z})^k$. So the polynomial $g(z,\bar{z})$ of the statement is well defined. If we define the polynomial

$$D(z,\bar{z}) = -\frac{\partial}{\partial\bar{z}}A + \frac{\partial}{\partial z}B = \sum_{k\neq l} d_{kl}z^k\bar{z}^l,$$

then

$$g(z,\bar{z})dH = \left(\sum_{k\neq l}\frac{2}{k-l}d_{kl}z^k\bar{z}^l\right)\frac{1}{2}(\bar{z}dz + zd\bar{z}),$$

where every monomial satisfies

$$d\left(\left(\frac{d_{kl}}{k-l}z^k\bar{z}^l\right)(\bar{z}dz + zd\bar{z})\right) = \frac{d_{kl}}{k-l}d\left(z^k\bar{z}^{l+1}dz + z^{k+1}\bar{z}^l d\bar{z}\right)$$
$$= d_{kl}z^k\bar{z}^l dz \wedge d\bar{z}.$$

Therefore $g(z,\bar{z})$ verifies

$$d\omega = d(g(z,\bar{z})dH),$$

because $d\omega = D(z,\bar{z})dz \wedge d\bar{z}$.

(ii) From the expressions of the function $H = \frac{1}{2}r^2$ and the 1-form $\omega = \alpha dr + \beta d\theta$, we obtain that $d\omega = (-\frac{\partial\alpha}{\partial\theta} + \frac{\partial\beta}{\partial r})dr \wedge d\theta$ and $d(g(r,\theta)rdr) = -r\frac{\partial g}{\partial\theta}dr \wedge d\theta$ which coincide in the case that the function g is defined as in the statement. $\square$

Proof of Lemma 4.2. Let γ_ε be the solution curve of $\omega_\varepsilon = 0$ defining $L(\rho, \varepsilon)$; γ_0 is the curve which defines $H = \rho$. So it is clear that over γ_ε the 1-form ω_ε vanishes, and we have

$$\int_{\gamma_\varepsilon} \omega_\varepsilon = 0.$$

Integration over γ_ε is taken in the same sense as for γ_0. Substituting in this equality the expression for ω_ε, we get

$$0 = \int_{\gamma_\varepsilon} \left(dH + \varepsilon \omega_1 + \varepsilon^2 \omega_2 + \ldots \right)$$

$$= H(L(\rho, \varepsilon)) - H(\rho) + \varepsilon \int_{\gamma_\varepsilon} \omega_1 + \varepsilon^2 \int_{\gamma_\varepsilon} \omega_2 + \ldots$$

$$= \left(\rho + L_1(\rho)\varepsilon + L_2(\rho)\varepsilon^2 + \ldots \right) - \rho + \varepsilon \int_{\gamma_0} (\omega_1 + O(\varepsilon))$$

$$+ \varepsilon^2 \int_{\gamma_0} (\omega_2 + O(\varepsilon)) + \ldots.$$

Comparing the terms in ε, we get the expression

$$L_1(\rho) = -\int_{H=\rho} \omega_1.$$

We suppose now that $L_1(\rho) \equiv 0$. Then from $\int_{H=\rho} \omega_1 \equiv 0$ (see Lemma 4.5) we get that there exist polynomials h_1 and S_1 such that $-\omega_1 = h_1 dH + dS_1$, or equivalently $d(\omega_1) = d(-h_1 dH)$. The existence of this function h_1 allows introduction of the 1-form

$$(1 + \varepsilon h_1)\omega_\varepsilon,$$

which also vanishes on γ_ε. From this form we get the following equalities:

$$0 = \int_{\gamma_\varepsilon} \left(dH + \varepsilon(\omega_1 + h_1 dH) + \varepsilon^2(\omega_2 + \omega_1 h_1) + \ldots \right)$$

$$= \int_{\gamma_\varepsilon} \left(d(H - \varepsilon S_1) + \varepsilon^2(\omega_2 + \omega_1 h_1) + \ldots \right)$$

$$= (H - \varepsilon S_1)(L(\rho, \varepsilon)) - (H - \varepsilon S_1)(\rho) + \varepsilon^2 \int_{\gamma_\varepsilon} (\omega_2 + \omega_1 h_1) + \ldots$$

$$= H(L(\rho, \varepsilon)) - \varepsilon S_1(L(\rho, \varepsilon)) - H(\rho) + \varepsilon S_1(\rho) + \varepsilon^2 \int_{\gamma_\varepsilon} (\omega_2 + \omega_1 h_1) + \ldots$$

$$= \rho + \varepsilon^2 L_2(\rho) + O(\varepsilon^3) - \varepsilon \left(S_1(\rho + O(\varepsilon^2)) - S_1(\rho) \right) - \rho$$

$$+ \varepsilon^2 \int_{\gamma_0} (\omega_2 + \omega_1 h_1) + O(\varepsilon^3).$$

Comparing the terms in ε^2, we get the expression of L_2,

$$L_2(\rho) = -\int_{H=\rho} (\omega_2 + \omega_1 h_1).$$

We suppose now that $L_k(\rho) \equiv 0$ for $k = 1, \ldots, m-1$, and that there exist polynomials h_k and S_k satisfying the relations

$$-\sum_{i=1}^{k} \omega_i h_{k-i} = h_k dH + dS_k, \text{ for } k = 1, \ldots, m-1.$$

In a manner similar to what we have done to obtain $L_2(\rho)$, we consider the 1-form $(1 + \varepsilon h_1 + \varepsilon^2 h_2 + \ldots + \varepsilon^{m-1} h_{m-1}) \omega_\varepsilon$, which vanishes on γ_ε. Integrating along γ_ε and considering the terms in ε^m we obtain the expression for L_m,

$$L_m(\rho) = -\int_{H=\rho} \sum_{i=1}^{m} \omega_i h_{m-i},$$

proving the theorem.   □

We prove now the main result.

Proof of Theorem 4.1. If we consider the rescaling

$$(x, y) = (\varepsilon X, \varepsilon Y), \tag{4.17}$$

then system (4.1) becomes

$$\dot{X} = \tfrac{1}{\varepsilon}(-\varepsilon Y + P(\varepsilon X, \varepsilon Y)) = -X + \sum_{k=2}^{\infty} \varepsilon^{k-1} P_k(X, Y),$$

$$\dot{Y} = \tfrac{1}{\varepsilon}(\varepsilon X + Q(\varepsilon X, \varepsilon Y)) = Y + \sum_{k=2}^{\infty} \varepsilon^{k-1} Q_k(X, Y). \tag{4.18}$$

For a point $(\rho, 0)$ in action-angle coordinates (ρ, θ), i.e., $\rho = H(X, Y)$, the return map of (4.18), $L(\rho, \varepsilon)$, is well defined. This point, with the change (4.17), becomes the point $(\varepsilon\sqrt{2\rho}, 0)$ in cartesian coordinates (x, y), such that the first return of this point, for system (4.3), is

$$\left(\varepsilon\sqrt{2\rho} + u_{2k+1}(\varepsilon\sqrt{2\rho})^{2k+1} + \ldots, 0\right).$$

Writing it in variables (X, Y), we obtain the point

$$\left(\sqrt{2\rho} + u_{2k+1}\varepsilon^{2k}(\sqrt{2\rho})^{2k+1} + \ldots, 0\right),$$

such that the value of $H(x, y)$ at this point is

$$\frac{1}{2}\left(\sqrt{2\rho} + u_{2k+1}\varepsilon^{2k}(\sqrt{2\rho})^{2k+1} + \dots\right)^2$$

$$= \frac{1}{2}\left(2\rho + 2\varepsilon^{2k}(\sqrt{2\rho})^{2k+2}u_{2k+1} + \dots\right)$$

$$= \rho + \varepsilon^{2k}2^{k+1}\rho^{k+1}u_{2k+1} + \dots,$$

and this gives the first return of $(\rho, 0)$, for system (4.18), that is to say $L(\rho, \varepsilon) = \rho + \varepsilon^{2k}L_{2k}(\rho) + \dots$. So we get

$$L_{2k}(\rho) = 2^{k+1}\rho^{k+1}u_{2k+1}.$$

Taking into account Theorem 4.2, that is the statement we wanted to prove.   □

4.5 The Algorithm

4.5.1 A Theoretical Description

Theorem 4.1 gives the theoretical basis for calculating the Lyapunov constants, V_{2n-1}, of system (4.3). In fact we use it to obtain an expression of these V_{2n-1}, in terms of words. This section has two objectives: to obtain alternative proofs of known properties about the fact that the Lyapunov constants are polynomials in the coefficients of system (4.3), and to create an effective algorithm for calculating them.

Before giving the version of Theorem 4.1 using words, we need to introduce a little notation. We first introduce the notion of word. Given a set $\{p_1, \dots, p_n\}$ of continuous functions on $[0, \omega]$, we define for $i = 1, \dots, n$ operators b_i : $C[0, \omega] \rightarrow C[0, \omega]$ by

$$(b_i f)(t) = \int_0^t p_i(s)f(s)ds,$$

where $C[0, \omega]$ is the space of the continuous maps from the interval $[0, \omega]$ into itself. We consider strings of these operators acting on the constant function $U(t) = 1$. So, for example,

$$(b_1 b_2 b_1 U)(t)) = \int_0^t p_1(s) \int_0^s p_2(r) \int_0^r p_1(u)\, du\, dr\, ds.$$

For convenience the function U is usually omitted. Thus we write $b_1 b_2 b_1$ for $b_1 b_2 b_1 U$.

These strings are regarded as *words* over the alphabet $A = \{b_1, \dots, b_n\}$. The b_i are the letters of A. The identity operator denoted by e acts as the empty word. The set of all words over A (including the empty word) is denoted by A^*. It is easy to provide to A^* with the structure of a vector space. Note that every word is an operator on the set $C[0, \omega]$.

We recall that system (4.3) in complex coordinates is written as (4.5)

Definition 4.6 *Given a polynomial $R \in \mathcal{P}$ of degree n, we can define for every integer $k \geq 2$ the functionals $\mathcal{F}_k$ and $\mathcal{H}_k$ as*

$$\mathcal{F}_k : \mathcal{P} \longrightarrow \mathcal{P}$$
$$R \longmapsto \mathcal{F}(R_k R),$$

$$\mathcal{H}_k : \mathcal{P} \longrightarrow \mathbb{R}$$
$$R \longmapsto -\frac{1}{(2\rho)^{\frac{n+1+k}{2}}} \int_{H=\rho} \mathrm{Im}\,(R_k R d\bar{z}),$$

where R_k is given in (4.5) and $\mathcal{F}$ in Definition 4.4.

We must note that the functionals $\mathcal{F}_k$ and $\mathcal{H}_k$ are defined only on certain subsets of $\mathcal{P}$, and so we will need to check that they are well defined every time that we use them.

In order to simplify the notation, we use the following notation:

$$\mathcal{H}_{m_0} \prod_m \mathcal{F}_{m_i} := \mathcal{H}_{m_0} \mathcal{F}_{m_1} \mathcal{F}_{m_2} \ldots \mathcal{F}_{m_s} := \mathcal{H}_{m_0} \left(\mathcal{F}_{m_1} \left(\mathcal{F}_{m_2} \left(\ldots \left(\mathcal{F}_{m_s} (1) \right) \right) \right) \right),$$

where $m = (m_1, \ldots, m_s)$, and if $m = (0)$ it is understood that $\prod_{(0)} \mathcal{F}_{m_i} = 1$.

Next lemma gathers together some properties of these operators that we will need later on.

Lemma 4.7 *The following statements hold.*

(i) *If we have a 1-form $\omega = \mathrm{Im}(f d\bar{z})$ such that $\int_{H=\rho} \omega = 0$, then the function $g(z, \bar{z})$ from Lemma 4.5 takes the form $\mathcal{F}(f)$.*
(ii) *The operators $\mathcal{F}$, $\mathcal{F}_k$, and $\mathcal{H}_k$ are real.*
(iii) *If R and R_k are polynomials of degree n and k, respectively, then the degree of $\mathcal{F}_k(R)$ is $n + k - 1$.*

Proof. If $\omega = \mathrm{Im}(f d\bar{z}) = (f d\bar{z} - \bar{f} dz)(2i)$, the function g given by Lemma 4.5 is

$$g = G\left(\frac{1}{2i} \left(\frac{\partial \bar{f}}{\partial \bar{z}} + \frac{\partial f}{\partial z} \right) \right) = \frac{1}{2i} \left(G\left(\frac{\partial \bar{f}}{\partial \bar{z}} \right) + G\left(\frac{\partial f}{\partial z} \right) \right)$$

$$= \frac{1}{2i} \left(-\overline{G\left(\frac{\partial f}{\partial z} \right)} + G\left(\frac{\partial f}{\partial z} \right) \right) = \mathrm{Im}\left(G\left(\frac{\partial f}{\partial z} \right) \right) = \mathcal{F}(f).$$

So statement *(i)* is proved. The other statements follow easily.   □

With this notation we can write Theorem 4.1 as:

Theorem 4.8 *An expression for the constant V_n for system (4.5) is*

$$V_n = \sum_{k=2}^{n} \mathcal{H}_k \left(\sum_{m \in S_{n-k}} \prod_m \mathcal{F}_{m_i} \right),$$

where $\mathcal{H}_k$ and $\mathcal{F}_m$ are those of Definition 4.6, and S_l is the set

$$S_l = \bigcup_{s \in \mathbb{N}^+} \left\{ m = (m_1, \ldots, m_s) \in \left(\mathbb{N}^+ \setminus \{1\} \right)^s : \sum_{i=1}^{s} (m_i - 1) = l \right\},$$

for $l \neq 0$ and $S_0 = \{(1)\}$.

Proof. From the expression of system (4.5), doing the same rescaling $(z \to \varepsilon z)$ as in the proof of Theorem 4.2, we get the 1-form

$$dH + \sum_{k=2}^{\infty} \varepsilon^{k-1} \operatorname{Im}(R_k(z, \bar{z}) d\bar{z}) = dH + \sum_{k=1}^{\infty} \varepsilon^k \omega_k.$$

From Theorem 4.2 and Definition 4.6 we know that

$$L_1(\rho) = -\int_{H=\rho} \omega_1 = -\int_{H=\rho} \operatorname{Im}(R_2 d\bar{z}) = (2\rho)^{3/2} \mathcal{H}_2(1) = (2\rho)^{3/2} \mathcal{H}_2,$$

and using Theorem 4.1, we obtain

$$V_2 = \frac{1}{(2\rho)^{3/2}} L_1(\rho) = \mathcal{H}_2.$$

When $L_1(\rho) \equiv V_2 \equiv 0$, from Lemma 4.7 the polynomial h_1 such that $d(-\omega_1) = d(h_1 dH)$ is $h_1 = \mathcal{F}_2$, and it has degree 1.

We consider now the 1-form $\omega_2 + \omega_1 h_1$. Arguing as in the previous step, we may write:

$$L_2 = -\int_{H=\rho} \omega_2 + \omega_1 h_1$$

$$= -\int_{H=\rho} (\operatorname{Im}(R_3 d\bar{z}) + \operatorname{Im}(R_2 d\bar{z}) \mathcal{F}_2)$$

$$= -\int_{H=\rho} (\operatorname{Im}(R_3 d\bar{z}) + \operatorname{Im}(R_2 \mathcal{F}_2 d\bar{z}))$$

$$= (2\rho)^{4/2} \mathcal{H}_3(1) + (2\rho)^{4/2} \mathcal{H}_2(\mathcal{F}_2),$$

and

$$V_3 = \frac{1}{(2\rho)^{4/2}} L_2 = \mathcal{H}_3 + \mathcal{H}_2 \mathcal{F}_2.$$

If it is true that $V_2 \equiv V_3 \equiv 0$, then Lemma 4.7 gives us that the polynomial h_2 which satisfies $-d(\omega_2 + \omega_1 h_1) = d(h_2 dH)$ is $h_2 = \mathcal{F}_3 + \mathcal{F}_2 \mathcal{F}_2$, and it has degree 2.

We now study the general case, that is, $V_2 \equiv V_3 \equiv \ldots \equiv V_{n-1} \equiv 0$. In order to calculate V_n, we suppose that for $l = 1, \ldots, n-1$, there exist polynomials h_l of degree l satisfying the equalities:

(i) $-d\left(\sum\limits_{k=0}^{l-1} \omega_{l-k} h_k\right) = d(h_l dH)$, and

(ii) $h_l = \sum\limits_{m \in S_l} \prod\limits_m \mathcal{F}_{m_i}$.

From Theorem 4.2 and Definition 4.6 we know that

$$
\begin{aligned}
L_n(\rho) &= -\int_{H=\rho} \left(\sum_{k=0}^{n-1} \omega_{n-k} h_k\right) \\
&= -\int_{H=\rho} \left(\sum_{k=0}^{n-1} \mathrm{Im}\,(R_{n-k+1} d\bar{z}) \sum_{m \in S_k} \prod_m \mathcal{F}_{m_i}\right) \\
&= -\int_{H=\rho} \left(\sum_{k=0}^{n-1} \mathrm{Im}\left(R_{n-k+1} \left(\sum_{m \in S_k} \prod_m \mathcal{F}_{m_i}\right) d\bar{z}\right)\right) \\
&= \sum_{k=0}^{n-1} -\int_{H=\rho} \left(\mathrm{Im}\left(R_{n-k+1} \left(\sum_{m \in S_k} \prod_m \mathcal{F}_{m_i}\right) d\bar{z}\right)\right) \\
&= \sum_{k=0}^{n-1} (2\rho)^{\frac{n-k+1+k+1}{2}} \mathcal{H}_{n-k+1} \left(\sum_{m \in S_k} \prod_m \mathcal{F}_{m_i}\right) \\
&= (2\rho)^{\frac{n+2}{2}} \sum_{k=2}^{n+1} \mathcal{H}_k \left(\sum_{m \in S_{n-k+1}} \prod_m \mathcal{F}_{m_i}\right),
\end{aligned}
$$

and using Theorem 4.1, we may say that

$$
V_{n+1} = \frac{1}{(2\rho)^{(n+2)/2}} L_n(\rho) = \sum_{k=2}^{n+1} \mathcal{H}_k \left(\sum_{m \in S_{n-k+1}} \prod_m \mathcal{F}_{m_i}\right),
$$

which is the expression of V_{n+1} given in the statement.

We must also find the expression of h_n in the case that $V_{n+1} \equiv 0$. Arguing as when we obtained L_n, and using Lemma 4.7, we get

$$
\begin{aligned}
h_n &= \sum_{k=0}^{n-1} \mathcal{F}_{n-k+1} \left(\sum_{m \in S_k} \prod_m \mathcal{F}_{m_i}\right) \\
&= \sum_{k=2}^{n+1} \mathcal{F}_k \left(\sum_{m \in S_{n-k+1}} \prod_m \mathcal{F}_{m_i}\right) = \sum_{m \in S_n} \prod_m \mathcal{F}_{m_i}. \qquad \square
\end{aligned}
$$

Consequently, we have a way to express the Lyapunov constants in the case of homogeneous nonlinear perturbations of degree n, which we will use later on.

Corollary 4.9 *The nonzero Lyapunov constant of system*

$$\dot{z} = iz + R_n(z, \bar{z}),$$

are

$$V_{n+j(n-1)} = \mathcal{H}_n(\mathcal{F}_n^j),$$

where j is a natural number if n is odd, and an odd natural number if n is even.

Proof. In this case, in the expression for V_j of the previous proposition we have only the operators $\mathcal{H}_n$ and $\mathcal{F}_n$, so the sum $\sum_{m \in S_{n-k}} \prod_m \mathcal{F}_{m_i}$ can have only the term $\mathcal{F}_m^i$. So we have from the definition of the set S_l nonzero constants for the values $i(n-1) = j - n$, and given that $j = n + i(n-1)$ has to be odd, as is proved in the next corollary, i is a natural number if n is odd, and must be odd if n is even.   □

From Corollary 4.9 it is very easy to give new proofs, shorter and unified, for all the known algebraic properties of the Lyapunov constants.

We say that a map f is a *quasihomogeneous map* with *weights* $a = (a_1, \ldots, a_n) \in \mathbb{N}^n$ and *weighted degree* $d = (d_1, \ldots, d_n) \in \mathbb{N}^n$ if

$$f_i(\lambda^{a_1} x_1, \lambda^{a_2} x_2, \ldots, \lambda^{a_n} x_n) = \lambda^{d_i} f_i(x_1, x_2, \ldots, x_n)$$

holds for every $i = 1, 2, \ldots, n$ and any $\lambda > 0$.

Corollary 4.10 (Algebraic properties of the Lyapunov constants)
Let $V_n := V_n(\{r_{k,l}, \bar{r}_{k,l}\})$ be a Lyapunov constant of system (4.5), where $R_m(z, \bar{z}) = \sum_{k+l=m} r_{k,l} z^k \bar{z}^l$. Then the following statements hold:

(i) $V_{2n} \equiv 0$ for every $n \geq 1$.
(ii) V_{2n+1} is a quasihomogeneous polynomial of weight 0 when we associate to each variable $r_{k,l}$ (respectively, $\bar{r}_{k,l}$) the weight $-k + l + 1$ (respectively, $k - l - 1$), i.e.,

$$V_{2n+1}(\lambda^{-k+l+1} r_{k,l}, \lambda^{k-l-1} \bar{r}_{k,l}) = V_{2n+1}(r_{k,l}, \bar{r}_{k,l}),$$

for every real λ.
(iii) V_{2n+1} is a quasihomogeneous polynomial of weighted degree $2n$ when we associate to each variable $r_{k,l}$ and $\bar{r}_{k,l}$ the weight $k + l - 1$, i.e.,

$$V_{2n+1}(\lambda^{k+l-1} r_{k,l}, \lambda^{k+l-1} \bar{r}_{k,l}) = \lambda^{2n} V_{2n+1}(r_{k,l}, \bar{r}_{k,l}),$$

for every real λ.

(iv) V_{2n+1} *can be written as* $\mathrm{Re}\left(V^o_{2n+1}\right) + \mathrm{Im}\left(V^e_{2n+1}\right)$, *where* V^e_{2n+1} *and* V^o_{2n+1} *are polynomials with real coefficients of degree even and odd, respectively, in the variables* $r_{k,l}$ *and* $\bar{r}_{k,l}$.

Before we prove these properties, we will prove some similar ones for the polynomial $h_N = \sum_{m \in S_N} \prod_m \mathcal{F}_{m_i}$, defined in the proof of Theorem 4.8.

Lemma 4.11 *In the hypothesis of Corollary 4.10, the following properties hold for the polynomial* $h_N = \sum_{m \in S_N} \prod_m \mathcal{F}_{m_i} = \sum_{k,l} f_{k,l} z^k \bar{z}^l$.

(i) h_N *is a homogeneous polynomial of degree N in $z, \bar{z}$.*

(ii) $f_{k,l}$ *is a quasihomogeneous polynomial of weight $-k + l$ when we associate to each variable $r_{k,l}$ (respectively, $\bar{r}_{k,l}$) the weight $-k + l + 1$ (respectively, $k - l - 1$).*

(iii) $f_{k,l}$ *is a quasihomogeneous polynomial of weighted degree $k + l$ when we associate to each variable $r_{k,l}$ and $\bar{r}_{k,l}$ the weight $k + l - 1$.*

(iv) h_N *can be written as* $\mathrm{Im}\left(h^o_N\right) + \mathrm{Re}\left(h^e_N\right)$, *where h^e_N and h^o_N are polynomials with coefficients in $\mathbb{Q}[z, \bar{z}]$ of even and odd degree, respectively, in the variables $r_{k,l}$ and $\bar{r}_{k,l}$.*

Proof. Given the linearity of the operators, it is enough to prove that each element of the sum $\sum_{m \in S_N} \prod_m \mathcal{F}_{m_i}$, has degree $\sum_{j=1}^n (m_j - 1) = N$, that they satisfy the properties of weight and weighted degree, and that they can be written as the imaginary or real part depending on the parity of the degree in $r_{k,l}$ and $\bar{r}_{k,l}$.

From the notation that we are using, if $R_{m_1} = \sum_{a+b=m_1} r_{a,b} z^a \bar{z}^b$, then

$$
\begin{aligned}
\mathcal{F}_{m_1} &= \mathcal{F}_{m_1}(1) = \mathcal{F}(R_{m_1}) \\
&= \mathcal{F}\Big(\sum_{a+b=m_1} r_{a,b} z^a \bar{z}^b \Big) = \sum_{a+b=m_1} \mathcal{F}(r_{a,b} z^a \bar{z}^b) \\
&= - \sum_{a+b=m_1} \mathrm{Im}\left(\mathcal{G}\left(a r_{a,b} z^{a-1} \bar{z}^b\right)\right) \\
&= - \sum_{a+b=m_1} \mathrm{Im}\left(\frac{2a}{a-b-1} r_{a,b} z^{a-1} \bar{z}^b \right) \\
&= -\frac{1}{2i} \sum_{a+b=m_1} \left(\frac{2a}{a-b-1} r_{a,b} z^{a-1} \bar{z}^b - \frac{2a}{a-b-1} \bar{r}_{a,b} \bar{z}^{a-1} z^b \right) \\
&= \sum_{a+b=m_1} \frac{1}{i} \frac{a}{a-b-1} \left(-r_{a,b} z^{a-1} \bar{z}^b + \bar{r}_{a,b} \bar{z}^{a-1} z^b \right).
\end{aligned}
$$

So $\mathcal{F}(R_{m_1})$ is a polynomial which has degree $m_1 - 1$, and the monomials $r_{a,b} z^{a-1} \bar{z}^b$ and $\bar{r}_{a,b} \bar{z}^{a-1} z^b$ are quasihomogeneous of weight $-a + b + 1 = -(a - 1) + b$ and $a - b - 1 = -b + (a-1)$, respectively, and weighted degree $a + b - 1$. Moreover, $\mathcal{F}(R_{m_1})$ can be written as

$$\mathrm{Im}\left(-\sum_{a+b=m_1}\left(\frac{2a}{a-b-1}r_{a,b}z^{a-1}\bar{z}^b\right)\right),$$

and so all statements hold for R_{m_1}.

Suppose now, as an induction hypothesis, that $\prod_{i=1}^{n}\mathcal{F}_{m_i} = \sum_{k,l} f_{k,l}z^k\bar{z}^l$ satisfies the statements of the lemma. Then we have that

$$\prod_{j=1}^{n+1}\mathcal{F}_{m_j} = \mathcal{F}_{m_{n+1}}\prod_{j=1}^{n}\mathcal{F}_{m_j} = \mathcal{F}\left(R_{m_{n+1}}\sum_{c+d=\sum_{j=1}^{n}(m_j-1)}f_{c,d}z^c\bar{z}^d\right)$$

$$= \mathcal{F}\left(\sum_{a+b=m_{n+1}}r_{a,b}z^a\bar{z}^b\sum_{c+d=\sum_{j=1}^{n}(m_j-1)}f_{c,d}z^c\bar{z}^d\right)$$

$$= \sum_{\substack{a+b=m_{n+1}\\c+d=\sum_{j=1}^{n}(m_j-1)}}\mathcal{F}(r_{a,b}f_{c,d}z^{a+c}\bar{z}^{b+d})$$

$$= -\sum_{\substack{a+b=m_{n+1}\\c+d=\sum_{j=1}^{n}(m_j-1)}}\mathrm{Im}\left(\mathcal{G}((a+c)r_{a,b}f_{c,d}z^{a+c-1}\bar{z}^{b+d})\right)$$

$$= -\sum_{\substack{a+b=m_{n+1}\\c+d=\sum_{j=1}^{n}(m_j-1)}}\mathrm{Im}\left(\frac{2(a+c)}{a+c-1-b-d}r_{a,b}f_{c,d}z^{a+c-1}\bar{z}^{b+d}\right)$$

$$= \sum_{\substack{a+b=m_{n+1}\\c+d=\sum_{j=1}^{n}(m_j-1)}}\frac{(a+c)}{a+c-1-b-d}\frac{1}{i}\left(\bar{r}_{a,b}\bar{f}_{c,d}\bar{z}^{a+c-1}z^{b+d}-r_{a,b}f_{c,d}z^{a+c-1}\bar{z}^{b+d}\right),$$

and so $\prod_{j=1}^{n+1}\mathcal{F}_{m_j}$ is a homogeneous polynomial of degree $m_{n+1}+\sum_{j=1}^{n}(m_j-1)-1 = \sum_{j=1}^{n+1}(m_j-1)$ in the variables $z,\bar{z}$ as statement (i) says.

The coefficients of each monomial in $z,\bar{z}$ $r_{a,b}f_{c,d}$ and $\bar{r}_{a,b}\bar{f}_{c,d}$, are quasihomogeneous of weight $-a+b+1-c+d = -(a+c-1)+(b+d)$ and $a-b-1+c-d = (a+c-1)-(b+d)$, respectively, and weighted degree $a+c-1+b+d = (a+b-1)+(c+d)$, as statements (ii) and (iii) say.

As can be seen in the set of equalities, the operator $\mathcal{F}_{m_n}$, changes with the parity of the degree of a polynomial in $r_{k,l}$ and $\bar{r}_{k,l}$, over the polynomial which we operate. From this fact, together with the equalities

$$\mathrm{Im}(A\,\mathrm{Im}(B)) = \mathrm{Re}\left(-\frac{AB-A\bar{B}}{2}\right),$$

$$\mathrm{Im}(A\,\mathrm{Re}(B)) = \mathrm{Im}\left(\frac{AB+A\bar{B}}{2}\right),$$

we get statement (iv).   □

Now we are ready to prove Corollary 4.10.

Proof of Corollary 4.10. From Theorem 4.8, the nth Lyapunov constant is

$$V_n = \sum_{k=2}^{n} \mathcal{H}_k \left(\sum_{m \in S_{n-k}} \prod_m \mathcal{F}_{m_i} \right).$$

So using Lemma 4.11, we only need to see how the operator $\mathcal{H}_k$ behaves with respect to the properties we wanted to prove.

Given the linearity of the operator $\mathcal{H}_k$, we consider only the case $R = f_{c,d} z^c \bar{z}^d$, with $c + d = n - k$.

From the definition of $\mathcal{H}_k$, we have that

$$\mathcal{H}_k(R) = -\frac{1}{(2\rho)^{\frac{c+d+1+k}{2}}} \int_{H=\rho} \mathrm{Im}\,(R_k R d\bar{z})$$

$$= -\frac{1}{(2\rho)^{\frac{n+1}{2}}} \int_{H=\rho} \mathrm{Im} \left(\sum_{a+b=k} r_{a,b} z^a \bar{z}^b f_{c,d} z^c \bar{z}^d dz \right)$$

$$= -\frac{1}{(2\rho)^{\frac{n+1}{2}}} \int_{H=\rho} \mathrm{Im} \left(\sum_{a+b=k} r_{a,b} f_{c,d} z^{a+c} \bar{z}^{b+d} dz \right)$$

$$= -\frac{1}{(2\rho)^{\frac{n+1}{2}}} \sum_{a+b=k} \int_{H=\rho} \frac{1}{2i} \left(r_{a,b} f_{c,d} z^{a+c} \bar{z}^{b+d} d\bar{z} - \right.$$

$$\left. - \bar{r}_{a,b} \bar{f}_{c,d} \bar{z}^{a+c} z^{b+d} dz \right)$$

$$= -\frac{1}{(2\rho)^{\frac{n+1}{2}}} \sum_{a+b=k} \frac{1}{2i} \left(r_{a,b} f_{c,d} \int_{H=\rho} z^{a+c} \bar{z}^{b+d} d\bar{z} - \right.$$

$$\left. - \bar{r}_{a,b} \bar{f}_{c,d} \int_{H=\rho} \bar{z}^{a+c} z^{b+d} dz \right)$$

$$= \frac{\pi}{(2\rho)^{\frac{n+1}{2}}} \left(\sum_{\substack{a+b=k \\ a+c-b-d-1=0}} r_{a,b} f_{c,d} (2\rho)^{\frac{a+c+b+d+1}{2}} + \right.$$

$$\left. + \sum_{\substack{a+b=k \\ b+d-a-c+1=0}} \bar{r}_{a,b} \bar{f}_{c,d} (2\rho)^{\frac{b+d+a+c+1}{2}} \right)$$

$$= \pi \sum_{\substack{a+b=k \\ a+c-b-d-1=0}} \left(r_{a,b} f_{c,d} + \bar{r}_{a,b} \bar{f}_{c,d} \right) = 2\pi \sum_{\substack{a+b=k \\ a+c-b-d-1=0}} \mathrm{Re}\,(r_{a,b} f_{c,d}).$$

From the relations $c + d = n - k$, $a + b = k$ and $a + c - b - d = 1$, we get that $a + c = \frac{n+1}{2}$ and that $b + d = \frac{n-1}{2}$. So if n is even, as a, b, c, d are integers, the sum of the previous expression has no terms. Therefore statement *(i)* is proved.

Using Lemma 4.11, we have that $\mathcal{H}_k(R)$ is a quasihomogeneous polynomial of weight $(-a+b+1)+(-c+d) = 0$, and weighted degree $(a+b-1)+(c+d) = (k-1)+(n-k) = n-1$, when the weights of statements *(ii)* and *(iii)* are associated, respectively. This proves these two statements.

In order to prove the last statement, we only need to use the next equalities,

$$\text{Re}(A\,\text{Im}(B)) = \text{Im}\left(-\frac{AB - A\overline{B}}{2}\right),$$

$$\text{Re}(A\,\text{Re}(B)) = \text{Re}\left(\frac{AB + A\overline{B}}{2}\right),$$

which together with the operator $\mathcal{H}_k$, they change the parity of the degrees of the polynomials in the variables $r_{k,l}$ and $\bar{r}_{k,l}$.   □

4.5.2 Practical Implementation

In this subsection we describe the algorithm based in Theorem 4.8, which allows the computation, in a simple way, of the Lyapunov constants for the system of differential (4.5). In order to show its speed and easiness of use, we have evaluated the first five Lyapunov constants for the general system (4.5). In Table 4.1, we give the measure and the computation time of the first four. The computer we have used (a personal computer with Pentium 120 MHz CPU and 32 Mb RAM) has been able to obtain the fifth constant in pieces, but if we try to get the complete expression, the computer crashes.

As far as we know, only the three first constants (V_3, V_5, V_7) for a general system had been found up to now. Moreover, as we will see later on, this is also an algorithm which allows us to get a new Lyapunov constant using the computations done to calculate the previous constants.

Applying this algorithm, the first two Lyapunov constants are:

$$V_3 = \mathcal{H}_3 + \mathcal{H}_2(\mathcal{F}_2),$$

$$V_5 = \mathcal{H}_5 + \mathcal{H}_4(\mathcal{F}_2) + \mathcal{H}_3(\mathcal{F}_2\mathcal{F}_2) + \mathcal{H}_2(\mathcal{F}_4 + \mathcal{F}_3\mathcal{F}_2 + \mathcal{F}_2\mathcal{F}_3 + \mathcal{F}_2\mathcal{F}_2\mathcal{F}_2),$$

where, in the computation of V_5 we can use the expression of $\mathcal{F}_2$, gotten to compute V_3.

Table 4.1. Computation time and measure of the constants of system (4.5)

constant	computation time	measure in bytes	number of monomials
3	0.07	80	4
5	0.48	1,753	54
7	11.73	20,864	526
9	544.70	17,9897	3,800

In order to compute V_7, we need to get the elements of S_k for $k = 0, \ldots, 5$ which are

$$S_0 = \{(1)\},$$
$$S_1 = \{(2)\},$$
$$S_2 = \{(2,2),(3)\},$$
$$S_3 = \{(2,2,2),(3,2),(2,3),(4)\},$$
$$S_4 = \{(2,2,2,2),(3,2,2),(2,3,2),(2,2,3),(3,3),(4,2),(2,4),(5)\},$$
$$\begin{aligned} S_5 = \{&(2,2,2,2,2),(3,2,2,2),(2,3,2,2),(2,2,3,2),\\ &(2,2,2,3),(3,3,2),(3,2,3),(2,3,3),(4,2,2),(2,4,2),\\ &(2,2,4),(4,3),(3,4),(5,2),(2,5),(6)\},\end{aligned}$$

and afterwards, to compute the expressions of $\prod_m \mathcal{F}_{m_i}$ for every $m \in S_k$. We must notice that with this method, we may use the expressions of

$$\mathcal{F}_2, \mathcal{F}_2\mathcal{F}_2, \mathcal{F}_3\mathcal{F}_2, \mathcal{F}_2\mathcal{F}_3, \mathcal{F}_2\mathcal{F}_2\mathcal{F}_2 \text{ and } \mathcal{F}_4,$$

computed when we got V_3 and V_5. This allows us to increase the speed of computation of the Lyapunov constants.

As we need a fast method for computing the constants, the best way to obtain the sets S_k is to use an inductive procedure which makes use of the previous computations. If we define the functions

$$\mathrm{op}(u) := \mathrm{op}((u_1, \ldots, u_n)) = u_1, \ldots, u_n, \text{ and}$$
$$\mathrm{add}(a, S) := \bigcup_{u \in S} \{(a, \mathrm{op}(u)), (\mathrm{op}(u), a)\},$$

then we can write

$$S_2 = \{(3)\} \bigcup \{\mathrm{add}(2, S_1)\},$$
$$S_3 = \{(4)\} \bigcup \{\mathrm{add}(2, S_2)\} \bigcup \{\mathrm{add}(3, S_1)\}, \text{ and}$$
$$S_n = \{(n)\} \bigcup_{k=1}^{n-1} \{\mathrm{add}(n-k+1, S_k)\}.$$

The computations done to test this algorithm have been done with MAPLE.

4.6 Applications

In this section we will apply the algorithm to several families of system (4.5). First we will apply it to some families widely studied in the past: quadratic systems, systems with linear plus cubic homogeneous parts, and others, just to check its efficiency. In Tables 4.2–4.5 we give the measure and the computation time, in seconds, of the Lyapunov constants for these families obtained using a personal computer. Finally, we will apply it to study the necessary conditions for existence of a center for the families (4.16), given in the introduction of this chapter. The objective for showing here these families is not to find all the centers they produce, but using them as effective applications of Theorem 4.8.

Given a system of the form

$$\dot{x} = \alpha x - y + p(x, y),$$
$$\dot{y} = x + \alpha y + q(x, y),$$
(4.19)

with $p(x, y)$ and $q(x, y)$ polynomial functions, which start with terms of order at least two, there is a weak focus at the origin if $\alpha = 0$. In this case, there exists an analytic function $V(x, y)$ defined in a neighborhood of the origin such that $\dot{V} = dV(x(t), y(t))/dt = \nu_2 r^2 + \nu_4 r^4 + \ldots$ where $r^2 = x^2 + y^2$. Bautin in [14] proves that the return map of system (4.19) is analytic and can be written as:

$$\Pi(x, \alpha, \lambda) = x + \sum_{k=1}^{\infty} V_k(\alpha, \lambda) x^k,$$

where the coefficients V_k are entire functions of (α, λ), the coefficients of system (4.19). Moreover, if $\alpha = 0$, these coefficients are polynomials of degree $k - 1$. Using these polynomials V_k, the *Bautin ideal* is defined as the ideal generated by the coefficients of the return map evaluated at $\alpha = 0$,

$$I = \langle V_1, V_2, \ldots, V_n, \ldots \rangle \subset \mathbb{R}[\lambda],$$

where $V_k = V_k(\lambda) = V_k(0, \lambda)$.

We define the *cyclicity* of a weak focus located at the origin as the maximum number of limit cycles that can bifurcate from the origin by means of a degenerate Hopf bifurcation.

The study of all these examples has brought to us the thought of the relation existing between the number of relevant Lyapunov constants, those which generate the Bautin's ideal, and the cyclicity of the weak focus.

4.6.1 Known Examples

The constants obtained to fill Tables 4.2 and 4.3 are enough to resolve the center problem for these systems .

As far as we know, the characterization of the centers for the systems studied in Tables 4.4 and 4.5 is still an open problem. The manipulator MAPLE, and the computer used to make the calculations, have allowed us to obtain only the results shown in the following tables.

In Table 4.6 we summarize the constants obtained for a case with non linear nonhomogeneous part.

Table 4.2. Computation time and measure of the constants of quadratic system $\dot{z} = iz + R_2(z, \bar{z})$. Obviously, we get that $v_9 = v_{11} = v_{13} = 0$ if v_3, v_5, and v_7 are zero

constant	computation time	measure in bytes	number of monomials
3	0.21	57	2
5	0.14	543	14
7	0.82	2,104	44
9	4.34	6,075	110
11	21.76	14,095	224
13	110.95	28,895	414

Table 4.3. Computation time and measure of the constants of cubic system $\dot{z} = iz + R_3(z, \bar{z})$. Obviously, we get that $v_{13} = v_{15} = v_{17} = 0$ if previous constants are also zero.

constant	computation time	measure in bytes	number of monomials
3	0.03	21	2
5	0.03	60	2
7	0.07	281	14
9	0.33	1,214	30
11	1.25	2,895	82
13	3.92	7,540	150
15	15.15	13,555	302
17	49.93	29,979	496

Table 4.4. Computation time and measure of the constants of quartic system $\dot{z} = iz + R_4(z, \bar{z})$.

constant	computation time	measure in bytes	number of monomials
7	0.24	114	4
13	1.04	2,778	64
19	33.57	22,796	404
25	1,220.26	12,0315	1,684

Finally, we see how far we can reach when we study the system

$$\dot{x} = -y + P_2(x, y) + P_3(x, y),$$
$$\dot{y} = \quad x + Q_2(x, y) + Q_3(x, y),$$

where the polynomials $P_i(x, y)$ and $Q_i(x, y)$ are homogeneous; see Table 4.7.

4.6.2 Kukles-Homogeneous Family

The centers of the homogeneous family

$$\dot{x} = -y + P_n(x, y),$$
$$\dot{y} = \quad x + Q_n(x, y),$$

Table 4.5. Computation time and measure of the constants of quintic system $\dot{z} = iz + R_5(z, \bar{z})$

constant	computation time	measure in bytes	number of monomials
5	0.04	21	2
9	0.05	120	4
13	0.33	837	32
17	2.02	4,636	106
21	14.62	14,019	358
25	148.75	49,927	870

Table 4.6. Computation time and measure of the constants of cubic system with degenerate infinity

Constant	computation time	measure in bytes	number of monomials
3	0.06	71	3
5	0.33	805	22
7	3.61	4,998	112
9	47.19	20,221	382
11	899.14	65,114	1,065

Table 4.7. Computation time and measure of the constants of cubic system

constant	computation time	measure in bytes	number of monomials
3	0.25	80	4
5	0.41	1,400	42
7	8.51	13,136	306
9	270.68	77,024	1,482
11	29,266.57	35,1832	5,694

have been widely studied. The problem of characterizing the centers of these families is still an open problem if $n \geq 4$. In order to show the difficulty of the problem, we consider a particular case of the homogeneous family, those named Kukles-homogeneous systems,

$$
\begin{aligned}
\dot{x} &= -y, \\
\dot{y} &= x + Q_n(x, y),
\end{aligned}
\tag{4.20}
$$

where $Q_n(x, y)$ is a homogeneous polynomial of degree n.

The algorithm described in Sect. 4.5 has allowed us to obtain the first Lyapunov constants, and in this way to obtain necessary conditions for (4.20) to have a center at the origin.

With the change $z = x + iy$, system (4.20) can be written as

$$
\dot{z} = iz + R_n(z, \bar{z}) = iz + \sum_{k+l=n} r_{k,l} z^k \bar{z}^l,
$$

Table 4.8. Bautin's number ($\mathbf{B}_n$) and relevant constants ($\mathbf{S}_n$) for system (4.20)

n	parameters	$\mathbf{B}_n$	$\mathbf{S}_n$
2	3	1	1
3	4	3	3
4	5	6	5
5	6	7	5

where $\bar{r}_{k,l} = r_{l,k}$. This way, we can apply Theorem 4.8 to calculate the Lyapunov constants.

In the case $n = 2$, there is only one center condition

$$v_3 = -2ir_{1,1}(r_{2,0} - r_{0,2}).$$

If $v_3 = 0$, in [14] it is proved that system (4.20) has a center at the origin.

In the case $n = 3$, there are three conditions for a center,

$$v_3 = -2r_{2,1} - 2r_{1,2},$$
$$v_5 = -4i(r_{3,0} + r_{0,3})r_{1,2},$$
$$v_7 = 3r_{3,0}r_{0,3}(r_{3,0} + r_{0,3}).$$

It is known that these three conditions are sufficient in order that system (4.20) have a center at the origin.

In the cases $n = 4, 5$, we will not reproduce here the expressions that we get for the Lyapunov constants because of their length. We must note that the method given by Theorem 4.8 has allowed us to obtain the first six constants in the case $n = 4$ and the first eight constants in the case $n = 5$. The characterization of the centers of these families needs the resolution of the system $\{v_4 = v_{10} = \ldots = 0\}$ in the case $n = 4$ and $\{v_5 = v_9 = \ldots = 0\}$ in the case $n = 5$. See Corollary 4.9 where the nonzero constants for the homogeneous family are given.

With the manipulator MAGMA, we have seen that

(i) The ideal generated by the first k Lyapunov constants is different from the one generated by the first $k - 1$, for $k = 2, \ldots, 6$, when $n = 4$ and for $k = 2, \ldots, 7$ when $n = 5$.

(ii) For $n = 5$, the ideal generated by the first eight constants is the same that the one generated by the first seven.

(iii) For $n = 4$, there exists a natural number k_1 such that $v_6^{k_1}$ belongs to the ideal generated by the first five constants.

(iv) For $n = 5$, there exist natural numbers k_2 and k_3, such that $v_6^{k_2}$ and $v_7^{k_3}$ belong to the ideal generated by the first five constants.

It is natural to think, from this list of properties, that the number of relevant Lyapunov constants for cases $n = 4$ and $n = 5$ will be 5. These facts are summarized in Table 4.8. In this table, $\mathbf{B}_n$ denotes the Bautin's number of the Kukles-homogeneous family of degree n, and the number $\mathbf{S}_n$ is a lower bound on the number of relevant Lyapunov constants.

Table 4.8 has led us to believe that we have enough Lyapunov constants to obtain all centers of the family defined by system (4.20), but the computer that we have used has not been able to solve the problem.

4.7 Bibliographical Comments

This chapter has essentially followed the exposition about a new algorithm for computing the Lyapunov constants due to Gasull and Torregrosa; see [71] and [157] (in catalan).

The study of the stability of a singular point of center-focus type having a non–degenerate linear part for an analytic vector field has generated a large number of papers during the twentieth century. As far back as 1893, Lyapunov already published a paper [105] where a partial solution to this problem was given. He defined the functions which determine the stability, that we now call Lyapunov constants. The major difficulty with these functions is their high complexity, and to find them explicitly becomes a computational problem. Another problem which has also produced a great quantity of results has been to determine how many limit cycles can bifurcate by perturbing the Hamiltonian $H = \frac{1}{2}(x^2 + y^2)$ using analytic functions. For a recent account on different methods that can be used we refer to [27]. Melnikov's theory based on calculating the derivative with respect to ε of a return map $L(\rho, \varepsilon)$, permits us to give a first answer to this problem. Françoise in [65] developed a new method in order for obtaining all Melnikov's functions for this system of differential equations. Later on, Gasull and Torregrosa [157] used the ideas of Françoise in order to get a new method for calculating the Lyapunov constants of the first problem by means of the Melnikov's functions of the second problem. This new method is good for obtaining both theoretical and practical results.

For a *monodromic* singular point of an analytic system of differential equations in the plane, there is the well known *center-focus problem* consisting in determining whether the singularity is a center or a focus. When the linear part of the system at the singular point has pure imaginary eigenvalues, the problem can be reduced to evaluation of its Lyapunov constants. Farr, Li, Labouriau, and Langford [62] give a summary of the different ways for finding them. Gasull, Guillamon, and Mañosa [69] and Pearson, Lloyd and Christopher [119] contributed to the computation of the Lyapunov constants and their application to the solution of the center-focus problem. In all these works, it is evident that serious computational problems appear, due to the amount of time needed to calculate and also to the fact that the complexity and length of these constants increase largely, not only when we increase the complexity of the differential system, but also for a simple systems when we try to find a Lyapunov constant of large order. In order to reduce these difficulties as much as possible, the method of Gasull and Torregrosa provides a great simplification in time and complexity, and it often allows us to give a simpler proof of known theoretical results. Their method can be generalized to study the center-focus problem.

One of the main ideas which has allowed the development of this algorithm comes from the method given by Françoise in [65], where the successive derivatives of the return map associated to the perturbations of some planar Hamiltonian systems are evaluated. In that paper and in the paper of Françoise and Pons [66], the authors apply their method to solve the center-focus problem, but only for systems having homogeneous nonlinear part, and without computing explicitly the Lyapunov constants; see also the work of Françoise and Yomdin [67]. This generalization has been described also by Iliev and Perko [87] and [86].

Using Theorem 4.1, it is not difficult to obtain the expression of the constant V_n in terms of words; see Theorem 4.8. For more details about this subject, the reader may look at the works of Devlin [47] and [48], where the author also uses words to study the Lyapunov constants of Abel's equations; see also Guillamon [78]. From this result, we can obtain the known algebraic properties of the Lyapunov constants, when they are considered as polynomials in the coefficients of system (4.5).

We have mentioned that Corollary 4.9 can be used to provide some new proofs, shorter and unified, of the known algebraic properties of the Lyapunov constants. Other proofs of these results can be found in the works of Cima, Gasull, Guillamon, Mañosa, and Mañosas [41], [70] and Żołądek [171].

As far as we know, only the three first constants (V_3, V_5, V_7) for a general system had been found up to now; see [69]

More information about the Bautin ideal can be found in Yakovenko [168].

See Roussarie [137] for additional results on the cyclicity of a weak focus.

The centers described in Sect. 4.6 have been studied by many authors; see for instance Bautin [14], Chavarriga, and Gine [28–30], Lloyd, Christopher, Devlin, Pearson, and Yasmin [104] and Sibirskii [147].

5

Poincaré and Poincaré–Lyapunov Compactification

In order to study the behavior of the trajectories of a planar differential system near infinity it is possible to use a compactification. One of the possible constructions relies on stereographic projection of the sphere onto the plane, in which case a single "point at infinity" is adjoined to the plane; see Bendixson [15]. A better approach for studying the behavior of trajectories near infinity is to use the so called Poincaré sphere, introduced by Poincaré [132]. It has the advantage that the singular points at infinity are spread out along the equator of the sphere and are therefore of a simpler nature than the singular points of the Bendixson sphere. However, some of the singular points at infinity on the Poincaré sphere may still be very complicated.

In order to improve the construction, we introduce the so called Poincaré–Lyapunov sphere, which is however based on a construction of a more abstract nature than the previous ones. The singularities are also spread along the equator but are in general simpler than for the Poincaré sphere. In the Poincaré–Lyapunov compactification we prefer to work on a hemisphere, calling it as the Poincaré–Lyapunov disk, and similarly talk about a Poincaré disk, if we restrict the Poincaré sphere to one of the hemispheres separated by the equator that represents the points at infinity.

5.1 Local Charts

In order to draw the phase portrait of a vector field, we would have to work over the complete real plane $\mathbb{R}^2$, which is not very practical. If the functions defining the vector field are polynomials, we can apply Poincaré compactification, which will tell us how to draw it in a finite region. Even more, it controls the orbits which tend to or come from infinity.

In this chapter we will use (x_1, x_2) as coordinates on the plane instead of (x, y).

Let $X = P\partial/\partial x_1 + Q\partial/\partial x_2$ be a polynomial vector field (the functions P and Q are polynomials of arbitrary degree in the variables x_1 and x_2), or in other words:

$$\begin{aligned} \dot{x}_1 &= P(x_1, x_2), \\ \dot{x}_2 &= Q(x_1, x_2). \end{aligned} \tag{5.1}$$

We recall that the degree of X is d if d is the maximum of the degrees of P and Q.

Poincaré compactification works as follows. First we consider $\mathbb{R}^2$ as the plane in $\mathbb{R}^3$ defined by $(y_1, y_2, y_3) = (x_1, x_2, 1)$. We consider the sphere $\mathbb{S}^2 = \{y \in \mathbb{R}^3 : y_1^2 + y_2^2 + y_3^2 = 1\}$ which we will call here *Poincaré sphere*; it is tangent to $\mathbb{R}^2$ at the point $(0, 0, 1)$. We may divide this sphere into $H_+ = \{y \in \mathbb{S}^2 : y_3 > 0\}$ (the northern hemisphere), $H_- = \{y \in \mathbb{S}^2 : y_3 < 0\}$ (the southern hemisphere) and $\mathbb{S}^1 = \{y \in \mathbb{S}^2 : y_3 = 0\}$ (the equator). Now we consider the projection of the vector field X from $\mathbb{R}^2$ to $\mathbb{S}^2$ given by the central projections $f^+ : \mathbb{R}^2 \to \mathbb{S}^2$ and $f^- : \mathbb{R}^2 \to \mathbb{S}^2$. More precisely, $f^+(x)$ (respectively, $f^-(x)$) is the intersection of the straight line passing through the point y and the origin with the northern (respectively, southern) hemisphere of $\mathbb{S}^2$:

$$f^+(x) = \left(\frac{x_1}{\Delta(x)}, \frac{x_2}{\Delta(x)}, \frac{1}{\Delta(x)} \right),$$

$$f^-(x) = \left(\frac{-x_1}{\Delta(x)}, \frac{-x_2}{\Delta(x)}, \frac{-1}{\Delta(x)} \right),$$

where

$$\Delta(x) = \sqrt{x_1^2 + x_2^2 + 1}.$$

In this way we obtain induced vector fields in each hemisphere. Of course, every induced vector field is analytically conjugate to X. The induced vector field on H_+ is $\overline{X}(y) = Df^+(x)X(x)$, where $y = f^+(x)$, and the one in H_- is $\overline{X}(y) = Df^-(x)X(x)$, where $y = f^-(x)$. We remark that $\overline{X}$ is a vector field on $\mathbb{S}^2 \setminus \mathbb{S}^1$ that is everywhere tangent to $\mathbb{S}^2$.

We notice that the points at infinity of $\mathbb{R}^2$ (two for each direction) are in bijective correspondence with the points of the equator of $\mathbb{S}^2$. Now we would like to extend the induced vector field $\overline{X}$ from $\mathbb{S}^2 \setminus \mathbb{S}^1$ to $\mathbb{S}^2$. Unfortunately it does not in general stay bounded as we get close to $\mathbb{S}^1$, obstructing the extension. It turns out, however, that if we multiply the vector field by the factor $\rho(x) = x_3^{d-1}$ then as we will check in a moment, the extension becomes possible. The extended vector field on $\mathbb{S}^2$ is called the *Poincaré compactification* of the vector field X on $\mathbb{R}^2$, and it is denoted by $p(X)$. On each hemisphere H_+ and H_- it is no longer C^ω-conjugate to X, but it remains C^ω-equivalent. The technique hence serves well to trace phase portraits, but precautions must be taken in making fine calculations, such as eigenvalues at singularities and periods of closed orbits.

We now work out the construction. As is usual in working with curved surfaces, we use charts to make calculations. For $\mathbb{S}^2$ we use the six local charts given by $U_k = \{y \in \mathbb{S}^2 : y_k > 0\}$, $V_k = \{y \in \mathbb{S}^2 : y_k < 0\}$ for $k = 1, 2, 3$. The corresponding local maps $\phi_k : U_k \to \mathbb{R}^2$ and $\psi_k : V_k \to \mathbb{R}^2$ are defined as $\phi_k(y) = -\psi_k(y) = (y_m/y_k, y_n/y_k)$ for $m < n$ and $m, n \neq k$. We denote by $z = (u, v)$ the value of $\phi_k(y)$ or $\psi_k(y)$ for any k, such that (u, v) will play different roles depending on the local chart we are considering. Geometrically the coordinates (u, v) can be expressed as in Fig. 5.1. The points of $\mathbb{S}^1$ in any chart have $v = 0$.

In what follows we make a detailed calculation of the expression of $p(X)$ only in the local chart U_1. We have $X(x) = (P(x_1, x_2), Q(x_1, x_2))$. Then $\overline{X}(y) = Df^+(x)X(x)$ with $y = f^+(x)$ and

$$D\phi_1(y)\overline{X}(y) = D\phi_1(y) \circ Df^+(x)X(x) = D(\phi_1 \circ f^+)(x)X(x).$$

Let $\overline{X}|_{U_1}$ denote the system defined as $D\phi_1(y)\overline{X}(y)$. Then since

$$(\phi_1 \circ f^+)(x) = \left(\frac{x_2}{x_1}, \frac{1}{x_1}\right) = (u, v),$$

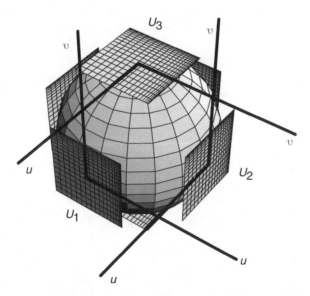

Fig. 5.1. The local charts (U_k, ϕ_k) for $k = 1, 2, 3$ of the Poincaré sphere

we have

$$
\overline{X}|_{U_1} = \begin{pmatrix} -\dfrac{x_2}{x_1^2} & \dfrac{1}{x_1} \\ -\dfrac{1}{x_1^2} & 0 \end{pmatrix} \begin{pmatrix} P(x_1, x_2) \\ Q(x_1, x_2) \end{pmatrix}
$$

$$
= \frac{1}{x_1^2} \left(-x_2 P\left(x_1, x_2\right) + Q\left(x_1, x_2\right), -P\left(x_1, x_2\right) \right)
$$

$$
= v^2 \left(-\frac{u}{v} P\left(\frac{1}{v}, \frac{u}{v}\right) + \frac{1}{v} Q\left(\frac{1}{v}, \frac{u}{v}\right), -P\left(\frac{1}{v}, \frac{u}{v}\right) \right).
$$

Now

$$
\rho(y) = y_3^{d-1} = \frac{1}{\Delta(x)^{d-1}} = \frac{v^{d-1}}{\Delta(z)^{d-1}} = v^{d-1} m(z),
$$

where $m(z) = (1 + u^2 + v^2)^{(1-d)/2}$. Consequently it follows that

$$
\rho(\overline{X}|_{U_1})(z) = v^{d+1} m(z) \left(-\frac{u}{v} P\left(\frac{1}{v}, \frac{u}{v}\right) + \frac{1}{v} Q\left(\frac{1}{v}, \frac{u}{v}\right), -P\left(\frac{1}{v}, \frac{u}{v}\right) \right).
$$

In order to prove that the extension of $\rho\overline{X}$ to $p(X)$ is defined on the whole of $\mathbb{S}^2$ we notice that while $\overline{X}|_{U_1}$ is not well defined when $v = 0$, $p(X)|_{U_1} = \rho\overline{X}|_{U_1}$ is well defined along $v = 0$, since the multiplying factor v^{d+1} cancels any factor of v which could appear in the denominator. Similar arguments can be applied to the rest of the local charts.

In order to simplify the extended vector field we also make a change in the time variable and remove the factor $m(z)$. We still keep a vector field on $\mathbb{S}^2$ which is C^ω-equivalent to X on any of the hemispheres H_+ and H_-.

The expression for $p(X)$ in local chart (U_1, ϕ_1) is given by

$$
\dot{u} = v^d \left[-uP\left(\frac{1}{v}, \frac{u}{v}\right) + Q\left(\frac{1}{v}, \frac{u}{v}\right) \right],
$$

$$
\dot{v} = -v^{d+1} P\left(\frac{1}{v}, \frac{u}{v}\right). \tag{5.2}
$$

The expression for (U_2, ϕ_2) is

$$
\dot{u} = v^d \left[P\left(\frac{u}{v}, \frac{1}{v}\right) - uQ\left(\frac{u}{v}, \frac{1}{v}\right) \right],
$$

$$
\dot{v} = -v^{d+1} Q\left(\frac{u}{v}, \frac{1}{v}\right), \tag{5.3}
$$

and for (U_3, ϕ_3) is

$$
\dot{u} = P(u, v),
$$

$$
\dot{v} = Q(u, v). \tag{5.4}
$$

The expression for $p(X)$ in the charts (V_k, ψ_k) is the same as for (U_k, ϕ_k) multiplied by $(-1)^{d-1}$, for $k = 1, 2, 3$.

To study X in the complete plane $\mathbb{R}^2$, including its behavior near infinity, it clearly suffices to work on $H_+ \cup \mathbb{S}^1$, which we call the Poincaré disk. All calculation can be done in the three charts (U_1, ϕ_1), (U_2, ϕ_2), and (U_3, ϕ_3) in which case the expressions are given by the formulas (5.2), (5.3), and (5.4).

It is clear that we do not need to go through the complete geometrical construction, as just presented, in order to get these expressions. The expressions in (5.4) clearly do not need any elaboration at all. To obtain (5.2) we start with (5.1) and introduce coordinates (u, v) by the formulas

$$(x_1, x_2) = (\frac{1}{v}, \frac{u}{v}). \tag{5.5}$$

This leads to a vector field $\bar{X}^u$ which we multiply by v^{d-1}. To obtain (5.3) we start with (5.1) and introduce coordinates (u, v) by the formulas

$$(x_1, x_2) = (\frac{u}{v}, \frac{1}{v}). \tag{5.6}$$

We again multiply the computed vector field $\bar{X}^v$ by v^{d-1}.

This more abstract way of constructing a vector field on a sphere (or better on a half-sphere) directly by means of charts, is not only the most practical way for precise calculations, but also will lead to a generalization (and improvement) of the construction, called the *Poincaré-Lyapunov disk*, as we will see in Sect. 5.3.

Up to C^ω-equivalence, the two operations (5.5) and (5.6) can be combined into a global construction at infinity:

$$(x_1, x_2) = (\frac{\cos\theta}{v}, \frac{\sin\theta}{v}), \tag{5.7}$$

taking $\theta \in \mathbb{S}^1$ and multiplying the vector field again by v^{d-1}. If we work out this construction we get

$$\dot{\theta} = v(\dot{x}_2 \cos\theta - \dot{x}_1 \sin\theta),$$
$$\dot{v} = -v^2(\dot{x}_1 \cos\theta + \dot{x}_2 \sin\theta).$$

This construction is in general preferred in cases where at infinity (on the equator) $p(X)$ has no singularities, implying that the set of points at infinity becomes a closed orbit. In order to study the behavior of $p(X)$ near that closed orbit, a global study of a neighborhood of it is needed. No such neighborhood is wholly contained in either chart (U_1, ϕ_1) or (U_2, ϕ_2).

We remark that in each local chart the local representative of $p(X)$ is a polynomial vector field.

We call *finite* (respectively, *infinite*) *singular points* of X or $p(X)$ the singular points of $p(X)$ which lie in $\mathbb{S}^2 \setminus \mathbb{S}^1$ (respectively, $\mathbb{S}^1$). We note that if $y \in \mathbb{S}^1$ is an infinite singular point, then $-y$ is also a singular point. Since the local behavior near $-y$ is the local behavior near y multiplied by $(-1)^{d-1}$, it follows that the orientation of the orbits changes when the degree is even. For example, if d is even and $y \in \mathbb{S}^1$ is a stable node of $p(X)$, then $-y$ is an unstable node. Due to the fact that infinite singular points appear in pairs of diametrally opposite points, it is enough to study half of them, and using the degree of the vector field one can determine the other half.

As we have already observed, the integral curves of $\mathbb{S}^2$ are symmetric with respect to the origin, such that it is sufficient to represent the flow of $p(X)$ only in the closed northern hemisphere, the so called *Poincaré disk* . In order to draw this, for practical purposes, as a disk in the plane we can project the points of the closed northern hemisphere onto the disk $\{(y_1, y_2, y_3) \in \mathbb{R}^3 \ : \ x_1^2 + x_2^2 \leq 1, x_3 = 0\}$. This could be done by projecting each point of the sphere onto the disk using a straight line parallel to the y_3-axis; we can however project using a family of straight lines passing through a point $(0, 0, y_3)$ with $y_3 < 0$. If y_3 is a value close to $-\infty$, we will get the same result, but if y_3 is close to zero then we might get a better representation of what is happening near infinity. In doing this we lose resolution in the regions close to the origin in (x_1, x_2)-plane.

5.2 Infinite Singular Points

We want to study the local phase portrait at infinite singular points. For this we choose an infinite singular point $(u, 0)$ and start by looking at the expression of the linear part of the field $p(X)$. We denote by P_i and Q_i the homogeneous polynomials of degree i for $i = 0, 1, \ldots, d$ such that $P = P_0 + P_1 + \ldots + P_d$ and $Q = Q_0 + Q_1 + \ldots + Q_d$. Then $(u, 0) \in \mathbb{S}^1 \cap (U_1 \cup V_1)$ is an infinite singular point of $p(X)$ if and only if

$$F(u) \equiv Q_d(1, u) - u P_d(1, u) = 0.$$

Similarly $(u, 0) \in \mathbb{S}^1 \cap (U_2 \cup V_2)$ is an infinite singular point of $p(X)$ if and only if

$$G(u) \equiv P_d(u, 1) - u Q_d(u, 1) = 0.$$

Also we have that the Jacobian of the vector field $p(X)$ at the point $(u, 0)$ is

$$\begin{pmatrix} F'(u) & Q_{d-1}(1, u) - u P_{d-1}(1, u) \\ 0 & -P_d(1, u) \end{pmatrix},$$

or

$$\begin{pmatrix} G'(u) & P_{d-1}(u, 1) - u Q_{d-1}(u, 1) \\ 0 & -Q_d(u, 1) \end{pmatrix},$$

if $(u, 0)$ belongs to $U_1 \cup V_1$ or $U_2 \cup V_2$, respectively.

We first remark that the equator of $\mathbb{S}^2$ can consist entirely of singularities, but in most cases the singularities are isolated. We confine our discussion to isolated singularities.

Among the hyperbolic singular points at infinity only nodes and saddles can appear. All the semi-hyperbolic singular points can appear at infinity.

From Chap. 2, we see that if one of these hyperbolic or semi-hyperbolic singularities at infinity is a (topological) saddle, then the straight line $\{v = 0\}$, representing the equator of $\mathbb{S}^2$, is necessarily a stable or unstable manifold, or a center manifold; see Fig. 5.2.

The same property also holds for semi-hyperbolic singularities of saddle-node type. They can hence have their hyperbolic sectors split in two different ways depending on the Jacobian of the system in the charts U_1 or U_2. The Jacobian can be either

$$\begin{pmatrix} \lambda & * \\ 0 & 0 \end{pmatrix},$$

or

$$\begin{pmatrix} 0 & * \\ 0 & \lambda \end{pmatrix},$$

with $\lambda \neq 0$. In the first case we say that the *saddle-node is of type SN1* and in the second case of *type SN2*. The two cases are represented in Fig. 5.3. The sense of the orbits can also be the opposite.

Among the nilpotent singular points at infinity all can appear with the exception of the cusp, the focus and the center.

The nilpotent points, as well as the singularities with zero linear part, have a behavior at infinity that is quite a bit more complicated than the hyperbolic and elementary singular points. Blow-up is needed to study them. It is not necessary for such singularities that the two orbits at infinity be separatrices.

Fig. 5.2. A hyperbolic or semi-hyperbolic saddle on the equator of $\mathbb{S}^2$

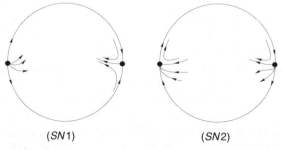

(SN1)   (SN2)

Fig. 5.3. Saddle-nodes of type SN1 and SN2 of $p(X)$ in the equator of $\mathbb{S}^2$

5.3 Poincaré–Lyapunov Compactification

Sometimes, the singularities at infinity in a Poincaré compactification are quite complicated, but there is a possibility of simplifying them by working with a so called *Poincaré–Lyapunov compactification*. The construction is very similar to the one expressed in (5.5) and (5.6) or in (5.7). We start with a generalization of (5.7), in the sense that we no longer keep the construction homogeneous, but make it quasihomogeneous. Thus we set

$$
\begin{aligned}
x_1 &= \cos\theta/s^\alpha, \\
x_2 &= \sin\theta/s^\beta,
\end{aligned}
\tag{5.8}
$$

for some well chosen powers $(\alpha, \beta) \in \mathbb{N} \times \mathbb{N}$ with $\alpha, \beta \geq 1$. If it is advantageous to do so, we can even replace the periodic functions $\cos\theta$ and $\sin\theta$ by the periodic functions $\mathrm{Cs}\,\theta$ and $\mathrm{Sn}\,\theta$, which are the unique solutions of the initial value problem

$$
\begin{aligned}
\frac{\mathrm{d}}{\mathrm{d}\theta}\,\mathrm{Cs}\,\theta &= -\,\mathrm{Sn}^{2\alpha-1}\theta, \\
\frac{\mathrm{d}}{\mathrm{d}\theta}\,\mathrm{Sn}\,\theta &= \mathrm{Cs}^{2\beta-1}\theta, \\
\mathrm{Cs}\,0 &= 1, \\
\mathrm{Sn}\,0 &= 0,
\end{aligned}
$$

which satisfy the relation $\beta\,\mathrm{Sn}^{2\alpha}\theta + \alpha\,\mathrm{Cs}^{2\beta}\theta = \alpha$. It is possible that for a system for which the usual Poincaré compactification has a nonelementary singular point at infinity, for well chosen α and β, the Poincaré–Lyapunov compactification has only elementary singular points at infinity, or even no singular points at infinity at all. For the calculations it is again sometimes better to work in different charts, comparable to (5.5) and (5.6), and this will be done in Sect. 9.1.

5.4 Bendixson Compactification

Let $X = P\partial/\partial x_1 + Q\partial/\partial x_2$ be a polynomial vector field (the functions P and Q are polynomials of maximum degree d). Its associated differential system is given by (5.1).

The construction of the Bendixson compactification is as follows (for more details see Chapt. 13 of [4]). We consider the sphere in $\mathbb{R}^3$ given by $y_1^2 + y_2^2 + y_3^2 = 1/4$, and call it the *Bendixson sphere*. We identify the x_1x_2-plane on which X is defined with the tangent plane to the sphere at the point $S = (0, 0, -1/2)$ given by the equation $y_3 = -1/2$. Let p_N be the stereographic projection from the north pole $N = (0, 0, 1/2)$ to the plane $y_3 = -1/2$. Thus p_N^{-1} defines an induced vector field X_N on $\mathbb{S}^2 \setminus N$. Clearly the infinity of the plane $y_3 = -1/2$ is transformed by p_N^{-1} into the north pole N.

Now we want to extend X_N to a vector field on $\mathbb{S}^2$. This extension is called the *Bendixson compactification* and the induced vector field on $\mathbb{S}^2$ which has the north pole as a singular point is denoted by $b(X)$.

We use two local charts to study the vector field $b(X)$, defined, respectively on $U_N = \mathbb{S}^2 \backslash N$ and $U_S = \mathbb{S}^2 \backslash S$. The corresponding coordinate maps are $p_N : U_N \to \mathbb{R}^2$ and $p_S : U_S \to \mathbb{R}^2$, where p_S is the stereographic projection of $\mathbb{S}^2$ from the south pole to the plane $y_3 = 1/2$.

The map $p_S \circ p_N^{-1}$ from the plane $y_3 = -1/2$ minus S to the plane $y_3 = 1/2$ minus N is given by

$$u = \frac{x_1}{x_1^2 + x_2^2}, \quad v = \frac{x_2}{x_1^2 + x_2^2} \;,$$

where (u, v) are the coordinates in the plane $y_3 = 1/2$. Hence the infinity of system (5.1) is transformed into the origin of the system

$$
\begin{aligned}
\dot{u} &= \frac{1}{(u^2 + v^2)^d} \widetilde{P}(u, v), \\
\dot{v} &= \frac{1}{(u^2 + v^2)^d} \widetilde{Q}(u, v) \;,
\end{aligned}
\tag{5.9}
$$

where $\widetilde{P}(u, v) = P(x_1(u, v), x_2(u, v))$ and $\widetilde{Q}(u, v) = Q(x_1(u, v), x_2(u, v))$, whose terms of lowest order are of degree at least $d + 2$. Finally we introduce a change of time scale $dt/d\tau = (u^2 + v^2)^d$ and system (5.9) becomes

$$
\begin{aligned}
\dot{u} &= \widetilde{P}(u, v), \\
\dot{v} &= \widetilde{Q}(u, v),
\end{aligned}
\tag{5.10}
$$

where $\dot{u} = du/d\tau$ and $\dot{v} = dv/d\tau$. Since $d + 2$ is the minimum of the degrees of the homogeneous parts of $\widetilde{P}(u, v)$ and $\widetilde{Q}(u, v)$ it is clear that $(0, 0)$ is a singular point of system (5.10).

5.5 Global Flow of a Planar Polynomial Vector Field

First, we determine the phase portrait on the Poincaré disk of the system

$$
\begin{aligned}
\dot{x} &= x, \\
\dot{y} &= -y.
\end{aligned}
\tag{5.11}
$$

This system has a unique finite singular point, the origin, which is a saddle. Let X be the vector field associated to system (5.11). Then the expression for $p(X)$ in the local chart U_1 is

$$
\begin{aligned}
\dot{u} &= -2u, \\
\dot{v} &= -v.
\end{aligned}
$$

Therefore there is a unique singular point in U_1, the origin, which is a stable node at infinity. Since the degree of X is odd, the origin of V_1 is also another stable node.

The expression for $p(X)$ in the local chart U_2 is

$$\dot{u} = 2u,$$

$$\dot{v} = v.$$

So at the origin of U_2 there is an unstable node. The same is true for the origin of V_2.

If we now draw the phase portrait of system (5.11) on the Poincaré disk we get Fig. 5.4. In this section we always project the northern hemisphere of the Poincaré sphere onto the Poincaré disk parallel to the y_3-axis. Note that the unique separatrices of this system are those of the saddle at the origin, which are contained in the invariant straight lines $x = 0$ and $y = 0$.

Second, we study the phase portrait on the Poincaré disk of the system

$$\dot{x} = -x - y^2,$$
$$\dot{y} = y + x^2. \tag{5.12}$$

This is a Hamiltonian system because it can be written as

$$\dot{x} = -\partial H/\partial y,$$

$$\dot{y} = \partial H/\partial x,$$

with Hamiltonian

$$H(x,y) = \frac{1}{3}(x^3 + y^3) + xy.$$

Therefore H is a first integral of system (5.12); i.e., H is constant on the solutions of (5.12), because on any solution $(x(t), y(t))$ of (5.12) we have

$$\frac{dH}{dt}(x(t), y(t)) = \frac{\partial H}{\partial x}(-x - y^2) + \frac{\partial H}{\partial y}(y + x^2)\,|_{(x,y)=(x(t),y(t))} = 0.$$

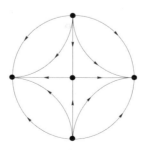

Fig. 5.4. The phase portrait in the Poincaré disk of system (5.11)

System (5.12) has two finite singular points, a saddle at $(0,0)$ and a linear center at $(-1,-1)$ with eigenvalues $\pm\sqrt{3}i$. Since the first integral H is well defined at $(-1,-1)$, this singular point is a center.

Now we shall compute the infinite singular points. Let X be the vector field associated to (5.12). Then the expression for $p(X)$ in the local chart U_1 is

$$\dot{u} = 1 + 2uv + u^3,$$
$$\dot{v} = v^2 + u^2 v.$$

Therefore on U_1 there is a unique singular point, $(-1,0)$ which is an unstable node at infinity. Since the degree of X is 2, the diametrically opposite point is a stable node in V_1.

The expression for $p(X)$ in the local chart U_2 is

$$\dot{u} = -1 - 2uv - u^3,$$
$$\dot{v} = -v^2 - u^2 v.$$

Since the origin of U_2 is not a singular point there do not exist additional infinite singular points.

The unique separatrices of system (5.12) are the separatrices of the saddle $(0,0)$. Since $H(0,0) = 0$, in order to locate such separatrices it is sufficient to draw the curve $H(x,y) = 0$. Hence the phase portrait of system (5.12) on the Poincaré disk as is given in Fig. 5.5.

Finally, we analyze the phase portrait on the Poincaré disk of the system

$$\dot{x} = y,$$
$$\dot{y} = -(1+y)(x+y). \tag{5.13}$$

This system has a unique finite singular point, the origin. Since this point has eigenvalues $(-1 \pm \sqrt{3}i)/2$, it is a stable focus.

Let X be the vector field associated to system (5.13). Then the expression for $p(X)$ in the local chart U_1 is

$$\dot{u} = -u - v - u^2 - uv - u^2 v,$$
$$\dot{v} = -uv^2.$$

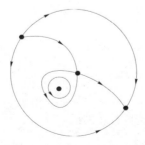

Fig. 5.5. The phase portrait in the Poincaré disk of system (5.12)

Therefore in U_1 there are two infinite singular points, $(0,0)$ and $(-1,0)$, both semi-hyperbolic. A computation shows that $(0,0)$ is a semi-hyperbolic saddle and that $(-1,0)$ is a semi-hyperbolic saddle-node. Looking at the linear part of the system at the saddle-node and checking that the eigenspace of the nonzero eigenvalue is transverse to the equator of $\mathbb{S}^2$, it is easy to deduce that the two hyperbolic sectors are in the chart V_1 while the nodal or parabolic sector is in the chart U_1. Of course, two of the separatrices of the saddle-node are on the equator of the Poincaré sphere.

The expression for $p(X)$ in the local chart U_2 is

$$
\begin{aligned}
\dot{u} &= u + v + u^2 + uv + u^2 v, \\
\dot{v} &= v + uv + v^2 + uv^2.
\end{aligned}
\tag{5.14}
$$

So the origin of U_2 is an unstable hyperbolic node. Since the degree of X is even, the diametrically opposite point is a stable node in V_2.

The fact that the straight line $y = -1$ is invariant under the flow of X tells us that the phase portrait of $p(X)$ on the Poincaré disk is the one described in Fig. 5.6. We may check that the center-unstable manifold of the infinite saddle-node spirals clock-wise around the focus $(0,0)$ since the flow on the straight half-line $\{(x,y) : x = 0, y > 0\}$ moves to the right.

We end this chapter with two examples in which it is clear that the singularities at infinity in a well chosen Poincaré–Lyapunov compactification can be much simpler than in a Poincaré compactification.

Example 5.1 We consider the system

$$
\begin{aligned}
\dot{x} &= y, \\
\dot{y} &= -x^3 - xy.
\end{aligned}
\tag{5.15}
$$

Performing the usual Poincaré compactification for cubic systems we get on the Poincaré disk a phase portrait like in Fig. 5.7a. On the circle at infinity we get two degenerate singularities.

If however we use a Poincaré–Lyapunov compactification of type $(1,2)$ then we get a phase portrait as in Fig. 5.7b of which the circle at infinity is a periodic orbit.

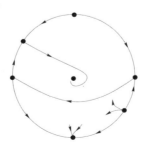

Fig. 5.6. The phase portrait in the Poincaré disk of system (5.13)

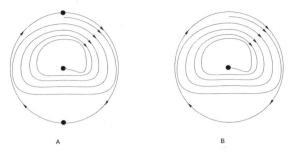

Fig. 5.7. Compactification of system (5.15)

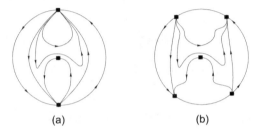

Fig. 5.8. Compactification of system (5.16)

Example 5.2 We consider the system

$$\begin{aligned}
\dot{x} &= y, \\
\dot{y} &= x^5 - xy.
\end{aligned} \tag{5.16}$$

Performing the Poincaré compactification for vector fields of degree 5 we get on the Poincaré disk a phase portrait like in Fig. 5.8a. On the circle at infinity we get two degenerate singularities.

If we use a Poincaré–Lyapunov compactification of type $(1,3)$ then we get a phase portrait as in Fig. 5.8b with only hyperbolic singularities on the circle at infinity.

In both examples we leave it as an exercise to make the necessary calculations at infinity, according to the procedures explained in this chapter.

5.6 Exercises

Exercise 5.1 Determine the phase portraits on the Poincaré disk of the following polynomial differential systems:

(i) $\dot{x} = x,\ \dot{y} = -y.$
(ii) $\dot{x} = x^2 + y^2 - 1,\ \dot{y} = 5(xy - 1).$

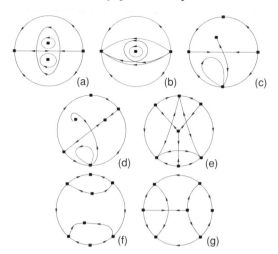

Fig. 5.9. Phase portraits of Exercise 5.2

(iii) $\dot{x} = -y(1+x)+\alpha x+(\alpha+1)x^2$, $\dot{y} = x(1+x)$, depending on the parameter $\alpha \in (0,1)$.

Exercise 5.2 Determine which of the global phase portraits shown in Fig. 5.9 correspond to the following quadratic systems (two of them do not correspond to any system presented).

(i) $\dot{x} = -4y + 2xy - 8$, $\dot{y} = 4y^2 - x^2$.
(ii) $\dot{x} = 2x - 2xy$, $\dot{y} = 2y - x^2 + y^2$.
(iii) $\dot{x} = -x^2 - y^2 + 1$, $\dot{y} = 2x$.
(iv) $\dot{x} = -x^2 - y^2 + 1$, $\dot{y} = 2xy$.
(v) $\dot{x} = x^2 - y^2 - 1$, $\dot{y} = 2y$.

Exercise 5.3 Consider a homogeneous quadratic differential system:

$$\dot{x} = ax^2 + bxy + cy^2,$$
$$\dot{y} = dx^2 + exy + fy^2.$$

Supposing that the singular point at the origin is isolated, hence unique, show that the phase portrait on the Poincaré disk must be one of the seven phase portraits given in Fig. 5.10.

5.7 Bibliographical Comments

For more information on Poincaré compactification; see the book of Andronov, Leontovich, Gordon and Maier [4], the book of Sotomayor [151] and the paper of Gonzalez [76].

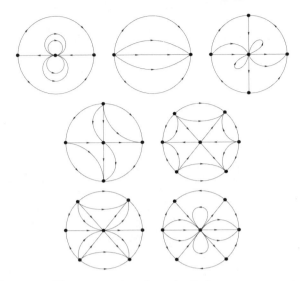

Fig. 5.10. Phase portraits of quadratic homogenous systems

The use of a quasihomogeneous compactification of polynomial planar vector fields and the related name Poincaré–Lyapunov compactification were introduced in papers of Dumortier at the end of the eighties. The inspiration came from the strongly related quasihomogeneous blow-up of singularities, as presented in Chap. 3. As one application among others, a systematic study by means of the technique has been made of all polynomial Liénard equations near infinity [55]

6

Indices of Planar Singular Points

In this chapter we study the index of a singular point of a vector field on $\mathbb{R}^2$ or on $\mathbb{S}^2$. For a given vector field X on $\mathbb{S}^2$ having finitely many singular points, we shall see that the sum of their indices is equal to 2 (the Euler characteristic of $\mathbb{S}^2$). This result is the famous Poincaré Index Theorem, which later on was extended by Hopf to vector fields on compact manifolds. We also prove the formula of Poincaré for computing the index of an isolated singularity. Finally we briefly discuss the relation between the index and the multiplicity of a singular point.

6.1 Index of a Closed Path Around a Point

A *path* in the plane $\mathbb{R}^2$ is a continuous map from the interval $I = [0, 1]$ to $\mathbb{R}^2$ ($\sigma : I \to \mathbb{R}^2$); that is, we assign to every $t \in [0, 1]$ the point $\sigma(t) = (\sigma_1(t), \sigma_2(t))$ in the plane, such that $\sigma_i : I \to \mathbb{R}$ are continuous maps.

The point $\sigma(0)$ is called the *origin* of the path σ and $\sigma(1)$ is called the *endpoint of the path*. We must not confuse a path with its image $\sigma(I) \subset \mathbb{R}^2$; $\sigma(I)$ is called a *curve*. For example, the paths $\sigma : \sigma(t) = (r \cos \pi t, r \sin \pi t)$ and $\tau : \tau(t) = (r(1 - 2t), 2r\sqrt{t(1 - t)})$, are obviously different, but the curves $\sigma(I)$ and $\tau(I)$ coincide; see Fig. 6.1.

Let q be a point of $\mathbb{R}^2$ which does not belong to $\sigma(I)$ and let r be a ray with origin at q. For every point $\sigma(t)$ we denote by $\bar{\varphi}(t)$ the angle formed by the rays r and $\overline{q\sigma(t)}$. The angle $\bar{\varphi}(t)$ is an element of the circle $\mathbb{R}/2\pi\mathbb{Z}$. The function $\bar{\varphi} : I \to \mathbb{R}/2\pi\mathbb{Z}$ is continuous with respect to the parameter t; see Fig. 6.2. We can cover it by a continuous mapping $\varphi : I \to \mathbb{R}$. From now on we shall work with such a continuous cover, with the extra condition that $\varphi(0) = 0$. We call φ an *angle function*.

A *closed path* is a path whose endpoint coincides with its origin (i.e., $\sigma(1) = \sigma(0)$). Equivalently, a closed path can be considered as a continuous map $\sigma : \mathbb{S}^1 \to \mathbb{R}^2$. Let q be a point of $\mathbb{R}^2$ which does not belong to $\sigma(I)$. Then

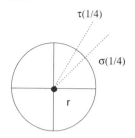

Fig. 6.1. Same image, different paths

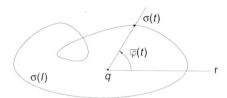

Fig. 6.2. Definition of $\varphi(t)$

the difference $\varphi(1) - \varphi(0)$ is a multiple of 2π, independent of the chosen ray r, and independent of the chosen cover φ. It is then acceptable to define the quotient

$$i(q,\sigma) = \frac{\varphi(1) - \varphi(0)}{2\pi},$$

which is an integer called the *index of the closed path σ around the point q*.

The index $i(q,\sigma)$ is an algebraic magnitude associated to a closed path and a point, which are topological concepts.

We consider some examples:

Example 6.1 Let $\sigma : I \to \mathbb{R}^2$ be the closed path defined by the expression $\sigma(t) = e^{2\pi i n t}$. We have that $\sigma(I) \equiv \mathbb{S}^1$. As the origin $q = (0,0) \notin \mathbb{S}^1$, we may consider the index $i(q,\sigma)$ using as ray r the positive x-axis. Then we have that $\varphi(t) = 2\pi n t$, and so

$$i(q,\sigma) = \frac{\varphi(1) - \varphi(0)}{2\pi} = n.$$

$\square$

Example 6.2 In Fig. 6.3 we have indicated the indices of different points with respect to the closed paths drawn. $\square$

Now we consider a method for calculating the index of a closed path around a point q. We choose as r a ray with origin at q which cuts $\sigma(I)$ in a finite number of points. Let $0 < t_1 < t_2 < \cdots < t_n < 1$ be the parameters for which

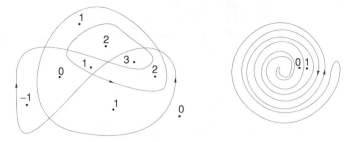

Fig. 6.3. Some examples of indices

$\sigma(t_i) \in r$. We assume that for every t_i there exists $\varepsilon_i > 0$ such that for every $t \neq t_i$ with $|t_i - t| < \varepsilon_i$, $\sigma(t) \notin r$, and so $\varphi(t) \neq \varphi(t_i)$. We may distinguish four cases:

(i) $\varphi(t) \leq \varphi(t_i)$ for every t such that $|t_i - t| < \varepsilon_i$;
(ii) $\varphi(t) \geq \varphi(t_i)$ for every t such that $|t_i - t| < \varepsilon_i$.

In these two cases, the ray r does not cross $\sigma(I)$ at the point $\sigma(t_i)$.

(iii) $\varphi(t) < \varphi(t_i)$ for $t_i - \varepsilon_i < t < t_i$ and $\varphi(t) > \varphi(t_i)$ for $t_i + \varepsilon_i > t > t_i$;
(iv) $\varphi(t) > \varphi(t_i)$ for $t_i - \varepsilon_i < t < t_i$ and $\varphi(t) < \varphi(t_i)$ for $t_i + \varepsilon_i > t > t_i$.

In these last two cases, the ray r crosses the path σ at the point $\sigma(t_i)$. We call case *(iii)* a *positive intersection point* and case *(iv)* a *negative intersection point*.

If we denote by $\psi(t) = [\varphi(t)/2\pi]$, the function representing the integer part of $\varphi(t)/2\pi$, then $\psi(t_i) = \varphi(t_i)/2\pi$. When we move from a $t < t_i$ to a $t > t_i$ (always in a neighborhood of $|t_i - t| < \varepsilon_i$), the value $\psi(t)$ increases by one in case *(iii)*, decreases by one in case *(iv)*, and does not change in the remaining cases. Then

$$\psi(1) - \psi(0) = \left[\frac{\varphi(1)}{2\pi}\right] - \left[\frac{\varphi(0)}{2\pi}\right] = \frac{\varphi(1) - \varphi(0)}{2\pi},$$

because $\varphi(0)/2\pi - [\varphi(0)/2\pi] = \varphi(1)/2\pi - [\varphi(1)/2\pi]$.

So if m represents the number of positive intersection points and n represents the number of negative intersection points, we have that $i(q, \sigma) = m - n$. Then the index of the closed path σ around the point q is equal to the number of positive intersection points minus the number of negative intersection points of σ with the ray r. In this construction it is possible most of the time to work with rays whose intersection with $\sigma(I)$ are all isolated crossing points.

The next two propositions give conditions in order that the index of a path around a point does not depend on the point (Proposition 6.3), or does not depend on the path (Proposition 6.5).

Proposition 6.3 *If the segment $\overline{q_1 q_2}$ does not cut the closed path σ, then* $i(q_1, \sigma) = i(q_2, \sigma)$.

Proof. Let $\sigma : I \to \mathbb{R}^2$ be a closed path, and q_1 and q_2 two points not belonging to $\sigma(I)$. Let φ_1 and φ_2 be angle functions with respect to q_1 and q_2. Then the function

$$\delta(t) = \frac{\varphi_1(t) - \varphi_2(t)}{\pi}$$

is continuous for every t, and we have that

$$\delta(1) - \delta(0) = \frac{\varphi_1(1) - \varphi_2(1)}{\pi} - \frac{\varphi_1(0) - \varphi_2(0)}{\pi}$$
$$= 2(i(q_1, \sigma) - i(q_2, \sigma)).$$

Suppose that $i(q_1, \sigma) \neq i(q_2, \sigma)$. Then we have that $|\delta(1) - \delta(0)| \geq 2$, and by the Intermediate Value Theorem, there exists $t_0 \in I$ such that $\delta(t_0)$ is odd, that is $\varphi_1(t_0) - \varphi_2(t_0) = n\pi$ with n odd.

But if we choose parallel rays with origins at q_1 and q_2 to define φ_1 and φ_2, respectively, in such a way that the ray starting at q_1 passes through q_2, then $\sigma(t_0) \in \overline{q_1 q_2}$, which contradicts the conditions of the proposition. $\quad\square$

Corollary 6.4 *If q_1 and q_2 belong to the same connected component of $\mathbb{R}^2 \setminus \sigma(I)$, then $i(q_1, \sigma) = i(q_2, \sigma)$.*

Proposition 6.5 *Let $\sigma_1, \sigma_2 : I \to \mathbb{R}^2$ be closed paths and $q \in \mathbb{R}^2$ such that for every $t \in I$ we have $q \notin \overline{\sigma_1(t)\sigma_2(t)}$. Then $i(q, \sigma_1) = i(q, \sigma_2)$.*

Proof. Let φ_1 and φ_2 be angle functions with respect to σ_1 and σ_2 and the point q. If we suppose that the indices are different, then by the same arguments as in Proposition 6.3, there exists a t_0 such that $\varphi_1(t_0) - \varphi_2(t_0) = n\pi$ with n odd.

But this implies that q belongs to a segment determined by $\sigma_1(t_0)$ and $\sigma_2(t_0)$, contradicting the hypothesis. $\quad\square$

Corollary 6.6 *Let $\sigma_1, \sigma_2 : I \to \mathbb{R}^2$ be closed paths and $q \in \mathbb{R}^2$ such that $q \notin \sigma_1(I) \cup \sigma_2(I)$. If for every $t \in I$ we have $\|\sigma_1(t) - \sigma_2(t)\| < \|q - \sigma_1(t)\|$, then $i(q, \sigma_1) = i(q, \sigma_2)$.*

Proof. Under the hypotheses of the corollary, $q \notin \overline{\sigma_1(t)\sigma_2(t)}$ for any $t \in I$. Then Proposition 6.5 applies. $\quad\square$

6.2 Deformations of Paths

Given a family of closed paths $\sigma_s : I \to \mathbb{R}^2$, with $s \in I$, we say that we have a *deformation of closed paths* if $\sigma(s, t) = \sigma_s(t)$ depend continuously on the variables $t \in I$ and $s \in I$.

Two closed paths which can be deformed one into the other, as we have described, are called *homotopic* closed paths; see Fig. 6.4.

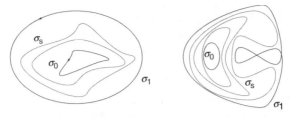

Fig. 6.4. Examples of homotopic closed paths

If the point $q \in \mathbb{R}^2$ does not belong to the image of σ, then we call it a *deformation* (or *homotopy*) in $\mathbb{R}^2 \setminus \{q\}$, and we say that the closed paths σ_0 and σ_1 are *homotopic* in $\mathbb{R}^2 \setminus \{q\}$. This implies that the point q does not belong to the image of any of the paths σ_s for $s \in I$.

Proposition 6.7 *If the closed paths σ_1 and σ_2 are homotopic in $\mathbb{R}^2 \setminus \{q\}$, then $i(q, \sigma_1) = i(q, \sigma_2)$.*

Proof. As $\sigma(I \times I)$ is compact, the distance d from the point $q \notin \sigma(I \times I)$ to $\sigma(I \times I)$ is positive.

Moreover, as every continuous map defined in a compact set is uniformly continuous, there exists $\delta > 0$ such that $\|\sigma(t,s) - \sigma(t',s')\| < d$ for every t, s, t', s' such that $|s - s'| < \delta$ and $|t - t'| < \delta$. Let $0 < s_0 < s_1 < \ldots < s_n < 1$ be such that $s_i - s_{i-1} < \delta$. Then for every $t \in I$ and every $i = 1, 2, \ldots, n - 1$ we have that

$$\|\sigma(t, s_i) - \sigma(t, s_{i-1})\| < d \leq \|q - \sigma(t, s_i)\|,$$

and we can apply Corollary 6.6 to prove that

$$i(q, \sigma_0) = i(q, \sigma_{s_1}) = i(q, \sigma_{s_2}) = \ldots = i(q, \sigma_{s_n}) = i(q, \sigma_1).$$

Thus the result is proved. $\qquad\qquad\qquad\qquad\qquad\qquad\qquad\qquad\qquad\square$

We say that a closed path σ is *contractible* if it is homotopic to a constant path, that is, a path whose image consists of a single point.

Corollary 6.8 *If a closed path σ is contractible in $\mathbb{R}^2 \setminus \{q\}$, then $i(q, \sigma) = 0$.*

Corollary 6.9 *If a closed path σ is homotopic in $\mathbb{R}^2 \setminus \{q\}$ to the closed path defined by $\sigma_n(t) = q + e^{2\pi i n t}$ with n an integer and $t \in I$, then $i(q, \sigma) = n$.*

The paths $\sigma_n(t) = q + e^{2\pi i n t}$ are especially important due to the next proposition, which is the converse of Corollary 6.9.

Proposition 6.10 *If σ is a closed path in $\mathbb{R}^2$ such that $i(q, \sigma) = n$ and $q \notin \sigma(I)$, then σ is homotopic in $\mathbb{R}^2 \setminus \{q\}$ to the closed path $\sigma_n(t) = q + e^{2\pi i n t}$.*

Proof. We choose as r the right horizontal ray passing through the point q. Let $\varphi(t)$ be an angle function for the path σ with respect to the point q. We set $d(t) = \|q - \sigma(t)\|$, and define then $\sigma : I \times I \to \mathbb{R}^2$ by

$$\sigma(t, s) = q + [d(t)(1 - s) + s]e^{(\varphi(t)(1-s)+2\pi nst)i}.$$

It is clear that σ is a continuous map, that $\sigma(t, 0) = \sigma(t)$, and that $\sigma(t, 1) = \sigma_n(t)$. Moreover, as $\varphi(1) = \varphi(0) + 2\pi n$, $\sigma(0, s) = \sigma(1, s)$ for every s. Finally, as $q \notin \sigma(I \times I)$, σ gives the required deformation. $\qquad\square$

This proposition can be completed to the next theorem.

Theorem 6.11 *Two closed paths are homotopic in $\mathbb{R}^2 \setminus \{q\}$ if and only if they have the same index around q.*

Corollary 6.12 *If $\sigma : I \to \mathbb{R}^2 \setminus \{q\}$ is a path with $i(q, \sigma) = 0$, then there exists a map $f : \mathbb{D}^2 \to \mathbb{R}^2 \setminus \{q\}$ with $f(e^{2\pi it}) = \sigma(t)$.*

6.3 Continuous Maps of the Closed Disk

A closed path $\sigma : I \to \mathbb{R}^2$ can also be considered as a continuous map $\tilde{\sigma} : \mathbb{S}^1 \to \mathbb{R}^2$ and conversely by the relation $\tilde{\sigma}(e^{2\pi it}) = \sigma(t)$. If no confusion is possible we also simply write σ instead of $\tilde{\sigma}$.

Let $\mathbb{D}^2$ be the closed disk of radius one centered at the origin and let $f : \mathbb{D}^2 \to \mathbb{R}^2$ be a continuous map expressed in polar coordinates as $f(r, \theta)$. The boundary of $\mathbb{D}^2$ is the circle $\mathbb{S}^1$. The restriction of f to $\mathbb{S}^1$ is the closed path $\sigma : \mathbb{S}^1 \to \mathbb{R}^2$. If $q \in \mathbb{R}^2$ does not belong to the image of f, then the map

$$\sigma : I \times I \longrightarrow \mathbb{R}^2$$
$$(t, s) \to f(1 - s, 2\pi t)$$

provides a deformation from the closed path σ, with $\sigma(t) = f(1, 2\pi t)$, to a constant path. So the closed path σ is contractible due to Corollary 6.8 and consequently $i(q, \sigma) = 0$. In short, we have proved the following result.

Theorem 6.13 *Let $f : \mathbb{D}^2 \to \mathbb{R}^2$ be a continuous map, and let q be a point of $\mathbb{R}^2$ which does not belong to $f(\mathbb{S}^1)$, and such that $i(q, \sigma) \neq 0$, with $\sigma(t) = f(1, 2\pi t)$ as above. Then $q \in f(\mathbb{D}^2)$.*

This theorem can be thought of as an extension of the Intermediate Value Theorem from the interval I to the disk $\mathbb{D}^2$.

6.4 Vector Fields Along the Unit Circle

We consider a continuous vector field X defined on the circle $\mathbb{S}^1$. Given a point $p = e^{2\pi it} \in \mathbb{S}^1$, we identify the vector $X(p) = (X_1(p), X_2(p))$ with the point $\sigma(t) \in \mathbb{R}^2$ such that $\overrightarrow{0\sigma(t)} = X(p)$. This allows us to define a continuous map

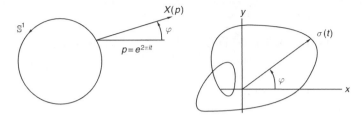

Fig. 6.5. Closed path associated to a vector field

$$\sigma : I \to \mathbb{R}^2$$

$$t \to \sigma(t)$$

such that $\sigma(0) = \sigma(1)$; then σ is a closed path in $\mathbb{R}^2$. We call it the *closed path associated to the vector field* $X|_{\mathbb{S}^1}$; see Fig. 6.5.

If $X(p) \neq 0$ for every $p \in \mathbb{S}^1$, that is, if the vector field does not vanish at any point of the circle, then the origin cannot belong to the image of the path σ. Then we can define the index $i(0, \sigma)$ which geometrically means the number of revolutions, taking into account their orientation, that the vector field X makes when moving along $\mathbb{S}^1$. We denote it by $i(X)$.

All that we have seen about closed paths and indices of paths can be translated to vector fields. In particular, we talk about a *deformation* of a vector field X_0 into a vector field X_1, when we have a continuous family of vector fields X_s, with $s \in I$, which depends continuously on a parameter s.

Proposition 6.14 *If the vector field X_0 can be deformed into the vector field X_1 in such a way that $X_s(p) \neq 0$ for every $p \in \mathbb{S}^1$ and every $s \in I$, then $i(X_0) = i(X_1)$.*

Proof. In this case, the closed paths σ_0 and σ_1 associated to X_0 and X_1 can be deformed continuously in $\mathbb{R}^2 \setminus \{0\}$ and so, by Proposition 6.7, $i(0, \sigma_0) = i(0, \sigma_1)$. □

Proposition 6.15 *Let X_0 and X_1 be two vector fields defined on $\mathbb{S}^1$ not vanishing at any point. If $X_0(p)$ and $X_1(p)$ never have opposite directions at any p, then $i(X_0) = i(X_1)$.*

Proof. The corresponding closed paths σ_0 and σ_1 satisfy the hypotheses of Proposition 6.5, so that $i(0, \sigma_0) = i(0, \sigma_1)$. □

Proposition 6.16 *Let X be a vector field defined on $\mathbb{D}^2$ such that $X(p) \neq 0$ for every $p \in \mathbb{S}^1$. If $i(X) \neq 0$, then there exists a point $p \in \mathbb{D}^2$ such that $X(p) = 0$.*

Proof. If σ is the closed path associated to the vector field X, then $i(0, \sigma) \neq 0$. Therefore, by Theorem 6.13, there exists a point $p \in \mathbb{D}^2$ such that $X(p) = 0$. □

We consider some examples.

Example 6.17 Let X be a vector field on $\mathbb{S}^1$ which does not vanish at any point and which is never pointing in the direction of the exterior normal. Then applying Proposition 6.15, the index $i(X)$ is equal to the number of turns of the field $Y = (-x, -y)$, and so it is 1. ☐

Example 6.18 Let X be a vector field on $\mathbb{R}^2$ which does not vanish at any point. Then by Proposition 6.16, the number of turns of X over the circle $\mathbb{S}^1$ has to be 0. So by the previous example, there is at least one point $p \in \mathbb{S}^1$ in which $X(p)$ points in the direction of the exterior normal. ☐

6.5 Index of Singularities of a Vector Field

Given an isolated singularity p of a vector field X defined on an open subset of $\mathbb{R}^2$, there is a neighborhood V of p on which there is no other singularity of X. Consider now a closed path $\sigma : \mathbb{S}^1 \to V \setminus \{p\}$. We define $i_{p,\sigma}(X) = i(X \circ \sigma)$; it is equal to the number of turns of $X \circ \sigma$ in a counter–clockwise sense.

Lemma 6.19 *Let p be an isolated singularity of the vector field X, and let σ and σ' be two homotopic closed paths in a punctured neighborhood $V \setminus \{p\}$ of p, on which X has no singularities. Then $i_{p,\sigma}(X) = i_{p,\sigma'}(X)$.*

Proof. Let $\sigma : I \times I \to V \setminus \{p\}$, given by $(t, s) \longmapsto \sigma(t, s)$, be a homotopy between $\sigma(t, 0) = \sigma(t)$ and $\sigma(t, 1) = \sigma'(t)$. Then $f : I \times I \to \mathbb{R}^2 \setminus \{(0, 0)\}$, given by $(t, s) \longmapsto X(\sigma(t, s))$, is a homotopy between the paths $X \circ \sigma$ and $X \circ \sigma'$, inducing the required result. ☐

For a situation like the one described in Lemma 6.19, we call $\sigma : I \to V \setminus \{p\}$ a *canonical closed path* if σ is homotopic in $V \setminus \{p\}$ to $\sigma_r : I \to V \setminus \{p\}$, where $\sigma_r(t) = p + re^{2\pi it}$ with $r > 0$ sufficiently small.

Let p be an isolated singularity of the vector field X. We define the *index* $i_p = i_p(X)$ *of the singularity p* of the vector field X as $i_{p,\sigma}(X)$ for a canonical closed path σ. By Lemma 6.19, this definition does not depend on the canonical closed path chosen, since they are mutually homotopic.

We now consider a method for computing the index $i_p(X)$, given a canonical closed path σ for p. For $t \in I$ let $u(t)$ denote the unit tangent vector to the curve $\sigma(I)$ at the point $\sigma(t)$, and let $v(t)$ denote the unit vector defined by $v(t) = X(\sigma(t))/\|X(\sigma(t))\|$. We work only with canonical closed paths σ for p having the following property: there exist finitely many values $0 < t_1 < \cdots < t_n < 1$ of the parameter t such that the vectors $u(t_i)$ and $v(t_i)$ coincide. Then for every t_i there exists $\varepsilon_i > 0$ such that for every $t_j \neq t_i$ we have $|t_j - t_i| > \varepsilon_i$, and for every $t \neq t_i$ with $|t - t_i| < \varepsilon_i$ the angle between the vectors $u(t)$ and $u(t_i)$ is smaller than $\pi/4$.

We may distinguish four cases:

(i) $u(t) \wedge v(t) \leq 0$ for every t such that $|t_i - t| < \varepsilon_i$;
(ii)$u(t) \wedge v(t) \geq 0$ for every t such that $|t_i - t| < \varepsilon_i$.

Here, as usual we define $u \wedge v$ as the real number $|u| \, |v| \sin \theta$, where θ is the angle going in the counter-clockwise sense from the vector u to the vector v. In these two cases the angle from the vector $u(t)$ to the vector $v(t)$, measured in the counter-clockwise sense, does not change the sign when we cross the point $\sigma(t_i)$ for increasing t.

(iii) $u(t) \wedge v(t) < 0$ for $t_i - \varepsilon_i < t < t_i$ and $u(t) \wedge v(t) > 0$ for $t_i - \varepsilon_i > t > t_i$;
(iv) $u(t) \wedge v(t) > 0$ for $t_i - \varepsilon_i < t < t_i$ and $u(t) \wedge v(t) < 0$ for $t_i - \varepsilon_i > t > t_i$.

In these last two cases the vectors $u(t)$ and $v(t)$ interchange their orientations when the path σ crosses the point $\sigma(t_i)$. We call case *(iii)* a *positive crossing point* and case *(iv)* a *negative crossing point*.

It is easy to check that if m represents the number of positive crossing points and n represents the number of negative crossing points, we have that $i_p(X) = 1 + m - n$. Then the index of an isolated singularity p of the vector field X with respect to a canonical closed path σ for p is equal to the number of positive crossing points minus the number of negative crossing points of $\sigma(I)$ plus 1, provided the number of crossing points is finite.

From Chap. 3 we know that an isolated singularity of an analytic vector field is either a center or a focus, or it has a finite sectorial decomposition as defined in Sect. 1.5. In both cases it is easy to see that we can find in a neighborhood of the singularity a C^∞ canonical curve with a finite number of crossing points.

Using this geometric interpretation of the index of an isolated singularity p of the vector field X with respect to a canonical closed path σ for p, it is now easy to prove the following interesting result:

Proposition 6.20 *If X at p and Y at q are locally C^1-conjugate, with p and q isolated singular points, then $i_p(X) = i_q(Y)$.*

We leave the proof as an exercise; it will suffice to mimic the argumentation that we will use in the proof of Proposition 6.32.

As a consequence of this proposition we see that the definition of the index $i_p(X)$ for an isolated singularity is independent of the chosen C^1-coordinates used to represent the vector field. This permits us to use this notion on surfaces, as we will do it in Sect. 6.6.

We consider some examples.

Example 6.21 *Null index, $i_p = 0$.* In this case we call the singularity *removable*; that is, we can modify the vector field X in the interior of a ball V of center p in such a way that the new field Y is continuous and is not zero in V.

In order to see this, let $f : V \to \mathbb{R}^2$ be the continuous map defined by the vector field X, and let σ be the closed path of $\mathbb{R}^2$ defined by the restriction of

X to the boundary ∂V of V. Then $i(p, \sigma) = 0$, and by Corollary 6.12, there is a continuous map $g : V \to \mathbb{R}^2$ which coincides with f over ∂V and whose image does not contain p. Now the map g allows us to define a continuous vector field Y which does not vanish at any point of V.   □

Example 6.22 *Index* 1. Typical examples are the vector fields $X = (x, y)$ and $X' = (-y, x)$, both having an isolated singularity of index $+1$ at the origin. In the first case, the solutions are rays emanating from the origin, and in the second the solutions are circles of radius r, $r \in (0, \infty)$. So in both cases, the number of turns of the vector field over the circle $\mathbb{S}^1$ is 1.   □

Example 6.23 *Index* -1. In this case the easiest example is the vector field $X = (x, -y)$ which has a singularity at the origin of index -1. The solutions of this vector field are the hyperbolas $xy = c$ with $c \in \mathbb{R} \setminus \{0\}$ and the rays in the x- and y-axes determined by the origin.

We may calculate the index using these solutions. The index of X at p is the number of turns of the vector field X on $\mathbb{S}^1$, which is the index of the closed path σ defined by X around p. But the angle α of Fig. 6.6 is exactly $\alpha = \arctan(-y/x) = -\varphi$, and now

$$i_p = i_p(X) = i(p, \sigma) = \frac{\alpha(1) - \alpha(0)}{2\pi} = \frac{-2\pi}{2\pi} = -1.$$

□

Example 6.24 *Index* 2. In this case we proceed differently. Instead of proposing a vector field, we look for the solutions that may give us a singularity of index 2, and afterward we give the corresponding vector field.

The variation of the angle α on the circle $\mathbb{S}^1$ has to be 4π. Then we take, for example, $\alpha = 2\varphi$. Then we have

$$\frac{dy}{dx} = \tan \alpha = \tan(2\varphi) = \frac{2 \tan \varphi}{1 - \tan^2 \varphi} = \frac{2y/x}{1 - (y/x)^2} = \frac{2xy}{x^2 - y^2}.$$

So a vector field that we can use is $X = (x^2 - y^2, 2xy)$ and the solutions are given by the curves $x^2 + y^2 + 2cy = 0$; see Fig. 6.7.   □

Fig. 6.6. Point of index -1

Fig. 6.7. Point of index 2

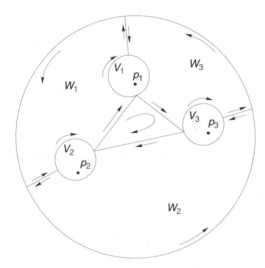

Fig. 6.8. Piecing of $\mathbb{D}^2$

Example 6.25 *Index n.* It is easy to see that if $\alpha = n\varphi$, then the index of the vector field at the origin is n. Then $dy/dx = \tan\alpha = \tan(n\varphi)$, and when we write $\tan(n\varphi)$ in terms of $\tan\varphi = y/x$ we get a vector field with a singularity at the origin having index n. □

Now for a vector field X defined on $\mathbb{D}^2$ we examine the relation between the number of turns of X on $\mathbb{S}^1$ and the indices of the singularities in $\mathbb{D}^2$.

Proposition 6.26 *Let X be a vector field defined on $\mathbb{D}^2$ such that $X(p) \neq 0$ for all $p \in \partial\mathbb{D}^2 = \mathbb{S}^1$. Suppose that the number of singularities of X in $\mathbb{D}^2$ is finite, and denote them by $p_1, p_2, \ldots, p_k$. Then $i(X|_{\mathbb{S}^1}) = \sum_{i=1}^{k} i_{p_i}(X)$.*

Proof. Let $V_1, V_2, \ldots, V_k$ be disks with centers $p_1, p_2, \ldots, p_k$, respectively, and with radius so small that each disk V_i is contained in the interior of $\mathbb{D}^2$ and that they are pairwise disjoint.

We decompose $\mathbb{D}^2 \setminus \cup V_i$ as in Fig. 6.8, that is, we use straight segments. We give an orientation coherent with the decomposition, as in Fig. 6.8. Each region W_i into which we have split $\mathbb{D}^2 \setminus (\cup_i V_i)$ is homeomorphic to the disk $\mathbb{D}^2$ and so we can talk about the number of turns $i(X, \partial W_i)$ of the vector field X in the boundary of each W_i.

If we now make the vector field X move over all the straight segments of the decomposition, each straight segment is traversed twice, but in opposite sense. So we have that

$$i(X, \partial W_1) + \ldots + i(X, \partial W_k) = i(X) - (i_{p_1} + \ldots + i_{p_k}).$$

But as we know that $i(X, \partial W_j) = 0$ for every j, by Proposition 6.16 we have the required result.   □

6.6 Vector Fields on the Sphere $\mathbb{S}^2$

Let X be a vector field on the unit sphere $\mathbb{S}^2 \subset \mathbb{R}^3$, that is to say, we assign a vector $X(p) = (X_1(p), X_2(p), X_3(p))$ to every point of $\mathbb{S}^2$ such that the components $X_1(p)$, $X_2(p)$, and $X_3(p)$, depend continuously on the point p. The vector field X is called a *tangent vector field to* $\mathbb{S}^2$ if for every $p \in \mathbb{S}^2$, $X(p)$ belongs to the tangent plane $T_p\mathbb{S}^2$ to $\mathbb{S}^2$ at the point p.

Let X be a tangent vector field on $\mathbb{S}^2$ and q an isolated singular point of X; that is, $X(q) = 0$, and there is no other singular point of X in a sufficiently small neighborhood of q. The stereographic projection of $\mathbb{S}^2$ from $-q$ to $E = T_q\mathbb{S}^2$, defines an analytic diffeomorphism from an open neighborhood of q in $\mathbb{S}^2$ to an open neighborhood U of q in E. This diffeomorphism transforms X into a planar vector field X' on U having q as an isolated singular point. We define the *index* i_q *of the singular point* q *of the tangent vector field* X as the index of q for the projected planar vector field X'. By Proposition 6.20 it is clear that any C^1 coordinate system around q can be used to define i_q, without altering the definition.

We consider some examples:

Example 6.27 For every point $p \in \mathbb{S}^2$, $X(p)$ is a unit tangent vector to the meridian passing through p. In this case, there are exactly two singular points, the north and south poles, each one with index 1 because they are nodes.   □

Example 6.28 For every point $p \in \mathbb{S}^2$, $X(p)$ is a unit tangent vector to the parallel passing through p. In this case, there are only two singular points, again the poles, each one with index 1, but now they are centers.   □

Example 6.29 Assume a vector field X' in the tangent plane E' to $\mathbb{S}^2$ at the south pole z' given by a constant vector; that is, the integral lines of X' are a set of parallel straight lines. By stereographic projection from the north pole z of the integral lines of X' onto $\mathbb{S}^2$ we get a family of integral lines on $\mathbb{S}^2$ which define a vector field X on $\mathbb{S}^2$. The unique point which is singular is the north pole z; see Fig. 6.9.

In order to evaluate the index of the vector field X at z we must project it in a normal way onto the tangent plane E to $\mathbb{S}^2$ at the point z. Projecting X onto E is just the inverse with respect to $\mathbb{S}^2$. It is easy to see then that the vector field X has a singular point of index 2 at z; see Fig. 6.10.   □

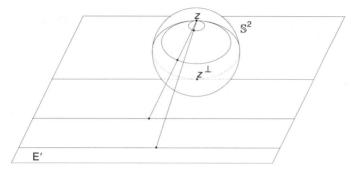

Fig. 6.9. Stereographic projection for example 6.29

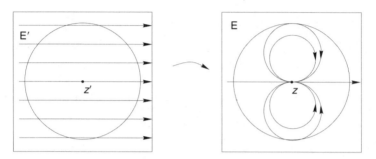

Fig. 6.10. Phase portraits for example 6.29

In each one of the previous examples, the sum of the indices of all the singular points for every vector field has been always 2. This is a general result for every tangent vector field defined on $\mathbb{S}^2$.

Theorem 6.30 (Poincaré–Hopf Theorem) *For every tangent vector field on* $\mathbb{S}^2$ *with a finite number of singular points, the sum of their indices is 2.*

Proof. Since the vector field X has finitely many singular points, we can always find a great circle $\mathbb{S}^1$ which contains no singular points. Let E^2 be the plane containing $\mathbb{S}^1$. We stereographically project the vector field X restricted to the southern hemisphere on E^2 from the north pole, and we get a vector field X' on E^2. We do a similar projection with the northern hemisphere and we get another projected vector field X'' on E^2.

If $q_1, q_2, \ldots, q_n$ are the singular points of X in the southern hemisphere with indices $i_1, i_2, \ldots, i_n$, and $q'_1, q'_2, \ldots, q'_m$ are the singular points of X in the northern hemisphere with indices $i'_1, i'_2, \ldots, i'_m$, then by Proposition 6.26, the vector fields X' and X'' have on E^2 a number of turns equal to $i(X') = i_1 + \ldots + i_n$ and $i(X'') = i'_1 + \ldots + i'_m$.

We will now compute $i(X') + i(X'')$ studying the relation between X' and X''.

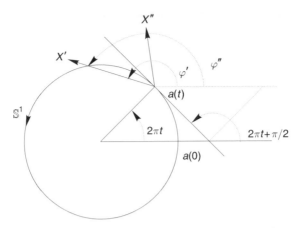

Fig. 6.11. The vector fields X' and X'' on $\mathbb{D}^2$

Let φ' and φ'' be the angle functions for X' and X'', respectively, over the points of $\mathbb{S}^1$. For each point $a(t) \in \mathbb{S}^1$ by construction the vector fields X' and X'' are symmetric with respect to the tangent line to $\mathbb{S}^1$ at $a(t)$; see Fig. 6.11. Then

$$\frac{\varphi'(t) + \varphi''(t)}{2} = 2\pi t + \pi/2,$$

consequently

$$\varphi'(t) + \varphi''(t) = 4\pi t + \pi.$$

Therefore

$$
\begin{aligned}
i(X') + i(X'') &= \frac{\varphi'(1) - \varphi'(0) + \varphi''(1) - \varphi''(0)}{2\pi} \\
&= \frac{\varphi'(1) + \varphi''(1) - (\varphi'(0) + \varphi''(0))}{2\pi} \\
&= \frac{4\pi + \pi - \pi}{2\pi} = 2,
\end{aligned}
$$

and the result is proved.   □

Corollary 6.31 *Every tangent vector field X on $\mathbb{S}^2$ has singular points.*

Proof. Considering any equator $\mathbb{S}^1 \subset \mathbb{S}^2$ and following the argumentation of the proof of Theorem 6.30, the absence of singular points implies that $i(X') = i(X'') = 0$. This contradicts the fact that $i(X') + i(X'') = 2$.   □

The definition of index on the unit sphere $\mathbb{S}^2 \subset \mathbb{R}^3$ can be transferred to an "abstract" sphere, meaning any two-dimensional oriented surface diffeomorphic to $\mathbb{S}^2$. As we have already observed, by Proposition 6.20, any local parametrization near the singular point can be used to calculate the index of a vector field near an isolated singular point. In fact, the index does not change even under a C^0-equivalence. The notion of index can also be defined for any C^0 vector fields on an arbitrary compact oriented surface without boundary.

The Poincaré–Hopf Theorem, as well as Corollary 6.31, clearly remain true on an "abstract" sphere. A similar theorem (with a different proof), can be proved on an arbitrary compact oriented surface without boundary. Thus, on a compact oriented surface without boundary S, it can be proved that the sum of the indices of the vector field with finitely many singular points is equal to the Euler characteristic of S. This theorem was due to Poincaré, and its extension to compact manifolds of arbitrary dimension is due to Hopf.

6.7 Poincaré Index Formula

In Sect. 1.5 we have defined parabolic, hyperbolic, and elliptic sectors, and the finite sectorial decomposition property for any isolated singular point different from a center or a focus. By definition we say that a center and a focus have neither elliptic, hyperbolic, nor parabolic sectors. It is easy to show that the index of a focus, as well as the index of a center, is $+1$.

In order to study the index of any of the other singular points we have the following result.

Proposition 6.32 (Poincaré Index Formula) *Let q be an isolated singular point having the finite sectorial decomposition property. Let e, h, and p denote the number of elliptic, hyperbolic, and parabolic sectors of q, respectively, and suppose that $e + h + p > 0$. Then $i_q = (e - h)/2 + 1$.*

Proof. Suppose that X is a continuous vector field defined in a neighborhood V of q and suppose that $X|V$ has the finite sectorial decomposition property defined in Sect. 1.5. If the sectorial decomposition is trivial, in the sense that it consists of one parabolic sector, then $i_q X = 1$ and $(e, h) = (0, 0)$ which clearly fits the statement. So we suppose that the decomposition is nontrivial, i.e., $(e, h) \neq (0, 0)$.

By definition ∂V is the image of a permissible parametrization, the vector field is nowhere tangent to ∂V except at e points $s_1, \ldots, s_e$ where it has an internal tangency (see Sect. 1.5) and h points $r_1, \ldots, r_h$ where it has an external tangency (see also Sect. 1.5). We can order these $e + h$ contact points as $q_1, \ldots, q_n$ with $n = e + h$, using the cyclic order on ∂V, and choose intermediate points $p_k \in (q_k, q_{k+1})$ for $k = 0, \ldots, n - 1$ and with $q_0 = q_n$. We can now choose a permissible parametrization $\rho : \mathbb{S}^1 \to \partial V$ such that $\rho(e^{2\pi i k/n}) = p_k$. We suppose that $X(p_0)$ is pointing outward. By construction, $n = e + h$ is even and the vectors $X(p_k)$ are pointing outward for k even and inward for k odd.

Let $N(e^{2\pi i t})$ denote the exterior normal related to ρ. We can consider the mapping

$$R : \mathbb{S}^1 \times (-\varepsilon, \varepsilon) \to \mathbb{R}^2$$
$$(e^{2\pi i t}, s) \to \rho(e^{2\pi i t}) + sN(e^{2\pi i t})$$

which is a C^1-diffeomorphism onto its image for $\varepsilon > 0$ sufficiently small. We define the continuous vector field Y along $\mathbb{S}^1$ by

$$Y(e^{2\pi it}, 0) = (dR_{(e^{2\pi it}, 0)})^{-1}(X(\rho(e^{2\pi it}))).$$

Y along $\mathbb{S}^1$ is nothing else but X along ∂V written in different C^1 coordinates. As such, $i(X, \partial V) = i(Y, \mathbb{S}^1)$.

Moreover, the tangency points of Y along $\mathbb{S}^1$ correspond to the tangency points of X along ∂V. We continue denoting them by $q_1, \dots, q_n$. We also write p_k for $e^{2\pi ik/n}$. The vectors $Y(p_k)$ are hence pointing outward for k even and inward for k odd. Since it does not change the index $i(Y, \mathbb{S}^1)$ we can suppose that $Y|\mathbb{S}^1$ consists of unit vectors, which can be achieved by considering $Y(p)/\|Y(p)\|$. We now fix neighborhoods $(p_k - \varepsilon, p_k + \varepsilon) \subset \mathbb{S}^1$, with $\varepsilon > 0$ small enough such that they do not contain the tangency points $q_1, \dots, q_n$. In each V_k separately we can now easily deform the vector field Y, not changing it at $\{p_k - \varepsilon, p_k + \varepsilon\}$ and keeping it pointing outward on V_k for k even and inward for k odd in such a way that, after deformation and still using the notation Y, $Y(p_k) = e^{2\pi ik/n}$ for k even and $Y(p_k) = -e^{2\pi ik/n}$ for k odd.

We can next deform the vector field Y on each $[p_k, p_{k+1}]$ separately, changing it into the field

$$Y(e^{2\pi it}) = e^{2\pi i(\frac{k}{n}(1-s)+s(\frac{k\pi}{n}+\pi))} \tag{6.1}$$

if there is a (unique) internal tangency in (p_k, p_{k+1}) and

$$Y(e^{2\pi it}) = e^{2\pi i(\frac{k}{n}(1-s)+s(\frac{k\pi}{n}-\pi))} \tag{6.2}$$

if there is a (unique) external tangency in (p_k, p_{k+1}).

In both cases we use $s = n(t - k/n)$ and the deformation can be taken in such a way that the beginning and ending values $Y(p_k)$ and $Y(p_{k+1})$ do not change. Thus, the different deformations on the respective $[p_k, p_{k+1}]$ can be joined to form a global deformation on $\mathbb{S}^1$. After this deformation we can suppose that the vector field Y is exactly given by (6.1) (respectively (6.2)) on the intervals $[p_k, p_{k+1}]$ in which there is a (unique) internal (respectively external) tangency.

In short, by means of a continuous change, the index of a singular point with (e, h, p) sectors is equal to the index of a singular point with exactly those same sectors, but each one of them being a triangular sector of angle $2\pi/m$ where $m = e + h + p$. Let σ be a circle of radius r with r small enough.

We can now consider the contribution to the index of each sector.

(i)   A triangular parabolic sector with angle $2\pi/m$ will start with an angle $\varphi(0) = \alpha$ and will end with an angle $\varphi(2\pi/m) = \alpha + 2\pi/m$, for a net gain of $2\pi/m$.

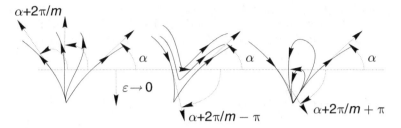

Fig. 6.12. Index given by sectors

(ii) A triangular hyperbolic sector with angle $2\pi/m$ will start with an angle $\varphi(0) = \alpha$ and will end with an angle $\varphi(2\pi/m) = \alpha + 2\pi/m - \pi$, for a net gain of $2\pi/m - \pi$.

(iii) A triangular elliptic sector with angle $2\pi/m$ will start with an angle $\varphi(0) = \alpha$ and will end with an angle $\varphi(2\pi/m) = \alpha + 2\pi/m + \pi$, for a net gain of $\pi + 2\pi/m$; see Fig. 6.12.

Now going through the entire closed curve σ we get

$$
i_q = i(q, \sigma) = \frac{\displaystyle\sum_{i=1}^{p} 2\pi/m + \sum_{i=1}^{h}(2\pi/m - \pi) + \sum_{i=1}^{e}(\pi + 2\pi/m)}{2\pi}
$$

$$
= \frac{(e - h)\pi}{2\pi} + \frac{(p + h + e)2\pi}{2m\pi} = \frac{e - h}{2} + 1.
$$

This completes the proof.

Corollary 6.33 *Suppose that X at p and Y at q are C^0-equivalent and that both satisfy the finite sectorial decomposition property. Then $i_p X = i_q Y$.*

Proof. Since X at p and Y at q have the same number of elliptic, hyperbolic, and parabolic sectors (at least in a minimal sectorial decomposition), Poincaré's formula gives the required result.   □

6.8 Relation Between Index and Multiplicity

In this section we want to discuss the relation between the notions of index and multiplicity for singular points of two-dimensional C^∞ differential systems. We shall present the results without proofs, providing references for them.

We deal with a differential system of the form

$$
\dot{x} = P(x, y), \quad \dot{y} = Q(x, y),
$$

where P and Q are C^∞ functions defined in a neighborhood of the origin. Suppose that the origin is a singular point of this system.

Here, the *germ* (P_0, Q_0) of (P, Q) at 0 is the equivalence class of C^∞ functions (F, G) defined in a neighborhood of 0 and equivalent to the function (P, Q) under the following equivalence relation. Two C^∞ functions (F, G) and (P, Q) are equivalent if there is some open neighborhood U of 0 such that for all $(x, y) \in U$, the identity $(F, G)(x, y) = (P, Q)(x, y)$ holds. All local properties of (P, Q) at 0 depend only on its germ (P_0, Q_0).

We consider the germ (P_0, Q_0) of (P, Q) at 0, and the local ring given by

$$C_0^\infty(\mathbb{R}^2)/ < P_0, Q_0 >$$

of (P_0, Q_0) at 0, where $C_0^\infty(\mathbb{R}^2)$ is the ring of germs at 0 of C^∞ real-valued functions on $\mathbb{R}^2$, and $< P_0, Q_0 >$ is the ideal generated by P_0 and Q_0. The *multiplicity* $\mu_0[P, Q]$ of (P, Q) at 0 is defined by $\mu_0[P, Q] = \dim_\mathbb{R}[C_0^\infty(\mathbb{R}^2)/ < P_0, Q_0 >]$ and we say that (P, Q) is a *finite map germ* if $\mu_0[P, Q] < \infty$. It is known that $\mu_0[P, Q]$ is the number of complex (P, Q)-preimages near 0 of a regular value of (P, Q) near 0. For more details and a proof of the following statements; see [4, 7].

Let $(P, Q) : (\mathbb{R}^2, 0) \to (\mathbb{R}^2, 0)$ be a finite map germ. Then the following statements hold.

(i) The multiplicity of (P, Q) at 0 does not depend on the chosen coordinates.

(ii) Let $P = P_k +$ higher order terms and $Q = Q_l +$ higher order terms, where P_k (respectively Q_l) denotes the homogeneous part of P (respectively Q) of degree k (respectively l). Then $\mu_0[P, Q] \geq kl$, and $\mu_0[P, Q] = kl$ if and only if the system $P_k = Q_l = 0$ only has the trivial solution $(0, 0)$ in $\mathbb{C}^2$.

(iii) If $P = P_1 P_2$, where $P_1(0) = P_2(0) = 0$, then $\mu_0[P, Q] = \mu_0[P_1, Q] + \mu_0[P_2, Q]$. A similar result holds for Q.

(iv) Let $(F, G) : (\mathbb{R}^2, 0) \to (\mathbb{R}^2, 0)$ also be a finite map germ. Then $\mu_0[(P, Q) \circ (F, G)] = \mu_0[P, Q]\mu_0[F, G]$.

(v) If $G = Q + AP$ with $A : (\mathbb{R}^2, 0) \to \mathbb{R}$ a C^∞ function, then $\mu_0[P, Q] = \mu_0[P, G]$. A similar result holds for P.

(vi) If $P = BF$ with $B : (\mathbb{R}^2, 0) \to \mathbb{R}$ a C^∞ function such that $B(0) \neq 0$, then $\mu_0[P, Q] = \mu_0[F, Q]$. A similar result holds for Q.

The next results, due to Eisenbud and Levine [60], show the relation between index and multiplicity of a singular point of a two-dimensional differential system.

Theorem 6.34 *Let the origin be a singular point of the C^∞ vector field $X = (P, Q)$. Then the following two statements hold.*

(i) $|i_0(X)| \leq \sqrt{\mu_0[X]}$.

(ii) $i_0(X) \equiv \mu_0[X] \pmod 2$.

The following natural question appears: Given the multiplicity $\mu = \mu_0[X]$, which values can the index $i_0(X)$ take? The next result due to Cima, Gasull, and Torregrosa [42] shows that the index $i_0(X)$ of a two-dimensional vector field is not subject to any restrictions other than the two conditions provided by Theorem 6.34.

Theorem 6.35 *For each $\mu \in \mathbb{N}$ with $\mu \geq 1$ and $i \in \mathbb{Z}$ satisfying $|i| \leq \sqrt{\mu}$ and $i \equiv \mu$ (mod 2), there exists a function germ $X = (P, Q) : (\mathbb{R}^2, 0) \to (\mathbb{R}^2, 0)$ such that $i_0(X) = i$ and $\mu_0[X] = \mu$.*

6.9 Exercises

Exercise 6.1 Prove the fundamental theorem of algebra. That is, let $p(z) = z^n + a_{n-1}z^{n-1} + \ldots + a_1 z + a_0$ be a polynomial with $n \geq 1$ and $a_i \in \mathbb{C}$ for $i = 0, 1, \ldots, n - 1$. Then show that there exists $z_0 \in \mathbb{C}$ such that $p(z_0) = 0$.
    Hint: Use Theorem 6.13 and Corollary 6.6.

Exercise 6.2 Let $\mathbb{S}^2 = \{x = (x_1, x_2, x_3) \in \mathbb{R}^3 : x_1^2 + x_2^2 + x_3^2 = 1\}$. A continuous map $f : \mathbb{S}^2 \to \mathbb{R}$ is odd, if for all $x \in \mathbb{S}^2$ we have that $f(x) = -f(x)$. Prove the Borsuk's Theorem: Two odd continuous maps $f, g : \mathbb{S}^2 \to \mathbb{R}$ always have a common zero.
    Hint: Use Theorem 6.13.

Exercise 6.3 Prove the Brouwer Fixed Point Theorem: A continuous map $f : \mathbb{D}^2 \to \mathbb{D}^2$ has at least one fixed point (i.e., there exists a point $p \in \mathbb{D}^2$ such that $f(p) = p$).
    Hint: Use Proposition 6.16 and Example 6.17.

Exercise 6.4 Let $f : \mathbb{S}^2 \to \mathbb{S}^2$ be a continuous map. Show that f has a point such that $f(p) = p$ or $f(p) = -p$. Here $\mathbb{S}^2 = \{x = (x_1, x_2, x_3) \in \mathbb{R}^3 : x_1^2 + x_2^2 + x_3^2 = 1\}$.
    Hint: Use Corollary 6.31.

6.10 Bibliographical Comments

The authors have used in the preparation of this chapter some notes of M. Castellet from the University Autònoma de Barcelona.

For advanced versions of the Poincaré–Hopf Theorem; see [85] and [111]. For additional results on index theory in dimension 2; see [169].

7

Limit Cycles and Structural Stability

In the qualitative theory of differential equations, research on limit cycles is an interesting and difficult topic. Limit cycles of planar vector fields were defined in the famous paper *Mémoire sur les courbes définies par une équation differentielle* [130,131]. At the end of the 1920s van der Pol [160], Liénard [99] and Andronov [3] proved that a periodic orbit of a self-sustained oscillation occurring in a vacuum tube circuit was a limit cycle as considered by Poincaré. After this observation, the existence and nonexistence, uniqueness, and other properties of limit cycles have been studied extensively by mathematicians and physicists, and more recently also by chemists, biologists, and economists.

After singular points, limit cycles are the main subject of study in the theory of two-dimensional differential systems. In this chapter we present the most basic results on limit cycles. In particular, we show that any topological configuration of a finite number of limit cycles is realizable by a suitable polynomial differential system. We define the multiplicity of a limit cycle, and we study the bifurcations of limit cycles for rotated families of vector fields. We present some results on structural stability.

7.1 Basic Results

We consider the system of differential equations

$$\dot{x} = \frac{dx}{dt} = P(x,y), \qquad \dot{y} = \frac{dy}{dt} = Q(x,y), \tag{7.1}$$

where x, y, and t are real variables, and P and Q are C^1 functions of x and y. Then the existence and uniqueness of its solutions is guaranteed.

We recall that if a solution $x = f(t)$, $y = g(t)$ of system (7.1) is a nonconstant periodic function of t, then $\gamma = \{(x,y) : x = f(t), y = g(t)\}$ is called a *periodic orbit* of system (7.1).

If for some arbitrarily small outer (inner) neighborhood of the periodic orbit γ there do not exist other periodic orbits, then γ is called an *externally (internally) limit cycle*.

If there is an arbitrarily small outer (inner) neighborhood of the periodic orbit γ filled with periodic orbits, then γ is called an *externally (internally) center-type periodic orbit*. Sometimes in the literature this is simply called an *externally (internally) periodic orbit*, omitting "center-type." We prefer to add center-type to avoid confusion.

If in any outer (inner) neighborhood of the periodic orbit γ there exist both nonperiodic and periodic orbits different from γ, then γ is called an *externally (internally) indefinite cycle*. If a periodic orbit is either externally or internally indefinite we call it an *indefinite cycle*. In the literature one can also find "compound" instead of "indefinite," but since "compound cycle" means something completely different in [6] we prefer to use "indefinite." Sometimes one can even find the word "compound limit cycles," although on at least one side no regular orbits tend to them, either in positive or negative time.

By the Poincaré–Bendixson Theorem, the following three propositions and the first theorem hold.

Proposition 7.1 *If γ is a periodic orbit of system (7.1) then there exists a sufficiently small neighborhood U of γ satisfying the following three statements.*

(i)   *U does not contain singular points.*

(ii) *Through any point $p \in \gamma$ there exists a sufficiently small transverse segment Σ_p for the flow of system (7.1); that is, the vector field associated to system (7.1) at the points of Σ_p is not tangent to it.*

(iii) *Any periodic orbit in U intersects Σ_p transversely at a unique point. Any nonperiodic orbit in U intersects Σ_p at an infinite number of points, which lie on the same side of γ.*

Proposition 7.2 *Any periodic orbit γ is either an external (internal) limit cycle, an external (internal) center-type periodic orbit or an external (internal) indefinite cycle. The first case can be divided into the following two subcases:*

(i) *There exists a sufficiently small outer (inner) neighborhood of γ such that all the orbits in it are nonperiodic, and they have γ as ω-limit set. Then γ is called an externally (internally) stable limit cycle. If γ is externally and internally stable, then we say that γ is a stable limit cycle.*

(ii) *There exists a sufficiently small outer (inner) neighborhood of γ such that all the orbits in it are nonperiodic, and they have γ as their α-limit set. Then γ is called an externally (internally) unstable limit cycle. If γ is externally and internally unstable, then we say that γ is an unstable limit cycle.*

If γ is externally stable and internally unstable or the opposite, then we say that γ is a semistable limit cycle.

Proposition 7.3 *If γ_1 and γ_2 are periodic orbits which form the boundary of an annular region A, and there are neither singular points nor other periodic orbits in A, then all the orbits in A have γ_1 as their ω-limit set and γ_2 as their α-limit set, or conversely. In other words, two adjacent periodic orbits (under the above conditions) possess different stability on the adjacent sides.*

Theorem 7.4 (Poincaré's Annular Region Theorem) *Let A be an annular region not containing singular points such that ∂A does not contain periodic orbits and such that every orbit crossing the boundary of A moves from the exterior to the interior (interior to the exterior). Then there exists at least one externally stable (unstable) limit cycle in A, and there exists at least one internally stable (unstable) limit cycle in A; it is possible that both limit cycles coincide, so as to be a single stable (unstable) limit cycle.*

For analytic systems some restrictions occur in the classification of periodic orbits.

Proposition 7.5 *If P and Q are analytic functions, then system (7.1) has no indefinite limit cycles.*

The proof easily follows from Propositions 7.1 and 7.2, observing that the Poincaré map on Σ_p is analytic, if we chose Σ_p to be an analytic regular arc with an analytic and regular parameter. For polynomial systems (7.1) there is a famous result, whose complicated proof has been given independently by Ecalle [59] and Il'yashenko [88].

Theorem 7.6 *When P and Q are polynomials then system (7.1) has at most finitely many limit cycles.*

The proof of the next result is straightforward.

Proposition 7.7 (Symmetry Principle) *Suppose that in equations (7.1)*

$$P(x,y) = P(-x,y), \qquad Q(-x,y) = -Q(x,y),$$

and the origin is the only singular point on the y-axis. If a trajectory γ starts from the positive y-axis and returns to the negative y-axis, then γ is a periodic orbit. If all the trajectories near the origin have this property, then the origin is a center.

Using the Intermediate Value Theorem the next result is easy to prove.

Proposition 7.8 *If the Poincaré map determined by the trajectory carries some closed segment into itself, then the equations must have a periodic orbit.*

Proposition 7.9 (Poincaré Method of Tangential Curves) *Consider a family of curves* $F(x,y) = C$*, where* $F(x,y)$ *is cont inuously differentiable. If in a region* R *the quantity*

$$\frac{dF}{dt} = \frac{\partial F}{\partial x}\frac{dx}{dt} + \frac{\partial F}{\partial y}\frac{dy}{dt} = P\frac{\partial F}{\partial x} + Q\frac{\partial F}{\partial y}$$

(which represents the rate of change of the function F *with respect to* t *along a trajectory of system (7.1)) has constant sign, and the curve*

$$P\frac{\partial F}{\partial x} + Q\frac{\partial F}{\partial y} = 0$$

(which represents the locus of points of contact between curves in the family and the trajectories of (7.1), and is called a tangential curve) does not contain a whole trajectory of (7.1) or any closed branch, then system (7.1) does not possess a periodic orbit which is entirely contained in R*.*

Proof. Suppose the hypotheses hold, but system (7.1) has a periodic orbit Γ that is wholly contained in R. If we integrate $P\partial F/\partial x + Q\partial F/\partial y$ along Γ in the direction of increasing t, we obtain

$$\int_\Gamma (P\frac{\partial F}{\partial x} + Q\frac{\partial F}{\partial y})dt = \int_\Gamma \frac{dF}{dt}\,dt.$$

Since Γ is periodic, we know that the right side of the equation above is equal to zero; however, on the other hand, the integrand on the left of this equality has a constant sign, but is not identically zero on Γ, and t monotonically increases along Γ; hence its value is different from zero, a contradiction. □

Theorem 7.10 (Bendixson's Theorem) *Assume that the divergence function* $\partial P/\partial x + \partial Q/\partial y$ *of system (7.1) has constant sign in a simply connected region* R*, and is not identically zero on any subregion of* R*. Then system (7.1) does not have a periodic orbit which lies entirely in* R*. (We assume that* P *and* Q *are* C^1*.)*

Proof. Suppose the hypotheses hold, but system (7.1) has a closed trajectory Γ. Since R is simply connected, Γ and its interior S lie entirely in R. From Green's formula we have

$$\int_\Gamma Pdy - Qdx = \iint_S \left(\frac{\partial P}{\partial x} + \frac{\partial Q}{\partial y}\right)dxdy. \qquad (7.2)$$

But $Pdy - Qdx = 0$ holds everywhere along Γ, since Γ is an orbit. Hence the left side of (7.2) is zero, while the integrand on the right side has constant sign but is not identically zero in S; hence the double integral is not zero, giving a contradiction. □

Remark 7.11 By the same arguments as in Theorem 7.10, i.e., the use of Green's formula, it follows that on any region bounded by periodic orbits, the integral of the divergence, as defined in (7.2) is equal to zero.

Theorem 7.12 (Dulac's Theorem) *If there exists a C^1 function $B(x,y)$ in a simply connected region R such that $\partial(BP)/\partial x + \partial(BQ)/\partial y$ has constant sign and is not identically zero in any subregion, then system (7.1) does not have a periodic orbit lying entirely in R. (We assume that P and Q are C^1.)*

Proof. We follow the proof of Theorem 7.10 but use BP and BQ in place of P and Q, respectively. □

We shall call $B(x,y)$ a *Dulac function*, and the method of proving the non–existence of a periodic orbit, as given in Theorem 7.10, is called the *method of Dulac functions*.

Theorems 7.10 and 7.12 can be generalized as follows.

Theorem 7.13 (Generalized Dulac's Theorem) *If we change the region R in Bendixson's Theorem or in Dulac's Theorem to be n-multiply connected (i.e., R has one or several outer boundary curves, and $n-1$ inner boundary curves), then system (7.1) has at most $n-1$ periodic orbits which lie entirely in R.*

Proof. From the proof of Theorem 7.10 we know that if there is a closed trajectory Γ of system (7.1) in R, then Γ should contain at least one inner boundary curve C of R in its interior. Similarly, we also know that if the interior of Γ also contains other closed trajectories $\Gamma_1, \ldots, \Gamma_k$, then the region in the interior of Γ but in the exterior of all the trajectories $\Gamma_1, \ldots, \Gamma_k$ also contains at least one inner boundary curve C, because of Remark 7.11. We say that C corresponds to Γ; we can see that for different Γ their corresponding curves C are also different. Hence if the number of closed trajectories in G is more than $n-1$, then the connectivity number of G must be greater than n. The theorem is proved. □

Proposition 7.14 *If P and Q of system (7.1) are C^1 on an open subset $U \subset \mathbb{R}^2$, and $\partial P/\partial x + \partial Q/\partial y \equiv 0$, i.e., the 1-form*

$$Pdy - Qdx \qquad (7.3)$$

is closed, then (7.1) does not have indefinite cycles in U, nor limit cycles nor one-sided limit cycles in R.

Proof. Suppose that U contains a closed orbit Γ of system X, expressed by (7.1) and consider a transverse section Σ consisting of a piece of orbit of the vector field $P\partial/\partial y - Q\partial/\partial x$, orthogonal to X. Consider now an open segment $\Sigma_0 \subset \Sigma$, containing $\Sigma \cap \Gamma$, on which a Poincaré map $P : \Sigma_0 \to \Sigma$ is defined. Take any $p \in \Sigma_0 \setminus \Gamma$, let $q = P(p)$ and denote by $A \subset U$ the annulus bounded by $\Gamma \cup \gamma_{pq} \cup \Sigma_{pq}$ where γ_{pq} is the piece of X-orbit between p and q, and Σ_{pq} is the segment in Σ between p and q. Since the divergence of X is equal to zero, it follows from Green's formula (7.2) on A that necessarily $p = q$ (we leave the details of the elaboration as an exercise). Hence all orbits in U sufficiently close to Γ have to be closed, proving our claim. $\qquad\square$

Let U be an open subset of $\mathbb{R}^2$. In this book we consider only C^1 maps V defined in U such that the set $\{(x, y) \in U : V(x, y) = 0\}$ is locally a one dimensional manifold except perhaps at finitely many points.

Let (P, Q) be a C^1 vector field defined in U. A standard method for determining an explicit solution of the vector field (P, Q) is to find a C^1 solution $V = V(x, y)$ of the linear partial differential equation

$$P\frac{\partial V}{\partial x} + Q\frac{\partial V}{\partial y} = RV, \qquad (x, y) \in U, \tag{7.4}$$

for some C^1 map $R = R(x, y)$. Thus the curve $V(x, y) = 0$ is formed by trajectories of the vector field (P, Q), because (7.4) shows that (P, Q) is tangent to the curve $V(x, y) = 0$.

In the next two results we use equation (7.4) to study the periodic solutions of the vector field (P, Q). Both results are proved in [72].

Theorem 7.15 *Let (P, Q) be a C^1 vector field defined on the open subset U of $\mathbb{R}^2$, let $(u(t), v(t))$ be a periodic solution of (P, Q) of period T, $R : U \to \mathbb{R}$ a C^1 map such that $\int_0^T R(u(t), v(t))\,dt \neq 0$, and $V = V(x, y)$ a C^1 solution of the linear partial differential equation (7.4). Then the closed trajectory $\gamma = \{(u(t), v(t)) \in U : t \in [0, T]\}$ is contained in $\Sigma = \{(x, y) \in U : V(x, y) = 0\}$, and γ is not contained in a period annulus of (P, Q). Moreover, if the vector field (P, Q) and the functions R and V are analytic, then γ is a limit cycle.*

We introduce a few new notions. We consider the C^1 vector field

$$X = P\frac{\partial}{\partial x} + Q\frac{\partial}{\partial y}$$

defined on the open subset U of $\mathbb{R}^2$. Then X is a *closed vector field* on U if

$$\frac{\partial P}{\partial x} = -\frac{\partial Q}{\partial y},$$

for all $(x, y) \in U$. Furthermore, if U is simply connected, then there exists a function $H : U \to \mathbb{R}$ satisfying

$$P = -\frac{\partial H}{\partial y}, \qquad Q = \frac{\partial H}{\partial x}.$$

The function H is the *Hamiltonian* of the *Hamiltonian vector field* X. Clearly the Hamiltonian function is a first integral of X.

A C^1 function $R : U \to \mathbb{R}$ such that

$$\frac{\partial(RP)}{\partial x} = -\frac{\partial(RQ)}{\partial y} \tag{7.5}$$

is an *integrating factor* of the vector field X. We know that R is an integrating factor of X in U if and only if R is a solution of the partial differential equation

$$P\frac{\partial R}{\partial x} + Q\frac{\partial R}{\partial y} = -\left(\frac{\partial P}{\partial x} + \frac{\partial Q}{\partial y}\right)R \tag{7.6}$$

in U.

A function $V : U \to \mathbb{R}$ is an *inverse integrating factor* of the vector field X if V verifies the partial differential equation

$$P\frac{\partial V}{\partial x} + Q\frac{\partial V}{\partial y} = \left(\frac{\partial P}{\partial x} + \frac{\partial Q}{\partial y}\right)V \tag{7.7}$$

in U. We note that V satisfies (7.7) in U if and only if $R = 1/V$ satisfies (7.6) in $U \setminus \{(x, y) \in U : V(x, y) = 0\}$.

Here we provide an easier and direct proof of the next theorem from [72].

Theorem 7.16 *Let X be a C^1 vector field defined on the open subset U of $\mathbb{R}^2$. Let $V : U \to \mathbb{R}$ be an inverse integrating factor of X. If γ is an indefinite cycle, a limit cycle or an one-sided limit cycle of X, then γ is contained in $S = \{(x, y) \in U : V(x, y) = 0\}$.*

Proof. Due to the existence of the inverse integrating factor V defined on U, we see that the vector field X/V is Hamiltonian in $U \setminus S$, hence its divergence is identically zero. The result now follows from Proposition 7.14.   □

Example 7.17 The system

$$\dot{x} = \lambda x - y + \lambda m_1 x^3 + (m_2 - m_1 + m_1 m_2)x^2 y + \lambda m_1 m_2 x y^2 + m_2 y^3,$$
$$\dot{y} = x + \lambda y - x^3 + \lambda m_1 x^2 y + (m_1 m_2 - m_1 - 1)x y^2 + \lambda m_1 m_2 y^3, \tag{7.8}$$

has the following inverse integrating factor

$$V(x, y) = (x^2 + y^2)(1 - m_1 x^2 + m_1 m_2 y^2).$$

Since V is defined in the whole plane, by Theorem 7.16 follows that (7.8) has at most one limit cycle, and when it exists it is algebraic and has the equation

$$1 - m_1 x^2 + m_1 m_2 y^2 = 0. \tag{7.9}$$

Note that such a limit cycle exists if and only if system (7.8) has no singular points on the curve (7.9).   □

7.2 Configuration of Limit Cycles and Algebraic Limit Cycles

By Theorem 7.6, a theorem proved independently by Écalle [59] and by Il'yashenko [88], we know that a polynomial planar system (7.1) can only have a finite number of limit cycles. It is not known, however, not even for quadratic systems, whether there exists a uniform upper bound, depending on the degree of the system. This is an old question, put forward by Hilbert in 1900 [82] as part of the 16th problem in his famous list of "problems for the 20th–century." Besides asking for such a uniform upper bound, Hilbert's 16th problem, or more precisely the second part of it (the first part dealing with ovals in a zero set of algebraic functions), also asks for a description of the possible configurations of limit cycles which polynomial systems can have.

We now deal with this problem, presenting the results obtained in [102].

A *configuration of closed curves* is a finite set $C = \{C_1, \ldots, C_n\}$ of disjoint simple closed curves of the plane such that $C_i \cap C_j = \emptyset$ for all $i \neq j$. We call it a *configuration of limit cycles* if the closed curves C_i are limit cycles, and a *configuration of algebraic limit cycles* if the closed curves C_i are algebraic limit cycles. A closed curve is called *algebraic* if it is a connected component of the zero set of some polynomial function.

Given a configuration of closed curves $C = \{C_1, \ldots, C_n\}$ the curve C_i is a *primary curve* if there is no curve C_j of C contained in the bounded region determined by C_i.

Two configurations of closed curves $C = \{C_1, \ldots, C_n\}$ and $C' = \{C'_1, \ldots, C'_m\}$ are *(topologically) equivalent* if there is a homeomorphism $h : \mathbb{R}^2 \to \mathbb{R}^2$ such that $h\left(\cup_{i=1}^n C_i\right) = \left(\cup_{i=1}^m C'_i\right)$. Of course, for equivalent configurations of closed curves C and C' we have that $n = m$.

We say that the vector field X *realizes* the configuration of closed curves C as a configuration of limit cycles if the set of all limit cycles of X is equivalent to C.

Theorem 7.18 *Let $C = \{C_1, \ldots, C_n\}$ be a configuration of closed curves, and let r be the number of its primary curves. Then the following statements hold.*

(i) The configuration C is realizable as a configuration of limit cycles by a polynomial vector field.

(ii) The configuration C is realizable as a configuration of algebraic limit cycles by a polynomial vector field of degree $\leq 2(n + r) - 1$.

In the proof of this theorem we provide an explicit expression for the polynomial differential system of degree at most $2(n+r) - 1$ satisfying statement *(ii)* of Theorem 7.18. Of course, statement *(i)* of Theorem 7.18 follows immediately from statement *(ii)*.

Proof. Let $C = \{C_1, \ldots, C_n\}$ be the configuration of closed curves given in the statement of Theorem 7.18. For every primary curve C_j we select a point p_j in the interior of the bounded component determined by C_j. Since we can work with an equivalent configuration of limit cycles, without loss of generality, we can assume that

(i) each curve C_i is a circle defined by

$$f_i(x, y) = (x - x_i)^2 + (y - y_i)^2 - r_i^2 = 0,$$

for $i = 1, \ldots, n$; and that

(ii) the primary curves of the configuration C are the curves C_j, and the selected points p_j have coordinates (x_j, y_j), for $j = 1, \ldots, r$.

For every selected point p_j we define

$$
\begin{aligned}
f_{n+2j-1}(x, y) &= (x - x_j) + i(y - y_j), \\
f_{n+2j}(x, y) &= (x - x_j) - i(y - y_j).
\end{aligned}
$$

Now we consider the function

$$\tilde{H} = f_1^{\lambda_1} \cdots f_n^{\lambda_n} f_{n+1}^{\lambda_{n+1}} f_{n+2}^{\lambda_{n+2}} \cdots f_{n+2r-1}^{\lambda_{n+2r-1}} f_{n+2r}^{\lambda_{n+2r}} = \prod_{k=1}^{n+2r} f_k^{\lambda_k},$$

with $\lambda_1 = \cdots = \lambda_n = 1$, and $\lambda_{n+2j-1} = 1 + i$ and $\lambda_{n+2j} = 1 - i$, for $j = 1, \ldots, r$. After an easy computation, we see that

$$\tilde{H}(x, y) = A(x, y)B(x, y)C(x, y),$$

where

$$A(x, y) = \prod_{i=1}^{n} \left[(x - x_i)^2 + (y - y_i)^2 - r_i^2 \right],$$

$$B(x, y) = \prod_{j=1}^{r} \left[(x - x_j)^2 + (y - y_j)^2 \right],$$

$$C(x, y) = \exp\left(-2 \sum_{j=1}^{r} \arg[(x - x_j) + i(y - y_j)] \right).$$

Clearly $\tilde{H}(x, y)$ is a real function. Therefore, the function

$$H = \log \tilde{H} = \sum_{k=1}^{n+2r} \lambda_k \log f_k$$

is also real.

We claim that the vector field

$$X = P(x,y)\frac{\partial}{\partial x} + Q(x,y)\frac{\partial}{\partial y}$$

$$= -\sum_{k=1}^{n+2r} \lambda_k \left(\prod_{\substack{l=1 \\ l \neq k}}^{n+2r} f_l \right) \frac{\partial f_k}{\partial y} \frac{\partial}{\partial x} + \sum_{k=1}^{n+2r} \lambda_k \left(\prod_{\substack{l=1 \\ l \neq k}}^{n+2r} f_l \right) \frac{\partial f_k}{\partial x} \frac{\partial}{\partial y},$$

satisfies the conclusion of statement *(ii)* of Theorem 7.18. Now we shall prove the claim.

First, we note that we have the equalities

$$\frac{\partial H}{\partial x} = \frac{Q}{\displaystyle\prod_{k=1}^{n+2r} f_k}, \qquad \frac{\partial H}{\partial y} = -\frac{P}{\displaystyle\prod_{k=1}^{n+2r} f_k}. \tag{7.10}$$

Therefore, since H and $\prod_{k=1}^{n+2r} f_k$ are real functions, we get that P, Q, and consequently X are real.

Clearly from the definition of X it follows that P and Q are polynomials of degree at most $n + 2r - 1$. So X is a real polynomial vector field of degree at most $n + 2r - 1$.

From (7.10) it follows that $V = \prod_{k=1}^{n+2r} f_k$ is a polynomial inverse integrating factor of X, and that H is a Hamiltonian for the Hamiltonian vector field

$$\frac{1}{V} X = \frac{P}{V} \frac{\partial}{\partial x} + \frac{Q}{V} \frac{\partial}{\partial y},$$

defined on $\mathbb{R}^2 \setminus \{V = 0\}$.

Since V is polynomial, V is defined on all of $\mathbb{R}^2$. Therefore, by Theorem 7.16 and since $V(x,y) = 0$ if and only if $(x,y) \in (\cup_{i=1}^n C_i) \cup \{p_1, \dots, p_r\}$, if the vector field X has limit cycles, these must be the circles C_i for $i = 1, \dots, n$. Now we shall prove that all these circles are limit cycles. Hence the polynomial vector field X will realize the configuration of limit cycles $\{C_1, \dots, C_n\}$ and the theorem will be proved.

We note that since $\tilde{H} = \exp(H)$ is a first integral of the vector field X on $\mathbb{R}^2 \setminus \{V = 0\}$, the circles are formed by solutions because they are contained in the level curve $\tilde{V} = 0$, where $\tilde{V}$ is the associated inverse integrating factor of the Hamiltonian $\tilde{H}$, and $V = 0$ is formed by solutions. Now we shall prove that on every circle C_i there are no singular points of X, hence C_i will be a periodic orbit. Assume to the contrary that (x_0, y_0) is a singular point of X on the circle C_i; i.e., $P(x_0, y_0) = Q(x_0, y_0) = f_i(x_0, y_0) = 0$. From the definition of P and Q we have that

$$P(x_0, y_0) = -\lambda_i \left(\prod_{\substack{l=1 \\ l \neq i}}^{n+2r} f_l(x_0, y_0) \right) \frac{\partial f_i}{\partial y}(x_0, y_0) = 0,$$

$$Q(x_0, y_0) = \lambda_i \left(\prod_{\substack{l=1 \\ l \neq i}}^{n+2r} f_l(x_0, y_0) \right) \frac{\partial f_i}{\partial x}(x_0, y_0) = 0.$$

Since $f_l(x_0, y_0) \neq 0$ for $l \neq i$, we obtain that $\frac{\partial f_i}{\partial x}(x_0, y_0) = 0$ and also that $\frac{\partial f_i}{\partial y}(x_0, y_0) = 0$. Therefore the point (x_0, y_0) is the center of the circle C_i, contradicting the fact that $f_i(x_0, y_0) = 0$. Hence every circle C_i is a periodic orbit of the vector field X. Now we shall prove that C_i is a limit cycle, and this will complete the proof of Theorem 7.18.

We note that all circles C_i and all points p_j are in the level $\tilde{H}(x, y) = 0$, and that they are the unique orbits of X at this level. Now suppose that C_i is not a limit cycle. Then there is a periodic orbit $\gamma = \{(x(t), y(t)) : t \in \mathbb{R}\}$ different from $C_1, \ldots, C_n$ and so close to C_i such that in the bounded component B determined by γ there are the same points of $\{p_1, \ldots, p_r\}$ as in the bounded component determined by C_i. Without loss of generality we can assume that these points are $p_1, \ldots, p_s$.

As γ is different from $C_1, \ldots, C_n$, there exists $h \neq 0$ such that

$$\tilde{H}(x(t), y(t)) = A(x(t), y(t))B(x(t), y(t)) \exp\left(-2 \sum_{j=1}^{r} \theta_j(t) \right) = h, \quad (7.11)$$

where $\theta_j(t) = \arg[(x(t) - x_j) + i(y(t) - y_j)]$. The function $A(x(t), y(t))$ $B(x(t), y(t))$ is bounded on γ. Clearly the angles $\theta_1(t), \ldots, \theta_s(t)$ all tend simultaneously (due to the definition) to either $+\infty$ or $-\infty$ as $t \to +\infty$, while the angles $\theta_{s+1}(t), \ldots, \theta_r(t)$ remain bounded when $t \to +\infty$. These facts are in contradiction to equality (7.11). Consequently, we have proved that C_i is a limit cycle. In short, Theorem 7.18 is proved. □

7.3 Multiplicity and Stability of Limit Cycles

We have already defined the stability of limit cycles for a system of equations (7.1). In applications only a stable limit cycle has practical significance, since every spiral sufficiently close to a limit cycle can approximately represent the limit cycle independent of initial conditions, and an unstable limit cycle is similar to an unstable equilibrium position in mechanics, which in reality cannot be seen. Hence the problem of distinguishing stability of limit cycles becomes a very important one.

From Sect. 1.6 we know that if along a closed trajectory γ of system (7.1)

$$\int_0^T \left(\frac{\partial P}{\partial x} + \frac{\partial Q}{\partial y} \right) dt < 0 \qquad (\text{respectively} > 0), \qquad (7.12)$$

then γ is a stable (respectively unstable) limit cycle. From this result it follows that for all other periodic orbits, we must have

$$\int_0^T \left(\frac{\partial P}{\partial x} + \frac{\partial Q}{\partial y}\right) dt = 0. \tag{7.13}$$

However, when γ is an ordinary stable or unstable limit cycle (7.13) may also hold since (7.12) is only a sufficient condition, not a necessary condition.

When condition (7.12) holds, we say that γ is a *hyperbolic limit cycle*; when condition (7.13) holds, we say that γ is a *multiple periodic orbit*, called a multiple limit cycle if it is a limit cycle, both externally and internally.

Let Σ be a transverse segment (parametrized by s) to the flow of (7.1) and let $p(s)$ be the Poincaré map over Σ. We define by $d(s) = p(s) - s$ the so called *displacement function*. We clearly have the following result, whose proof we leave as an exercise.

Theorem 7.19 *If, for a given closed trajectory γ,*

$$d'(0) = d''(0) = \ldots = d^{(k-1)}(0) = 0 \text{ and } d^{(k)}(0) < 0 \text{ (respectively } > 0), \tag{7.14}$$

for k odd, then γ is a stable (respectively unstable) limit cycle. If

$$d'(0) = d''(0) = \ldots = d^{(k-1)}(0) = 0 \text{ and } d^{(k)}(0) \neq 0, \tag{7.15}$$

for k even, then γ is a semistable limit cycle.

Any γ satisfying condition (7.14) or (7.15) is called a *k-multiple limit cycle*.

7.4 Rotated Vector Fields

Consider the system of differential equations

$$\frac{dx}{dt} = P(x, y, \alpha), \quad \frac{dy}{dt} = Q(x, y, \alpha), \tag{7.16}$$

with parameter α. Until now we have not yet considered how an orbit or how the phase portrait changes as a parameter α varies. Such problems are interesting but can also be very complicated. Suppose that as the parameter α is perturbed slightly near α_0, the topological structure of the phase portrait of $(7.16)_{\alpha_0}$ is unchanged; then α_0 is called a *regular value* of α, and the system $(7.16)_{\alpha_0}$ is called *structurally stable with respect to perturbations of α*. If for arbitrarily small perturbations α near α_0, the topological structure of the phase portrait for system $(7.16)_\alpha$ is changed, then we say that α_0 is a *bifurcation value* and the change in topological structure is called a *bifurcation*. For example, as the parameter α changes, a limit cycle may appear or disappear near a singular point, or one limit cycle may split into several ones;

these are called *bifurcation phenomena*. Bifurcation theory is an active area in differential equations; the reader is referred to [35].

In this section, we concentrate on the dependence of limit cycles on the parameter α. The changes are very complicated for general 1-parameter families; hence, we will restrict our attention to a special situation: rotated vector fields. The method of rotated vector fields is important in the sense that it provides tractable 1-parameter families of polynomial vector fields of a fixed degree, although it is definitely useful in a much more general context. We discuss the changes of limit cycles in rotated planar vector fields as the parameter varies. That is, we study the appearance and disappearance of limit cycles as the parameter varies. In the special situation under consideration, the changes are very systematic.

In this section we assume that the vector field (7.16) has only isolated singular points, and that $P(x, y, \alpha)$ and $Q(x, y, \alpha)$ are C^1 functions on $R \times I$, where $I : 0 \leq \alpha \leq T$ or $-\infty < \alpha < +\infty$ and $R \subset \mathbb{R}^2$ is an open region.

Suppose that, as α varies in $[0, T]$, the singular points of the vector field (7.16) are unchanged, and at all the regular points

$$\begin{vmatrix} P & Q \\ \dfrac{\partial P}{\partial \alpha} & \dfrac{\partial Q}{\partial \alpha} \end{vmatrix} > 0; \tag{7.17}$$

moreover

$$\begin{aligned} P(x, y, \alpha + T) &= -\, P(x, y, \alpha), \\ Q(x, y, \alpha + T) &= -\, Q(x, y, \alpha). \end{aligned} \tag{7.18}$$

Then (7.16) is said to form a *complete family of rotated vector fields*, for $0 \leq \alpha \leq T$.

From (7.18), it follows that $P(x, y, \alpha)$ and $Q(x, y, \alpha)$ can be extended as periodic functions of α with period $2T$.

Let θ be the angle between the vector field (P, Q) and the x-axis; then we have

$$\frac{\partial \theta}{\partial \alpha} = \frac{\partial}{\partial \alpha} \operatorname{Arctan} \frac{Q}{P} = \frac{1}{P^2 + Q^2} \begin{vmatrix} P & Q \\ \dfrac{\partial P}{\partial \alpha} & \dfrac{\partial Q}{\partial \alpha} \end{vmatrix}.$$

From condition (7.17), it follows that at all regular points $p = (x, y)$, when the parameter α increases, the vector field (7.16) rotates counter-clockwise at the point p. From condition (7.18), when the parameter changes from α to $\alpha + T$, the vector (P, Q) rotates exactly π radians counter-clockwise at the point p, and the length of the vector remains the same. Thus, when α changes to $\alpha + 2T$, the vector field (P, Q) rotates 2π radians counter-clockwise to its original position. This is the geometric meaning of "rotation and complete" in Duff's definition; see for more details [50]. As the parameter α varies, the change on limit cycles in rotated vector fields is relatively systematic; however, the restrictions in the definition of complete family of rotated vector fields are quite strong. These restrictions can be substantially reduced if we retain only the fundamental requirements. For example, Chen Xiang–Yan introduced generalized rotated vector fields, where limit cycles vary in the same systematic way as the parameter varies; see [165–167]. This will be our approach.

Suppose that as α varies in (a, b), the singular points of the vector fields (7.16) remain unchanged, and for any given point $p = (x, y)$ and any parameters $\alpha_1 < \alpha_2$ in (a, b), we have

$$\begin{vmatrix} P(x, y, \alpha_1) & Q(x, y, \alpha_1) \\ P(x, y, \alpha_2) & Q(x, y, \alpha_2) \end{vmatrix} \geq 0 \text{ (or } \leq 0), \tag{7.19}$$

where equality cannot hold on an entire periodic orbit of $(7.16)_{\alpha_i}$, $i = 1, 2$. Then (7.16) are called *generalized rotated vector fields*. Here, the interval (a, b) can be either bounded or unbounded.

If for a regular point (x_0, y_0) and parameter α_0, there exists $\delta(x_0, y_0, \alpha_0)$ positive such that for all $\alpha \in [\alpha_0 - \delta, \alpha_0 + \delta]$, equality holds in (7.19), then α_0 is called a *stopping point* for (x_0, y_0); otherwise, α_0 is called a *rotating point*. Stopping points are allowed in generalized rotated vector fields. Moreover, generalized rotated vector fields do not necessarily depend on α periodically; in particular, condition (7.18) is not required.

The geometric meaning of condition (7.19) is that, at any point $p = (x, y)$, the oriented angle between $(P(x, y, \alpha_1), Q(x, y, \alpha_1))$ and $(P(x, y, \alpha_2), Q(x, y, \alpha_2))$ has the same (or opposite) sign as $\operatorname{sgn}(\alpha_2 - \alpha_1)$. That is, at any point $p = (x, y)$, as the parameter α increases, the vector $(P(x, y, \alpha), Q(x, y, \alpha))$ can only rotate in one direction; moreover, the angle of rotation cannot exceed π.

In the following, we describe two examples of rotated vector fields.

Example 7.20 Consider the system of differential equations

$$\frac{dx}{dt} = P(x, y), \quad \frac{dy}{dt} = Q(x, y)$$

where P, Q are of class C^1. Consider the system of differential equations containing parameter α

$$\frac{dx}{dt} = P \cos \alpha - Q \sin \alpha, \quad \frac{dy}{dt} = P \sin \alpha + Q \cos \alpha. \tag{7.20}$$

It is not difficult to verify that equations (7.20) satisfy conditions (7.17) and (7.18) and thus form a complete family of rotated vector fields. In $0 < \alpha \leq \pi$ (7.20) represents generalized rotated vector fields, but not in $0 < \alpha \leq 2\pi$. We remark that (7.20) is analytic when P and Q are, while the vector fields in (7.20) are all polynomial of degree n if (P, Q) represents a polynomial vector field of degree n. □

In fact (7.20) can be regarded as a formula for axis rotation. It rotates the original vector field by an angle of α, and keeps the vector lengths unchanged. Thus (7.20) are called *uniformly rotated vector fields*.

Example 7.21 Consider the system of differential equations

$$\frac{dx}{dt} = -\alpha y, \quad \frac{dy}{dt} = \alpha x - \alpha y f(\alpha x), \tag{7.21}$$

where $0 < \alpha < +\infty$, and $f(x)$ is C^1 and monotonically increasing as $|x|$ increases. It can be verified by condition (7.19) that (7.21) are generalized rotated vector fields; however, it is not a complete family of rotated vector fields. □

In the following, we present a few important results concerning bifurcation of limit cycles for generalized rotated vector fields. Naturally, they will also apply to complete families of rotated vector fields. We leave the proofs to the reader.

Theorem 7.22 (Nonintersection Theorem) *Suppose that (7.16) are generalized rotated vector fields. Then for distinct α_1 and α_2, the periodic orbits of system $(7.16)_{\alpha_1}$ and $(7.16)_{\alpha_2}$ cannot intersect each other.*

In the following, we discuss the changes in the limit cycles as the parameter α changes in system (7.16).

(i) Let $(P(x, y, \alpha), Q(x, y, \alpha))$ be generalized rotated vector fields, satisfying inequality (7.19) with a determinant that is ≥ 0. Suppose that for $\alpha = \alpha_0$, γ_{α_0} is an externally stable limit cycle for system $(7.16)_{\alpha_0}$, turning counterclockwise (clockwise). Then for an arbitrarily small positive number ε, there exists $\alpha_1 < \alpha_0$ (or $\alpha_0 < \alpha_1$) such that for any $\alpha \in (\alpha_1, \alpha_0)$ (or $\alpha \in (\alpha_0, \alpha_1)$), there is at least one externally stable limit cycle γ_α and one internally stable limit cycle $\overline{\gamma}_\alpha$ for system $(7.16)_\alpha$ in an exterior ε-neighborhood of γ_α. (Here, γ_α may coincide with $\overline{\gamma}_\alpha$). Moreover, there is an exterior δ-neighborhood of γ_α (with $\delta \leq \varepsilon$), such that the neighborhood is filled with periodic orbits $\{\gamma_\alpha\}$ of $(7.16)_\alpha$, for different $\alpha \in (\alpha_1, \alpha_0)$ (or $\alpha \in (\alpha_0, \alpha_1)$). When $\alpha > \alpha_0$ (or $\alpha < \alpha_0$), there is no periodic orbit of $(7.16)_\alpha$ in the exterior δ-neighborhood of γ_{α_0}.

(ii) Let $(P(x, y, \alpha), Q(x, y, \alpha))$ be generalized rotated vector fields, satisfying inequality (7.19) with a determinant that is ≥ 0. Suppose that for $\alpha = \alpha_0$, γ_{α_0} is an internally stable limit cycle for system $(7.16)_{\alpha_0}$, turning counterclockwise (clockwise). Then for an arbitrarily small positive number ε, there exists $\alpha_2 > \alpha_0$ (or $\alpha_2 < \alpha_0$) such that for any $\alpha \in (\alpha_0, \alpha_2)$ (or $\alpha \in (\alpha_2, \alpha_0)$), there is at least one externally stable limit cycle γ_α and one internally stable limit cycle $\overline{\gamma}_\alpha$ for system $(7.16)_\alpha$ in an interior ε-neighborhood of γ_{α_0}. (Here, γ_{α_0} may coincide with $\overline{\gamma}_\alpha$). Moreover, there is an interior δ-neighborhood of γ_{α_0} (with $\delta \leq \varepsilon$), such that the neighborhood is filled with periodic orbits $\{\gamma_\alpha\}$ of $(7.16)_\alpha$, for different $\alpha \in (\alpha_0, \alpha_2)$ (or $\alpha \in (\alpha_2, \alpha_0)$). When $\alpha < \alpha_0$ (or $\alpha > \alpha_0$), there is no periodic orbit of $(7.16)_\alpha$ in the interior δ-neighborhood of γ_{α_0}.

For an unstable limit cycle γ_{α_0}, there are two results analogous to (i) and (ii). However, for a fixed orientation of γ_{α_0}, the parameter α should be taken in the opposite direction from that in the statements above.

From the statements above, we see that for generalized rotated vector fields, the evolution of stable or unstable limit cycles is fairly systematic. When the parameter changes monotonically, the limit cycle will not disappear; it will expand or contract. When the generalized rotated vector fields satisfy inequality (7.19) for a determinant that is positive, we now tabulate in Table (7.1) the evolution of the stable or unstable limit cycle γ_α for system $(7.16)_\alpha$ as α increases.

As to semistable limit cycles, we have the following statement concerning their evolution as the parameter in rotated vector fields varies.

(iii) Let $(P(x, y, \alpha), Q(x, y, \alpha))$ be generalized rotated vector fields, and let γ_{α_0} be a semistable limit cycle for system $(7.16)_{\alpha_0}$. When the parameter varies in a suitable direction, γ_{α_0} will bifurcate into at least one stable and one unstable limit cycle. They will lie on distinct sides of γ_{α_0}, one on the inside and one on the outside. When α varies in the opposite direction γ_{α_0} disappears.

We represent in Table 7.2 the situation for the changes of semistable limit cycles γ_α according to the direction of their movement, when the parameter α varies.

Now when α varies monotonically in a rotated vector field, are the expansions and contractions of stable or unstable limit cycles γ_α monotonic? In order to understand this, we must clarify whether γ_α can bifurcate into several limit cycles

(iv) Let $(P(x, y, \alpha), Q(x, y, \alpha))$ be generalized rotated vector fields. Then a simple limit cycle of system (7.16) cannot split nor disappear as the parameter α_0 varies monotonically. Moreover, the cycle will expand or contract monotonically.

Table 7.1. Behavior of γ_α as α varies

direction	counter-clockwise	counter-clockwise	clockwise	clockwise
stability	stable	unstable	stable	unstable
evolution	contract	expand	expand	contract

Table 7.2. Behavior of semistable γ_α as α varies

direction	counter-clockwise	counter-clockwise	clockwise	clockwise
stability	externally stable, internally unstable	externally unstable, internally stable	externally stable, internally unstable	externally unstable, internally stable
α increases	disappears	splits into two or more cycles	splits into two or more cycles	disappears
α decreases	splits into two or more cycles	disappears	disappears	splits into two or more cycles

7.5 Structural Stability

Since in Sect. 7.4 we have touched on the notion of structural stability, we believe it is interesting to say more about it. After all, it is one of the crucial notions in the qualitative theory of differential equations (see [118]). We first give a precise and rather general definition. We need some space $\mathcal{X}$ of vector fields on some subset $U \subset \mathbb{R}^2$ or $\mathbb{S}^2$ like C^r vector fields on $\mathbb{S}^2$, or polynomial vector fields of some degree on $\mathbb{R}^2$; we endow $\mathcal{X}$ with some topology $\mathcal{T}$ and then $X \in \mathcal{X}$ is called $\mathcal{T}$-structurally stable if there exists a $\mathcal{T}$-neighborhood V of X such that for all $Y \in V$ there exists a topological equivalence h_Y defined on U, between Y and X.

We remark that some authors require in the definition of structural stability that the equivalence h_Y depend continuously on Y and have the property that $h_X =$ Id. In the results that we are going to state, this would not make a difference, but in general one has to pay attention to it.

To fix ideas and to be able to state nice and simple results we will limit our discussion to $U = \mathbb{S}^2$ or $U = \mathbb{D}^2$, i.e., the closed northern hemisphere of the unit sphere $\mathbb{S}^2 = \{(x, y, z) \in \mathbb{R}^3; \; x^2 + y^2 + z^2 = 1\}$.

For $\mathbb{S}^2$ we consider $\mathcal{X}^r(\mathbb{S}^2)$, with $r = 1, 2, \ldots, \infty, \omega$, the space of C^r vector fields on $\mathbb{S}^2$, and we endow it with a C^s-topology $\mathcal{T}_s$ where $s \in \mathbb{N}$ with $1 \leq s \leq r$. A system $X \in \mathcal{X}^r(\mathbb{S}^2)$ is called s-structurally stable if it is structurally stable for the topology $\mathcal{T}_s$.

We can consider similar spaces $(\mathcal{X}^r(\mathbb{D}^2), \mathcal{T}_s)$ requiring for a vector field $X \in \mathcal{X}^r(\mathbb{D}^2)$ that $\mathbb{S}^1 = \partial(\mathbb{D}^2)$ be an invariant set (i.e., consist of orbits).

Theorems 7.23 and 7.24 below follow from results of Peixoto ([120–123]) and an adaptation for the analytic situation by Perelló ([126]). Before stating these theorems, we introduce the notion of a saddle-connection.

A *saddle-connection* is an orbit that is both a stable separatrix at a hyperbolic saddle p and an unstable separatrix at a hyperbolic saddle q ($q = p$ is permitted). If $p \neq q$ we call it a *heteroclinic connection*; if $p = q$ we call it a *homoclinic connection*.

Theorem 7.23 *Consider $(\mathcal{X}^r(\mathbb{S}^2), \mathcal{T}_s)$ with $r = 1, 2, \ldots, \infty, \omega$ and $s \in \mathbb{N}$ with $1 \leq s \leq r$. Then $X \in \mathcal{X}^r(\mathbb{S}^2)$ is s-structurally stable if and only if*

(i)   the singularities of X are hyperbolic,
(ii)   the periodic orbits of X are hyperbolic limit cycles, and
(iii) there are no saddle-connections.

Moreover, the structurally stable systems form an open and dense subset of $(\mathcal{X}^r(\mathbb{S}^2), \mathcal{T}_s)$.

We will not give a proof of this theorem, but will present some ingredients in the exercises.

Theorem 7.24 *Consider $(\mathcal{X}^r(\mathbb{D}^2), \mathcal{T}_s)$ with $r = 1, 2, \ldots, \infty, \omega$ and $s \in \mathbb{N}$ with $1 \leq s \leq r$. Then $X \in \mathcal{X}^r(\mathbb{D}^2)$ is s-structurally stable if and only if*

(i)   the singularities of X are hyperbolic,

(ii) the periodic orbits of X are hyperbolic limit cycles, and
(iii) there are no saddle-connections in $\mathbb{D}^2 \setminus \mathbb{S}^1$.

Moreover, the structurally stable systems form an open and dense subset of $(\mathcal{X}^r(\mathbb{D}^2), \mathcal{T}_s)$.

Remark 7.25 Structurally stable systems in $(\mathcal{X}^r(\mathbb{D}^2), \mathcal{T}_s)$ can of course have stable saddle connections inside $\mathbb{S}^1$ since $\mathbb{S}^1$ is supposed invariant for all $X \in \mathcal{X}^r(\mathbb{D}^2)$.

Remark 7.26 In both Theorems 7.23 and 7.24 we see that the characterization of the structurally stable systems does not depend on the chosen topology $\mathcal{T}_s$. The openness of the subset of the structurally stable systems is a trivial consequence of the definition. As already said, we do not prove density but we will present some steps of the proof in the exercises.

Given our interest in polynomial systems, we say a few words about their structural stability. To avoid complication near infinity (see e.g., [146]) we prefer to work with their Poincaré compactification.

We denote by $P_n(\mathbb{D}^2)$ the space of analytic vector fields on $\mathbb{D}^2$ that are obtained from polynomial vector fields of degree at most n by the Poincaré compactification explained in Chap. 5. We denote by $P_n(\mathbb{R}^2)$ the vector space of polynomial vector fields of degree at most n.

It seems quite natural to consider on $P_n(\mathbb{R}^2)$ the *coefficient topology*, i.e., such that $P_n(\mathbb{R}^2)$ is isomorphic to $\mathbb{R}^M$ where the isomorphism expresses a vector field by means of the M-tuple $(a_{00}, a_{10}, \ldots, a_{nn}, b_{00}, b_{10}, \ldots, b_{nn})$ of its coefficients. We simply denote $P_n(\mathbb{R}^2)$, endowed with this coefficient topology, by $(P_n(\mathbb{R}^2), \mathcal{T}^n)$. The topology is generated by the Euclidean metric on $\mathbb{R}^M$. Using the mapping

$$P_c : P_n(\mathbb{R}^2) \to P_n(\mathbb{D}^2)$$

which to each $X \in P_n(\mathbb{R}^2)$ assigns its nth-degree Poincaré compactification, we can transport the Euclidean metric to $P_n(\mathbb{D}^2)$ and we denote the associated topology equally by $\mathcal{T}^n$. We are now interested in structural stability inside $(P_n(\mathbb{D}^2), \mathcal{T}^n)$.

From Pugh [134] and dos Santos [49], we can state the following theorem:

Theorem 7.27 $(P_n(\mathbb{D}^2), \mathcal{T}^n)$ *contains an open and dense subset S_n in which the vector fields X are characterized by the following properties:*

(i)   the singularities of X are hyperbolic,
(ii) the periodic orbits of X are hyperbolic limit cycles, and
(iii) there are no saddle connections in $\mathbb{D}^2 \setminus \mathbb{S}^1$.

The vector fields in S_n are structurally stable.

In fact, on $P_n(\mathbb{D}^2)$ the topology $\mathcal{T}^n$ is the same as the topology $\mathcal{T}_s$ for any $s \in \mathbb{N}$ with $s \geq 1$, even for $s = 0$. We leave it as an exercise to prove this.

It might seem logical to expect that all the structurally stable systems of $(P_n(\mathbb{D}^2), \mathcal{T}^n)$ belong to S_n, but this is still an open problem, even for $n = 2$.

What is known is that structurally stable systems in $(P_n(\mathbb{D}^2), \mathcal{T}^n)$ necessarily only have hyperbolic singularities, have no saddle connections in $\mathbb{D}^2 \setminus \mathbb{S}^1$, and have only isolated periodic orbits, hence limit cycles, whose multiplicity cannot be even. Unfortunately, it is not clear how to perturb a limit cycle γ with multiplicity $2n + 1$, for $n \geq 1$, in a way that there are nearby systems which in the neighborhood of γ exhibit more than one limit cycle.

In any case, all possible phase portraits of structurally stable systems in $(P_n(\mathbb{D}^2), \mathcal{T}^n)$ can be realized by systems in S_n.

For the moment it is beyond the possibilities of the theory to classify all structurally stable systems in $(P_n(\mathbb{D}^2), \mathcal{T}^n)$, even for $n = 2$. The solution of this problem would at least require a solution of Hilbert's 16th problem. It makes sense, however, to ask for a classification of the structurally stable vector fields without closed curves. This has been solved in [10] for $n = 2$. There are exactly 44 distinct phase portraits of such vector fields that are structurally stable in $(P_2(\mathbb{D}^2), \mathcal{T}^2)$. In [10] it has also been proven that the structurally stable vector fields have the same extended skeleton as one of the list of 44, or in other words, they can be obtained out of the 44 listed phase portraits by changing one or two of the singularities of index 1 (also called antisaddles) by a small disk, whose boundary is a limit cycle and inside which one can have a finite number of other limit cycles, all surrounding a unique strong focus.

Another problem that is yet unsolved consists in determining how many limit cycles can exist for each one of the 44 extended skeletons presented in [10]. For more information we refer to that paper and to [11].

A number of limit cycles surrounding a unique singularity is called a *nest of limit cycles*. It is well known that a quadratic vector field can only have two nests of limit cycles, and each nest necessarily surrounds a single focus, which is strong if the vector field is structurally stable.

In [129] it has been proved that when two nests occur, for a quadratic vector field, then necessarily one of the nests contains exactly one limit cycle, which is hyperbolic.

Such results are typical for quadratic vector fields. For cubic vector fields there can be more than two nests. And moreover there can occur limit cycles that surround more than one singularity, and even if a nest surrounds just one singularity it needs not be a focus; it can also be a node, for example.

In Figs. 7.1 and 7.2 we present two phase portraits that can appear in cubic systems and that cannot be realized by a quadratic vector field. The proof of the occurrence of such phase portraits for cubic systems can be found in [56] for the first example and in [58] as well as in [77] for the second.

Another question that has been considered concerns the structure of $P_n(\mathbb{D}^2) \setminus S_n$, with S_n as in Theorem 7.27. It at first seemed quite reasonable that this set might be described by polynomial equations, as had been

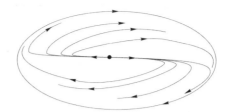

Fig. 7.1. A limit cycle with a node inside

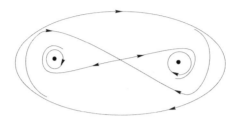

Fig. 7.2. A limit cycle surrounding a saddle, two antisaddles, and two limit cycles in different nests

put forward as a problem in [45] for $n = 2$. In fact it has been proved in [54] that already for $n = 2$, the set $P_n(\mathbb{D}^2) \setminus S_n$ is not even semi-analytic (see e.g., [18] for the definition of semi-analyticity).

7.6 Exercises

Exercise 7.1 Let X_i, $i = 1, 2$ be C^∞ vector fields, defined on open domains $U_i \subset \mathbb{R}^2$. Suppose that $\gamma_i \subset U_i$ is a periodic orbit for X_i. Take segments Σ_i transversely cutting γ_i at p_i, that are regularly parametrized by a parameter which is 0 at p_i. Consider the associated Poincaré maps P_{X_i} with respect to Σ_i. Then P_{X_1} and P_{X_2} are C^0-conjugate at 0 if and only if X_1 at γ_1 is C^0-equivalent to X_2 at γ_2, in the sense that there exist neighborhoods V_1 and V_2 of γ_1 and γ_2, respectively, and a homeomorphism $h : V_1 \to V_2$ with $h(\gamma_1) = \gamma_2$ and such that h is a C^0-equivalence between $X_1|_{V_1}$ and $X_2|_{V_2}$.

Exercise 7.2 Consider the family of equations

$$\dot{x} = f(x, y, \lambda_1, \ldots, \lambda_n)$$
$$\dot{y} = g(x, y, \lambda_1, \ldots, \lambda_n),$$

with f and g smooth; $\lambda = (\lambda_1, \ldots, \lambda_n)$ represents an n dimensional real parameter. Suppose that for some $p_0 = (x_0, y_0)$ and some $\lambda^0 = (\lambda_1^0, \ldots, \lambda_n^0)$, we have that $X_{\lambda^0}(p_0) = 0$, and that $D(X_{\lambda^0})(p_0)$ is non–degenerate, in the sense that $\det D(X_{\lambda^0})(p_0) \neq 0$. Show that for all λ sufficiently close to λ^0, there exists $p = p(\lambda)$ with $X_\lambda(p(\lambda)) = 0$ and $p(\lambda^0) = p_0$. Show also that $p(\lambda)$ is a C^∞ function. If f and g are analytic, then $p(\lambda)$ is analytic, too.

Exercise 7.3 Consider the same system as in the previous exercise, but suppose that $D(X_{\lambda^0})(p_0)$ is hyperbolic. Show that for $\lambda \approx \lambda^0$:

(i) $D(X_\lambda)(p_\lambda)$ is hyperbolic, too.

(ii) X_λ at $p(\lambda)$ is C^0-conjugate to X_{λ^0} at p_0. (We say that X_{λ^0} at p_0 is locally structurally stable).

Exercise 7.4 Let $f : U \subset \mathbb{R}^m \to \mathbb{R}^n$ be a C^∞ function with U open. Sard's Theorem states that almost all $b \in \mathbb{R}^n$ (with respect to Lebesgue measure) are regular values of f, meaning that either $b \notin f(U)$ or at all $x \in f^{-1}(b)$ we have that $Df_x : \mathbb{R}^m \to \mathbb{R}^n$ is surjective.

(i) Let $b \in f(U)$ be a regular value of f; show that $f^{-1}(b) \subset U$ is a $(m - n)$ dimensional submanifold of U.

(ii) Use Sard's Theorem and point (i) to show that for any C^∞ vector field X on $U \subset \mathbb{R}^2$ with U open, and $X(0) = 0$, there exists a vector field having an expression $X + \varepsilon$ with $\varepsilon \in \mathbb{R}^2$, and ε arbitrarily close to 0, such that all singularities of $X + \varepsilon$ in U are non-degenerate.

Exercise 7.5 Let $X : U \subset \mathbb{R}^2 \to \mathbb{R}^2$ be a C^∞ vector field $X = X_1 \frac{\partial}{\partial x} + X_2 \frac{\partial}{\partial y}$ having at 0 a non-degenerate singularity. Consider the following family of vector fields $X(a, b, c, d)$ depending on the four parameters (a, b, c, d):

$$\dot{x} = ax + by + X_1(x, y)$$
$$\dot{y} = cx + dy + X_2(x, y).$$

Show that there exist (a, b, c, d) arbitrarily close to $(0, 0, 0, 0)$ such that $X_{a,b,c,d}$ has a hyperbolic singularity at $(x, y) = (0, 0)$.

Exercise 7.6 Let X be a C^∞ vector field on some open domain $U \subset \mathbb{R}^2$ and suppose that at $p \in U$, X has a semi-hyperbolic singularity which is a generic saddle-node, i.e., the restriction of X to any center manifold W^c starts with quadratic terms with respect to a regular parametrization of W^c. Show the existence of a 1-parameter family X_ε, a neighborhood $V \subset U$ of p and an $\varepsilon_0 > 0$ such that for all $\varepsilon \in (0, \varepsilon_0)$, X_ε has no singularities in V, while for all $\varepsilon \in (-\varepsilon_0, 0)$, X_ε has two singularities in V.

   Hint: Use a normal form.

Exercise 7.7 Consider the previous exercise again, replacing the generic saddle-node singularity by a generic nilpotent cusp point.

Exercise 7.8 Let $X_\lambda = X_{\lambda_1,\dots,\lambda_n}$ be a smooth n-parameter family of vector fields on some open subset $U \subset \mathbb{R}^2$ with $\lambda \in B(\lambda^0, \varepsilon)$. Suppose that X_{λ^0} has a hyperbolic limit cycle γ_0. Show that there exists a $0 < \delta < \varepsilon$ and a neighborhood V of γ_0 in U such that for all $\lambda \in B(\lambda^0, \delta)$, X_λ has a unique periodic orbit in V. Moreover the periodic orbit is hyperbolic and $X_\lambda|_V$ is C^0-equivalent to $X_{\lambda^0}|_V$. Extra question: Can we be sure that $X_\lambda|_V$ is C^0-conjugate to $X_{\lambda^0}|_V$? If so, give a proof, if not, give a counterexample.

   Hint: Use a Poincaré map.

Exercise 7.9 Let X be a smooth vector field defined on some open subset of $\mathbb{R}^2$ and let X_a, for $a \in \mathbb{R}$ and $a \approx 0$ be a smooth rotated family of vector fields. Suppose that X has γ as a limit cycle of even multiplicity (γ is semistable). Show that for $a \neq 0$ and $a \approx 0$, X_a has either no closed curves, or at least two.

Hint: Use the Poincaré–Bendixson Theorem.

Exercise 7.10 Consider the ε-family of smooth Liénard system X_ε

$$\dot{x} = y,$$
$$\dot{y} = g(x) + yf(x) + \varepsilon y,$$

such that for $\varepsilon = 0$ the system X_0 has a hyperbolic saddle at both $p_1 = (x_1, 0)$ and $p_2 = (x_2, 0)$ together with a heteroclinic connection between them. Show that for $\varepsilon \neq 0$, $\varepsilon \approx 0$, system X_ε has no heteroclinic connection between p_1 and p_2.

Hint: Use the property of rotated vector fields.

Exercise 7.11 Let f_1, f_2 be C^2 functions on $\mathbb{R}^2$. Given $a > 0$ prove that a necessary condition for the system

$$\dot{x} = y + \mu f_1(x, y),$$
$$\dot{y} = -x + \mu f_2(x, y),$$
$$(7.22)$$

to have a periodic solution $\varphi(t, a, \mu)$ of period $\tau(\mu)$ for every μ sufficiently small is that $\varphi_a = \varphi(t, q, 0) = a(\cos t, -\sin t)$ and $\tau(\mu)$ is differentiable with $\tau(0) = 2\pi$, and that

$$\beta(a) = \int_{\varphi_a} f_1 dy + f_2 dx = 0.$$

Prove that if $\beta(a) = 0$ and $\beta'(a) \neq 0$, then (7.22) in fact has the properties described above.

Hint: Use polar coordinates $x = r\cos\theta$, $y = r\sin\theta$ transforming (7.22) into

$$\dot{r} = \mu R_1(r, \theta, \mu),$$
$$\dot{\theta} = 1 + R_2(r, \theta, \mu),$$

which is equivalent to an equation of the type

$$\frac{dr}{d\theta} = \mu R(r, \theta, \mu). \qquad (7.23)$$

Prove that a solution of (7.23) is of the form $\rho(r, \theta, \mu) = r + \mu\beta(r) + \varepsilon(r, t)\mu$.

Exercise 7.12 Use the previous exercise to prove that the equation of van der Pol

$$x'' = -x + \varepsilon x'(1 - x^2)$$

has, for every $\varepsilon > 0$ sufficiently small, a unique limit cycle stable in a neighborhood of the circle $x^2 + (x')^2 = 4$. Prove also that as $\varepsilon \to 0$, this limit cycle tends to the given circle.

Exercise 7.13 Let $X = (X_1, X_2)$ be a vector field on $\mathbb{R}^2$, where

$$X_1 = y + x(1 - x^2 - y^2)$$
$$X_2 = -x + y(1 - x^2 - y^2).$$

Prove that this vector field has a unique periodic orbit γ. Find a Poincaré map π associated to γ and prove that $\pi' \neq 1$ at γ.

Hint: In polar coordinates the system above becomes

$$\dot{r} = r(1 - r^2),$$
$$\dot{\theta} = -1.$$

Show that

$$\int \frac{dr}{r(1 - r^2)} = \frac{1}{2} \log\left(\frac{r^2}{1 - r^2}\right),$$

and then show that $\pi : \mathbb{R}^+ \to \mathbb{R}^+$ (where $\mathbb{R}^+$ denotes the positive x–axis) is given by

$$\pi(r) = \frac{re^\pi}{\sqrt{1 - r^2 + r^2 e^{2\pi}}}.$$

Exercise 7.14 Let γ be a stable periodic orbit of $X = (X_1, X_2)$. Let

$$X_\theta = \begin{pmatrix} \cos\theta & \sin\theta \\ -\sin\theta & \cos\theta \end{pmatrix} \begin{pmatrix} X_1 \\ X_2 \end{pmatrix}.$$

This is the vector field obtained from X after doing a rotation of angle θ.

(i)   Prove that there exists $\varepsilon > 0$ such that X_θ with $|\theta| < \varepsilon$ has a periodic orbit γ_θ such that $\gamma_\theta \to \gamma$ when $\theta \to 0$.
(ii) Prove that the γ_θ are all disjoint, that is

$$\gamma_{\theta_1} \cap \gamma_{\theta_2} = \emptyset \text{ if } \theta_1 \neq \theta_2,$$

and prove that $\bigcup_{|\theta| \leq \varepsilon} \gamma_\theta$ is an annular region of the plane.
(iii) If γ is an unstable periodic orbit prove an analogous version of (i) and (ii).
(iv) If γ is semistable prove that for θ with appropriate sign (positive or negative depending on the case), there exist two periodic orbits γ_{θ_1} and γ_{θ_2} with $\gamma_{\theta_i} \to \gamma$ when $\theta \to 0$, for $i = 1, 2$.

Exercise 7.15 Show that the systems

$$x'' + (5x^4 - 9x^2)x' + x^5 = 0,$$
$$x'' + (x^6 - x^2)x' + x = 0,$$

have a periodic orbit.

Exercise 7.16 Prove Poincaré's Theorem: A planar analytic system $\dot{x} = f(x)$, cannot have an infinite number of limit cycles which accumulate on a limit cycle of the system.

Exercise 7.17 Show that $\gamma(t) = (2\cos 2t, \sin 2t)$ is a periodic solution of the system

$$\dot{x} = -4y + x\left(1 - \frac{x^2}{4} - y^2\right),$$

$$\dot{y} = x + y\left(1 - \frac{x^2}{4} - y^2\right),$$

that lies in the ellipse $(x/2)^2 + y^2 = 1$; i.e., $\gamma(t)$ represents a cycle Γ of this system. Then show that Γ is a stable limit cycle.

Exercise 7.18 Show that the system

$$\dot{x} = -y + x(1 - x^2 - y^2)^2,$$

$$\dot{y} = x + y(1 - x^2 - y^2)^2,$$

has a limit cycle Γ represented by $\gamma(t) = (\cos t, \sin t)$. Show that Γ is a multiple limit cycle. Since Γ is a semistable limit cycle, we know that the multiplicity k of Γ is even. Can you show that $k = 2$?

Exercise 7.19 Use the Poincaré–Bendixson Theorem and the fact that the planar system

$$\dot{x} = x - y - x^3,$$

$$\dot{y} = x + y - y^3,$$

has only a single singular point at the origin to show that this system has a periodic orbit in the annular region $A = \{x \in \mathbb{R}^2 : 1 \le |x| \le \sqrt{2}\}$.

   Hint: Convert to polar coordinates and show that for all $\varepsilon > 0$, $\dot{r} < 0$ on the circle $r = \sqrt{2} + \varepsilon$ and $\dot{r} > 0$ on $r = 1 - \varepsilon$.

Exercise 7.20 Following exercise 7.20 show that there is at least one stable limit cycle in A (In fact this system has exactly one limit cycle in A and it is stable.)

Exercise 7.21 Show that

$$\dot{x} = y,$$

$$\dot{y} = -x + (1 - x^2 - y^2)y,$$

has a unique stable limit cycle which is the ω-limit set of every trajectory except the singular point at the origin.

   Hint: Compute $\dot{r}$.

Exercise 7.22 Use the Dulac function $B(x,y) = be^{-2\beta x}$ to show that the system

$$\dot{x} = y,$$
$$\dot{y} = -ax - by + \alpha x^2 + \beta y^2,$$

has no limit cycle in $\mathbb{R}^2$.

Exercise 7.23 Show that the system

$$\dot{x} = \frac{y}{1+x^2},$$
$$\dot{y} = \frac{-x + y(1+x^2+x^4)}{1+x^2},$$

has no limit cycle in $\mathbb{R}^2$.

Exercise 7.24 Prove that the system

$$\dot{x} = \delta x - y + x^2 + mxy + ny^2,$$
$$\dot{y} = x + bxy,$$

with $b = [(1+n)^2 + \delta(m+mn+\delta n)]/(\delta^2 n)$, $\delta n(1+n) \neq 0$ and $n > 0$, has no periodic orbits.

Hint: Use the fact that this system has an inverse polynomial integrating factor of degree 2.

Exercise 7.25 Let $\dot{x} = P(x,y)$ and $\dot{y} = Q(x,y)$ be a quadratic polynomial differential equation. Prove that if the pencil of conics $P + \lambda Q$ contains an imaginary conic, a real conic reduced to a single point, or a double straight line, then the system has no periodic orbits.

Hint: Use Theorem 7.15 with $V = e^{x+\lambda_0 y}$, where $P + \lambda_0 Q$ is one of the conics mentioned.

Exercise 7.26 Consider the system

$$\dot{x} = -y + ax(x^2+y^2-1),$$
$$\dot{y} = x + by(x^2+y^2-1).$$

Let $\gamma = \{(x,y) \in \mathbb{R}^2 : x^2+y^2 = 1\}$. Prove:

(i) if $a+b=0$ and $ab \neq 0$, then γ is the unique limit cycle of the system.
   Hint: Use Theorem 7.16.
(ii) if $a+b \neq 0$, then γ is a hyperbolic limit cycle, stable if $a+b < 0$ and unstable if $a+b > 0$.

Exercise 7.27 Consider the system

$$\dot{x} = y - (a_1 x + a_2 x^2 + a_3 x^3 + x^4),$$
$$\dot{y} = -x.$$

For fixed $a_2, a_3 \in \mathbb{R}$, show that this system defines a family of generalized rotated vector fields with respect to the parameter a_1.

Exercise 7.28 Consider the system

$$\dot{x} = -x + y^2,$$
$$\dot{y} = -\mu x + y + \mu y^2 - xy.$$

Using the fact that μ is a generalized rotation parameter for this system, describe the phase portrait of the system in the Poincaré disk as μ varies in $\mathbb{R}$. Note that as μ is changed this system has a simultaneous Hopf bifurcation of limit cycles at the foci $(1, \pm 1)$ after $\mu = 1/2$ and that both limit cycles end simultaneously in two homoclinic loops of the saddle $(0,0)$ at $\mu \approx 0.52$. Note also that this system is invariant under the transformation $(x, y, \mu) \rightarrow (x, -y, -\mu)$.

7.7 Bibliographical Comments

This chapter follows partly the book of Ye Yan Qian et al. (see [135]). Sect. 7.2 follows the paper [102]. In 1900 Hilbert [82], in the second part of his 16th problem, proposed finding an uniform upper bound on the number of limit cycles of all polynomial vector fields of a given degree, and also studying their distribution or configuration in the plane. This has been one of the main problems in the qualitative theory of planar differential equations in the 20th century. The contributions of Bamon [13], Golitsina [73] and Kotova [97] for the particular case of quadratic vector fields, and mainly of Écalle [59] and Il'yashenko [88] in proving that any polynomial vector field has but finitely many limit cycles have been the best results in this area. But until now it has not been proved that there exists an uniform upper bound depending only on the degree. This problem remains open even for quadratic polynomial vector fields.

People interested in 16th Hilbert's problem, concerning the maximum number of limit cycles which polynomial vector fields of a given degree can have, can also have a look at [89].

The problem of the realization of a given configuration of closed curves by a polynomial differential system has been studied by several authors. For C^r vector fields the problem was solved by Al'mukhamedov [1], Balibrea and Jimenez [12] and Valeeva [158]. Statement (i) of Theorem 7.18 was solved by Schecter and Singer [140] and Sverdlove [155], but they did not provide an explicit polynomial vector field satisfying the given configuration of limit cycles. The result presented in statement (ii) of Theorem 7.18 appears in [102], and its proof provides simultaneously the shortest and easiest proof of statement (i) of Theorem 7.18.

There are relatively complete results on the rotated vector field theory. The earliest work can be found in the paper [50] of Duff in 1953. Later, Seifert [145],

Perko [127] and Chen Xiang-Yan [165–167], successively improved the work of Duff. We should especially note that Chen Xiang-Yan introduced the concept of generalized rotated vector fields, which greatly weakens the conditions of Duff, and leads to important applications of rotated vector fields.

For additional information about general results on limit cycles in planar polynomial differential systems see [169].

Integrability and Algebraic Solutions in Polynomial Vector Fields

In contrast to earlier chapters, throughout this chapter we will work with complex two-dimensional differential systems, although our main aim still remains in the study of real planar differential equations.

For a two-dimensional vector field the existence of a first integral completely determines its phase portrait. The simplest planar vector fields having a first integral are the Hamiltonian ones. The integrable planar vector fields which are not Hamiltonian are, in general, very difficult to detect. In this chapter we study the existence of first integrals for planar polynomial vector fields through the Darbouxian theory of integrability. This kind of integrability provides a link between the integrability of polynomial vector fields and the number of invariant algebraic curves that they have.

8.1 Introduction

By definition a two-dimensional *complex planar polynomial differential system* or simply a *polynomial system* will be a differential system of the form

$$\frac{dx}{dt} = \dot{x} = P(x, y), \qquad \frac{dy}{dt} = \dot{y} = Q(x, y), \tag{8.1}$$

where the dependent variables x and y, and the independent one (the time) t are complex, and P and Q are polynomials in the variables x and y with complex coefficients. Throughout this chapter $m = \max\{\deg P, \deg Q\}$ denotes the degree of the polynomial system, and we always assume that the polynomials P and Q are relatively prime in the ring of complex polynomials in the variables x and y.

We want to show the fascinating relationships between integrability (a topological phenomenon) and the existence of exact algebraic solutions for a polynomial system.

8.2 First Integrals and Invariants

We denote by $\mathbb{F}$ either the real field $\mathbb{R}$ or the complex field $\mathbb{C}$; and by an $\mathbb{F}$-*polynomial system* the polynomial system (8.1) with the variables x, y and the coefficients of the polynomials P and Q in $\mathbb{F}$. We also denote by $\mathbb{F}[x, y]$ the ring of polynomials in the variables x and y and coefficients in $\mathbb{F}$.

The vector field X associated to system (8.1) is defined by

$$X = P\frac{\partial}{\partial x} + Q\frac{\partial}{\partial y}.$$

The $\mathbb{F}$-polynomial system (8.1) is *integrable* on an open subset U of $\mathbb{F}^2$ if there exists a nonconstant analytic function $H : U \to \mathbb{F}$, called a *first integral* of the system on U, which is constant on all solution curves $(x(t), y(t))$ of system (8.1) contained in U; i.e., $H(x(t), y(t)) = $ constant for all values of t for which the solution $(x(t), y(t))$ is defined and contained in U. Clearly H is a first integral of system (8.1) on U if and only if $XH = PH_x + QH_y \equiv 0$ on U.

Let $U \subset \mathbb{F}^2$ be an open set. We say that an analytic function $H(x, y, t) : U \times \mathbb{F} \to \mathbb{F}$ is an *invariant* of the polynomial vector field X on U if $H(x, y, t) = $ constant for all values of t for which the solution $(x(t), y(t))$ is defined and contained in U. If an invariant H is independent of t then of course it is a first integral.

The knowledge provided by an invariant is weaker than the one provided by a first integral. The invariant, in general, gives information only about either the α- or the ω-limit set of the orbits of the system.

8.3 Integrating Factors

Let U be an open subset of $\mathbb{F}^2$ and let $R : U \to \mathbb{F}$ be an analytic function which is not identically zero on U. The function R is an *integrating factor* of the $\mathbb{F}$-polynomial system (8.1) on U if one of the following three equivalent conditions holds on U:

$$\frac{\partial(RP)}{\partial x} = -\frac{\partial(RQ)}{\partial y}, \qquad \mathrm{div}(RP, RQ) = 0, \qquad XR = -R\,\mathrm{div}(P, Q).$$

As usual the divergence of the vector field X is defined by

$$\mathrm{div}(X) = \mathrm{div}(P, Q) = \frac{\partial P}{\partial x} + \frac{\partial Q}{\partial y}.$$

The *first integral H associated to the integrating factor R* is given by

$$H(x, y) = \int R(x, y)P(x, y)\, dy + h(x), \tag{8.2}$$

where h is chosen such that $\frac{\partial H}{\partial x} = -RQ$. Then

$$\dot{x} = RP = \frac{\partial H}{\partial y}, \qquad \dot{y} = RQ = -\frac{\partial H}{\partial x}. \tag{8.3}$$

In (8.2) we suppose that the domain of integration U is well adapted to the specific expression.

Conversely, given a first integral H of system (8.1) we always can find an integrating factor R for which (8.3) holds.

Proposition 8.1 *If the $\mathbb{F}$-polynomial system (8.1) has two integrating factors R_1 and R_2 on the open subset U of $\mathbb{F}^2$, then in the open set $U \setminus \{R_2 = 0\}$ the function R_1/R_2 is a first integral, provided R_1/R_2 is non-constant.*

Proof. Since R_i is an integrating factor, it satisfies $X R_i = -R_i \mathrm{div}(P, Q)$ for $i = 1, 2$. Therefore the proposition follows immediately from the computation

$$X\left(\frac{R_1}{R_2}\right) = \frac{(X R_1) R_2 - R_1 (X R_2)}{R_2^2} = 0. \qquad \square$$

8.4 Invariant Algebraic Curves

Let $f \in \mathbb{C}[x, y]$, f not identically zero. The algebraic curve $f(x, y) = 0$ is an *invariant algebraic curve* of the $\mathbb{F}$-polynomial system (8.1) if for some polynomial $K \in \mathbb{C}[x, y]$ we have

$$Xf = P\frac{\partial f}{\partial x} + Q\frac{\partial f}{\partial y} = Kf. \tag{8.4}$$

The polynomial K is called the *cofactor* of the invariant algebraic curve $f = 0$. We note that, since the polynomial system has degree m, any cofactor has degree at most $m - 1$.

On the points of the algebraic curve $f = 0$ the gradient $(\partial f/\partial x, \partial f/\partial y)$ of f is orthogonal to the vector field $X = (P, Q)$ (see (8.4)). Hence at every point of $f = 0$ the vector field X is tangent to the curve $f = 0$, so the curve $f = 0$ is formed by trajectories of the vector field X. This justifies the name "invariant algebraic curve" since it is invariant under the flow defined by X.

We remark that in the definition of invariant algebraic curve $f = 0$ we always allow this curve to be complex; that is $f \in \mathbb{C}[x, y]$ even in the case of a real polynomial system. As we will see this is due to the fact that sometimes for real polynomial systems the existence of a real first integral can be forced by the existence of complex invariant algebraic curves. Of course when we look for a complex invariant algebraic curve of a real polynomial system we are thinking of the real polynomial system as a complex one.

In the next proposition and throughout the whole chapter (complex) conjugation stands for conjugation of the coefficients of the polynomials only. This amounts of course to generic conjugation if we restrict the variables to $(x, y) \in \mathbb{R}^2$

Proposition 8.2 *For a real polynomial system (8.1), $f = 0$ is an invariant algebraic curve with cofactor K if and only if $\bar{f} = 0$ is an invariant algebraic curve with cofactor $\bar{K}$.*

Proof. We assume that $f = 0$ is an invariant algebraic curve with cofactor K of the real polynomial system (8.1). Then equality (8.4) holds. Since P and Q are real polynomials conjugating equality (8.4) we obtain

$$P\frac{\partial \bar{f}}{\partial x} + Q\frac{\partial \bar{f}}{\partial y} = \bar{K}\bar{f}.$$

Consequently, $\bar{f} = 0$ is an invariant algebraic curve with cofactor $\bar{K}$ of system (8.1). The proof in the converse is similar.   □

Lemma 8.3 *Let $f, g \in \mathbb{C}[x, y]$. We assume that f and g are relatively prime in the ring $\mathbb{C}[x, y]$. Then for a polynomial system (8.1), $fg = 0$ is an invariant algebraic curve with cofactor K_{fg} if and only if $f = 0$ and $g = 0$ are invariant algebraic curves with cofactors K_f and K_g, respectively. Moreover, $K_{fg} = K_f + K_g$.*

Proof. It is clear that

$$X(fg) = (Xf)g + f(Xg). \tag{8.5}$$

We assume that $fg = 0$ is an invariant algebraic curve of system (8.1) with cofactor K_{fg}. Then $X(fg) = K_{fg}fg$ and from equality (8.5) we get $K_{fg}fg = (Xf)g + fXg$. Therefore, since f and g are relatively prime, we obtain that f divides Xf, and g divides Xg. If we denote by K_f the quotient Xf/f and by K_g the quotient Xg/g, then $f = 0$ and $g = 0$ are invariant algebraic curves of system (8.1) with cofactors K_f and K_g, respectively, and $K_{fg} = K_f + K_g$.

The proof of the converse follows in a similar way, again using equality (8.5).   □

Proposition 8.4 *Suppose $f \in \mathbb{C}[x, y]$ and let $f = f_1^{n_1} \ldots f_r^{n_r}$ be its factorization into irreducible factors over $\mathbb{C}[x, y]$. Then for a polynomial system (8.1), $f = 0$ is an invariant algebraic curve with cofactor K_f if and only if $f_i = 0$ is an invariant algebraic curve for each $i = 1, \ldots, r$ with cofactor K_{f_i}. Moreover $K_f = n_1 K_{f_1} + \ldots + n_r K_{f_r}$.*

Proof. From Lemma 8.3, we know that $f = 0$ is an invariant algebraic curve with cofactor K_f if and only if $f_i^{n_i} = 0$ is an invariant algebraic curve for each $i = 1, \ldots, r$ with cofactor $K_{f_i^{n_i}}$; furthermore $K_f = K_{f_1^{n_1}} + \ldots + K_{f_r^{n_r}}$.

Now to prove the proposition it is sufficient to show, for each $i = 1, \ldots, r$, that $f_i^{n_i} = 0$ is an invariant algebraic curve with cofactor $K_{f_i^{n_i}}$ if and only if $f_i = 0$ is an invariant algebraic curve with cofactor K_{f_i}, and that $K_{f_i^{n_i}} = n_i K_{f_i}$. We assume that $f_i^{n_i} = 0$ is an invariant algebraic curve with cofactor $K_{f_i^{n_i}}$. Then

$$K_{f_i^{n_i}} f_i^{n_i} = X\left(f_i^{n_i}\right) = n_i f_i^{n_i-1} X(f_i),$$

or equivalently

$$X(f_i) = \frac{1}{n_i} K_{f_i^{n_i}} f_i.$$

So defining $K_{f_i} = K_{f_i^{n_i}}/n_i$ we obtain that $f_i = 0$ is an invariant algebraic curve with cofactor K_{f_i} such that $K_{f_i^{n_i}} = n_i K_{f_i}$. The proof of the converse follows in a similar way.   □

An *irreducible invariant algebraic curve* $f = 0$ will be an invariant algebraic curve such that f is an irreducible polynomial in the ring $\mathbb{C}[x, y]$.

8.5 Exponential Factors

There is another object, the so-called exponential factor, that plays the same role as the invariant algebraic curves in obtaining a first integral of the polynomial system (8.1). Before defining it formally, we explain how the notation arises naturally. Suppose we have invariant algebraic curves $h_\varepsilon = h + \varepsilon g + O(\varepsilon^2) = 0$ with cofactors K_{h_ε} for $\varepsilon \in [0, \varepsilon_0]$ with ε_0 sufficiently small. Using the fact that $X(h_\varepsilon) = K_{h_\varepsilon} h_\varepsilon$, if we expand the cofactor K_{h_ε} as a power series in ε we obtain that $K_{h_\varepsilon} = K_h + \varepsilon K + O(\varepsilon^2)$, where K is some polynomial of degree at most $m - 1$. We can now make a local study near a point where h is not zero.

Since

$$
\begin{aligned}
X\left(\frac{h_\varepsilon}{h}\right) &= \frac{X(h_\varepsilon)h - (h_\varepsilon)Xh}{h^2} \\
&= \frac{K_{h_\varepsilon}(h_\varepsilon)h - (h_\varepsilon)K_h h}{h^2} \\
&= \frac{(K_h + \varepsilon K + O(\varepsilon^2))(h + \varepsilon g + O(\varepsilon^2))h - (h + \varepsilon g + O(\varepsilon^2))K_h h}{h^2} \\
&= \varepsilon K + O(\varepsilon^2),
\end{aligned}
$$

we have

$$
\begin{aligned}
X\left(\left(\frac{h_\varepsilon}{h}\right)^{\frac{1}{\varepsilon}}\right) &= \frac{1}{\varepsilon}\left(\frac{h_\varepsilon}{h}\right)^{\frac{1}{\varepsilon}}\left(\frac{h_\varepsilon}{h}\right)^{-1} X\left(\frac{h_\varepsilon}{h}\right) \\
&= \frac{1}{\varepsilon}\left(\frac{h_\varepsilon}{h}\right)^{\frac{1}{\varepsilon}}(1 + O(\varepsilon))\left(\varepsilon K + O(\varepsilon^2)\right) \qquad (8.6) \\
&= (K + O(\varepsilon))\left(\frac{h_\varepsilon}{h}\right)^{\frac{1}{\varepsilon}}.
\end{aligned}
$$

Therefore the function

$$\left(\frac{h + \varepsilon g + O(\varepsilon^2)}{h}\right)^{\frac{1}{\varepsilon}}$$

has cofactor $K + O(\varepsilon)$. As ε tends to zero, the expression above tends to

$$\exp\left(\frac{g}{h}\right), \qquad (8.7)$$

and from (8.6) we obtain that

$$X\left(\exp\left(\frac{g}{h}\right)\right) = K \exp\left(\frac{g}{h}\right). \qquad (8.8)$$

Therefore, function (8.7) satisfies the same equation (8.4) as do the invariant algebraic curves, with a cofactor of degree at most $m - 1$.

Let $h, g \in \mathbb{C}[x, y]$ and assume that h and g are relatively prime in the ring $\mathbb{C}[x, y]$ or that $h \equiv 1$. Then the function $\exp(g/h)$ is called an *exponential factor* of the $\mathbb{F}$-polynomial system (8.1) if for some polynomial $K \in \mathbb{C}[x, y]$ of degree at most $m - 1$ it satisfies equation (8.8). As before we say that K is the *cofactor* of the exponential factor $\exp(g/h)$.

As we will see, from the point of view of the integrability of polynomial systems (8.1) the importance of the exponential factors is twofold. On the one hand, they satisfy equation (8.8), and on the other hand, their cofactors are polynomials of degree at most $m - 1$. These two facts mean that they play the same role as the invariant algebraic curves in the integrability of a polynomial system (8.1). We note that the exponential factors do not define invariant curves for the flow of system (8.1), because they are never zero.

We remark that in the definition of exponential factor $\exp(g/h)$ we always allow that this function be complex; that is $h, g \in \mathbb{C}[x, y]$ even in the case of a real polynomial system. The reason is the same as in the case of invariant algebraic curves. That is, sometimes for real polynomial systems the existence of a real first integral can be forced by the existence of complex exponential factors. Again, in looking for a complex exponential factor of a real polynomial system, we consider the real polynomial system as being on $\mathbb{C}^2$.

Proposition 8.5 *For a real polynomial system (8.1) the function $\exp(g/h)$ is an exponential factor with cofactor K if and only if the function $\exp(\bar{g}/\bar{h})$ is an exponential factor with cofactor $\bar{K}$.*

Proof. We assume that $\exp(g/h)$ is an exponential factor of the real polynomial system (8.1) with cofactor K. Then equality (8.8) holds. Since P and Q are real polynomials, conjugating equality (8.8) we obtain that

$$P\frac{\partial \exp(\bar{g}/\bar{h})}{\partial x} + Q\frac{\partial \exp(\bar{g}/\bar{h})}{\partial y} = \bar{K} \exp(\bar{g}/\bar{h}).$$

Consequently, $\exp(\bar{g}/\bar{h})$ is an exponential factor of system (8.1) with cofactor $\bar{K}$. The proof of the converse is similar. $\qquad \square$

Proposition 8.6 *If $F = \exp(g/h)$ is an exponential factor for the polynomial system (8.1), then $h = 0$ is an invariant algebraic curve, and g satisfies the equation*

$$Xg = gK_h + hK_F,$$

where K_h and K_F are the cofactors of h and F, respectively.

Proof. Since $F = \exp(g/h)$ is an exponential factor with cofactor K_F, we have

$$K_F \exp\left(\frac{g}{h}\right) = X\left(\exp\left(\frac{g}{h}\right)\right)$$

$$= \exp\left(\frac{g}{h}\right) X\left(\frac{g}{h}\right) = \exp\left(\frac{g}{h}\right) \frac{(Xg)h - g(Xh)}{h^2},$$

or equivalently

$$(Xg)h - g(Xh) = h^2 K_F.$$

Hence since h and g are relatively prime, we obtain that h divides Xh. So $h = 0$ is an invariant algebraic curve with cofactor $K_h = Xh/h$. Now replacing Xh by $K_h h$ in the last equality, we have that $Xg = gK_h + hK_F$. $\qquad\square$

In fact, the way that the exponential factor $\exp(g/h)$ has been introduced at the beginning of this section implies that it appears when the invariant algebraic curve $h = 0$ has geometric multiplicity higher than 1.

8.6 The Method of Darboux

Before stating the main results of the Darboux theory we need some definitions. If $S(x, y) = \sum_{i+j=0}^{m-1} a_{ij} x^i y^j$ is a polynomial of degree at most $m - 1$ with $m(m + 1)/2$ coefficients in $\mathbb{F}$, then we write $S \in \mathbb{F}_{m-1}[x, y]$. We identify the linear vector space $\mathbb{F}_{m-1}[x, y]$ with $\mathbb{F}^{m(m+1)/2}$ through the isomorphism

$$S \rightarrow (a_{00}, a_{10}, a_{01}, \ldots, a_{m-1,0}, a_{m-2,1}, \ldots, a_{0,m-1}).$$

We say that r points $(x_k, y_k) \in \mathbb{F}^2$, $k = 1, \ldots, r$, are *independent* with respect to $\mathbb{F}_{m-1}[x, y]$ if the intersection of the r hyperplanes

$$\sum_{i+j=0}^{m-1} x_k^i y_k^j a_{ij} = 0, \qquad k = 1, \ldots, r,$$

in $\mathbb{F}^{m(m+1)/2}$ is a linear subspace of dimension $[m(m + 1)/2] - r$.

We remark that the maximum number of isolated singular points of the polynomial system (8.1) is m^2 (by Bézout's Theorem), that the maximum number of independent isolated singular points of the system is $m(m + 1)/2$, and that $m(m + 1)/2 < m^2$ for $m \geq 2$.

A singular point (x_0, y_0) of system (8.1) is called *weak* if the divergence, $\text{div}(P, Q)$, of system (8.1) at (x_0, y_0) is zero.

Theorem 8.7 *[Darboux Theory of integrability for complex polynomial systems] Suppose that a $\mathbb{C}$-polynomial system (8.1) of degree m admits p irreducible invariant algebraic curves $f_i = 0$ with cofactors K_i for $i = 1, \ldots, p$, q exponential factors $\exp(g_j/h_j)$ with cofactors L_j for $j = 1, \ldots, q$, and r independent singular points $(x_k, y_k) \in \mathbb{C}^2$ such that $f_i(x_k, y_k) \neq 0$ for $i = 1, \ldots, p$ and for $k = 1, \ldots, r$.*

(i)   There exist $\lambda_i, \mu_j \in \mathbb{C}$ not all zero such that $\sum_{i=1}^{p} \lambda_i K_i + \sum_{j=1}^{q} \mu_j L_j = 0$, if and only if the (multivalued) function

$$f_1^{\lambda_1} \ldots f_p^{\lambda_p} \left(\exp \left(\frac{g_1}{h_1} \right) \right)^{\mu_1} \ldots \left(\exp \left(\frac{g_q}{h_q} \right) \right)^{\mu_q} \tag{8.9}$$

is a first integral of system (8.1).

(ii) If $p + q + r \geq [m(m+1)/2] + 1$, then there exist $\lambda_i, \mu_j \in \mathbb{C}$ not all zero such that $\sum_{i=1}^{p} \lambda_i K_i + \sum_{j=1}^{q} \mu_j L_j = 0$.

(iii) If $p + q + r \geq [m(m+1)/2] + 2$, then system (8.1) has a rational first integral, and consequently all trajectories of the system are contained in invariant algebraic curves.

(iv) There exist $\lambda_i, \mu_j \in \mathbb{C}$ not all zero such that $\sum_{i=1}^{p} \lambda_i K_i + \sum_{j=1}^{q} \mu_j L_j = -\operatorname{div}(P, Q)$, if and only if function (8.9) is an integrating factor of system (8.1).

(v)   If $p + q + r = m(m+1)/2$ and the r independent singular points are weak, then function (8.9) is a first integral if $\sum_{i=1}^{p} \lambda_i K_i + \sum_{j=1}^{q} \mu_j L_j = 0$, or an integrating factor if $\sum_{i=1}^{p} \lambda_i K_i + \sum_{j=1}^{q} \mu_j L_j = -\operatorname{div}(P, Q)$, under the condition that not all $\lambda_i, \mu_j \in \mathbb{C}$ are zero.

(vi) If there exist $\lambda_i, \mu_j \in \mathbb{C}$ not all zero such that $\sum_{i=1}^{p} \lambda_i K_i + \sum_{j=1}^{q} \mu_j L_j = -s$ for some $s \in \mathbb{C} \setminus \{0\}$, then the (multivalued) function

$$f_1^{\lambda_1} \ldots f_p^{\lambda_p} \left(\exp \left(\frac{g_1}{h_1} \right) \right)^{\mu_1} \ldots \left(\exp \left(\frac{g_q}{h_q} \right) \right)^{\mu_q} \exp(st)$$

is an invariant of system (8.1).

Of course, each irreducible factors of each h_j is one of the f_i's.

Proof of Theorem 8.7: We write $F_j = \exp(g_j/h_j)$. By hypothesis we have p invariant algebraic curves $f_i = 0$ with cofactors K_i, and q exponential factors F_j with cofactors L_j. That is, the f_is satisfy $X f_i = K_i f_i$, and the F_js satisfy $X F_j = L_j F_j$.

(i) Clearly statement (i) follows from the fact that

$$X \left(f_1^{\lambda_1} \ldots f_p^{\lambda_p} F_1^{\mu_1} \ldots F_q^{\mu_q} \right) =$$

$$\left(f_1^{\lambda_1} \ldots f_p^{\lambda_p} F_1^{\mu_1} \ldots F_q^{\mu_q} \right) \left(\sum_{i=1}^{p} \lambda_i \frac{X f_i}{f_i} + \sum_{j=1}^{q} \mu_j \frac{X F_j}{F_j} \right) =$$

$$\left(f_1^{\lambda_1} \ldots f_p^{\lambda_p} F_1^{\mu_1} \ldots F_q^{\mu_q} \right), \left(\sum_{i=1}^{p} \lambda_i K_i + \sum_{j=1}^{q} \mu_j L_j \right) = 0.$$

(ii) Since the cofactors K_i and L_j are polynomials of degree $m - 1$, we have that $K_i, L_j \in \mathbb{C}_{m-1}[x, y]$. We note that the dimension of $\mathbb{C}_{m-1}[x, y]$ as a vector space over $\mathbb{C}$ is $m(m+1)/2$.

Since (x_k, y_k) is a singular point of system (8.1), $P(x_k, y_k) = Q(x_k, y_k) = 0$. Then from $Df_i = P(\partial f_i/\partial x) + Q(\partial f_i/\partial y) = K_i f_i$, it follows that $K_i(x_k, y_k) f_i(x_k, y_k) = 0$. As we have assumed that $f_i(x_k, y_k) \neq 0$, therefore $K_i(x_k, y_k) = 0$ for $i = 1, \ldots, p$. Again, from $DF_j = P(\partial F_j/\partial x) + Q(\partial F_j/\partial y) = L_j F_j$, it follows that $L_j(x_k, y_k) F_j(x_k, y_k) = 0$. Since $F_j = \exp(g_j/h_j)$ does not vanish, $L_j(x_k, y_k) = 0$ for $j = 1, \ldots, q$. Consequently, since the r singular points are independent, all the polynomials K_i and L_j belong to a linear subspace S of $\mathbb{C}_{m-1}[x, y]$ of dimension $[m(m+1)/2] - r$. We have $p+q$ polynomials K_i and L_j and since from the assumptions $p+q > [m(m+1)/2] - r$, we obtain that the $p + q$ polynomials must be linearly dependent in S. So there are $\lambda_i, \mu_j \in \mathbb{C}$ not all zero such that $\sum_{i=1}^{p} \lambda_i K_i + \sum_{j=1}^{q} \mu_j L_j = 0$. Hence statement (ii) is proved.

(iii) Since the number of independent singular points $r \leq m(m + 1)/2$, it follows that $p + q \geq 2$. Under the assumptions of statement (iii) we apply statement (ii) to two subsets of $p + q - 1 > 0$ functions defining invariant algebraic curves or exponential factors. Thus we get two linear dependencies between the corresponding cofactors, which after some linear algebra and relabeling, we can write in the following form

$$M_1 + \alpha_3 M_3 + \ldots + \alpha_{p+q-1} M_{p+q-1} = 0, \quad M_2 + \beta_3 M_3 + \ldots + \beta_{p+q-1} M_{p+q-1} = 0,$$

where M_l are the cofactors K_i and L_j, and the α_l and β_l are complex numbers. Then by statement (i), it follows that the two functions

$$G_1 G_3^{\alpha_3} \ldots G_{p+q-1}^{\alpha_{p+q-1}}, \quad G_2 G_3^{\beta_3} \ldots G_{p+q-1}^{\beta_{p+q-1}},$$

are first integrals of system (8.1), where G_l is the polynomial defining an invariant algebraic curve or the exponential factor having cofactor M_l for $l = 1, \ldots, p + q - 1$. Then taking logarithms of the two first integrals above, we obtain that

$$H_1 = \log(G_1) + \alpha_3 \log(G_3) + \ldots + \alpha_{p+q-1} \log(G_{p+q-1}),$$
$$H_2 = \log(G_2) + \beta_3 \log(G_3) + \ldots + \beta_{p+q-1} \log(G_{p+q-1}),$$

are first integrals of system (8.1) on their domain of definition. Each provides an integrating factor R_i such that

$$P = R_i \frac{\partial H_i}{\partial y}, \quad Q = -R_i \frac{\partial H_i}{\partial x}.$$

Therefore, we obtain that

$$\frac{R_1}{R_2} = \frac{\partial H_2}{\partial x} \Big/ \frac{\partial H_1}{\partial x}.$$

Since the functions G_l are polynomials or exponentials of a quotient of polynomials, it follows that the functions $\partial H_i/\partial x$ are rational for $i = 1, 2$. So from the last equality, we get that the quotient of the two integrating factors R_1/R_2 is a rational function. Proposition 8.1 implies statement (iii).

(iv) Since the equality $\sum_{i=1}^{p} \lambda_i K_i + \sum_{j=1}^{q} \mu_j L_j = -\operatorname{div}(P,Q)$ is equivalent to the equality

$$X \left(f_1^{\lambda_1} \ldots f_p^{\lambda_p} F_1^{\mu_1} \ldots F_q^{\mu_q} \right) =$$

$$\left(f_1^{\lambda_1} \ldots f_p^{\lambda_p} F_1^{\mu_1} \ldots F_q^{\mu_q} \right) \left(\sum_{i=1}^{p} \lambda_i K_i + \sum_{j=1}^{q} \mu_j L_j \right) =$$

$$- \left(f_1^{\lambda_1} \ldots f_p^{\lambda_p} F_1^{\mu_1} \ldots F_q^{\mu_q} \right) \operatorname{div}(P,Q),$$

statement (iv) follows.

(v) Let $K = \operatorname{div}(P,Q)$; clearly $K \in \mathbb{C}_{m-1}[x,y]$. By assumption the r singular points (x_k, y_k) are weak, therefore $K(x_k, y_k) = 0$ for $k = 1, \ldots, r$. So K belongs to the linear subspace S of the proof of statement (ii).

On the other hand, since $\dim S = p + q = [m(m+1)/2] - r \geq 0$ and we have $p + q + 1$ polynomials $K_1, \ldots, K_p, L_1, \ldots, L_q, K$ in S (we are using the same arguments as in the proof of statement (ii)), it follows that these polynomials are linearly dependent in S. Therefore, we obtain $\lambda_i, \mu_j, \alpha \in \mathbb{C}$ not all zero such that

$$\left(\sum_{i=1}^{p} \lambda_i K_i \right) + \left(\sum_{j=1}^{q} \mu_j L_j \right) + \alpha K = 0. \tag{8.10}$$

If $\alpha = 0$ then as in the proof of statement (i), we obtain that function (8.9) is a first integral of system (8.1).

We assume now that $\alpha \neq 0$. Dividing the equality (8.10) by α (if necessary), we can assume without loss of generality that $\alpha = 1$. So we have that

$$K = - \left(\sum_{i=1}^{p} \lambda_i K_i \right) - \left(\sum_{j=1}^{q} \mu_j L_j \right).$$

Therefore, statement (v) follows from:

$$X \left(f_1^{\lambda_1} \ldots f_p^{\lambda_p} F_1^{\mu_1} \ldots F_q^{\mu_q} \right) =$$

$$\left(f_1^{\lambda_1} \ldots f_p^{\lambda_p} F_1^{\mu_1} \ldots F_q^{\mu_q} \right) \left(\sum_{i=1}^{p} \lambda_i K_i + \sum_{j=1}^{q} \mu_j L_j \right) =$$

$$- \left(f_1^{\lambda_1} \ldots f_p^{\lambda_p} F_1^{\mu_1} \ldots F_q^{\mu_q} \right) K =$$

$$- \left(f_1^{\lambda_1} \ldots f_p^{\lambda_p} F_1^{\mu_1} \ldots F_q^{\mu_q} \right) \operatorname{div}(P,Q).$$

(vi) We have $\lambda_i, \mu_j \in \mathbb{C}$ not all zero such that $\sum_{i=1}^{p} \lambda_i K_i + \sum_{j=1}^{q} \mu_j L_j = -s$. Then from

$$X\left(f_1^{\lambda_1} \ldots f_p^{\lambda_p} F_1^{\mu_1} \ldots F_q^{\mu_q} e^{st}\right) =$$

$$\left(f_1^{\lambda_1} \ldots f_p^{\lambda_p} F_1^{\mu_1} \ldots F_q^{\mu_q} e^{st}\right) \left(\sum_{i=1}^{p} \lambda_i K_i + \sum_{j=1}^{q} \mu_j L_j + s\right) = 0,$$

statement *(iv)* follows.   □

A (multivalued) function of the form (8.9) is called a *Darbouxian function*.

Now we shall see that if the polynomial differential system is real, then the first integral provided by the Darboux theory of integrability is also real. This follows from the following fact. Since the polynomial differential system (8.1) is real, it is well known that if a complex invariant algebraic curve or exponential factor appears, then its conjugate must appear simultaneously (see Propositions 8.2 and 8.5). If among the invariant algebraic curves of the real system (8.1) a complex conjugate pair $f = 0$ and $\overline{f} = 0$ occurs, the function (8.9) has a real factor of the form $f^{\lambda} \overline{f}^{\overline{\lambda}}$, which is the multivalued real function

$$\left[(\operatorname{Re} f)^2 + (\operatorname{Im} f)^2\right]^{\operatorname{Re} \lambda} \exp\left(-2 \operatorname{Im} \lambda \, \arg\left(\operatorname{Re} f + i \operatorname{Im} f\right)\right),$$

if $\operatorname{Im} \lambda \operatorname{Im} f \not\equiv 0$. If among the exponential factors of the real system (8.1) a complex conjugate pair $F = \exp(h/g)$ and $\overline{F} = \exp(\overline{h}/\overline{g})$ occurs, the first integral (8.9) has a real factor of the form

$$\left(\exp\left(\frac{h}{g}\right)\right)^{\mu} \left(\exp\left(\frac{\overline{h}}{\overline{g}}\right)\right)^{\overline{\mu}} = \exp\left(2 \operatorname{Re}\left(\mu \frac{h}{g}\right)\right).$$

In short, the function (8.9) is real when the polynomial differential system (8.1) is real.

8.7 Some Applications of the Darboux Theory

In what follows we present applications of each statement of Darboux's theorem to quadratic systems. For quadratic systems $m(m+1)/2 = 3$

Example 8.8 If $a \neq 0$ the quadratic system

$$\dot{x} = -y(ay + b) - (x^2 + y^2 - 1), \qquad \dot{y} = x(ay + b), \qquad (8.11)$$

has the algebraic solutions $f_1 = ay + b = 0$ with cofactor $K_1 = ax$, and $f_2 = x^2 + y^2 - 1 = 0$ with cofactor $K_2 = -2x$. Since $2K_1 + aK_2 = 0$, by Theorem 8.7*(i)* we have that $H = (ay + b)^2 (x^2 + y^2 - 1)^a$ is a first integral of system (8.11).   □

Example 8.9 If $abc \neq 0$ then the real quadratic system

$$\dot{x} = x(ax + c), \qquad \dot{y} = y(2ax + by + c), \qquad (8.12)$$

has exactly the following five invariant straight lines (i.e. algebraic solutions of degree 1): $f_1 = x = 0$, $f_2 = ax + c = 0$, $f_3 = y = 0$, $f_4 = ax + by = 0$, $f_5 = ax + by + c = 0$. Then by Theorem 8.7(ii) we know that system (8.5) must have a first integral of the form $H = f_1^{\lambda_1} f_2^{\lambda_2} f_3^{\lambda_3} f_4^{\lambda_4} f_5^{\lambda_5}$ with $\lambda_i \in \mathbb{F}$ satisfying $\sum_{i=1}^{5} \lambda_i K_i = 0$, where K_i is the cofactor of f_i. An easy computation shows that $K_1 = ax + c$, $K_2 = ax$, $K_3 = 2ax + by + c$, $K_4 = ax + by + c$ and $K_5 = ax + by$. Then a solution of $\sum_{i=1}^{5} \lambda_i K_i = 0$ is $\lambda_1 = \lambda_5 = -1$, $\lambda_2 = \lambda_4 = 1$ and $\lambda_3 = 0$. Therefore a first integral of system (8.12) is

$$H = \frac{(ax + c)(ax + by)}{x(ax + by + c)}.$$

We note that since this system has 5 invariant algebraic curves, by Theorem 8.7(iii) it must have a rational first integral, as we have found.   □

Example 8.10 The real quadratic system

$$\dot{x} = -y - b(x^2 + y^2) = P, \qquad \dot{y} = x = Q, \qquad (8.13)$$

has the invariant algebraic curve $f_1 = x^2 + y^2$ with cofactor $K_1 = -2bx$. Since $K_1 = \mathrm{div}(P, Q)$, from Theorem 8.7(iv), it follows that f_1^{-1} is an integrating factor. Then an easy computation shows that $H = \exp(2by)(x^2 + y^2)$ is a first integral of system (8.13).   □

Example 8.11 If $a_{02} \neq 0$ then the real quadratic system

$$\dot{x} = x^2 - 1 = P, \quad \dot{y} = a_{00} + a_{10}x + a_{01}y + a_{20}x^2 + a_{11}xy + a_{02}y^2 = Q,$$

with $a_{00} = (2a_{11} + a_{01}^2 - 1)/(4a_{02})$, $a_{10} = a_{01}a_{11}/(2a_{02})$ and $a_{20} = a_{11}(a_{11} - 2)/(4a_{02})$, has the following three algebraic solutions: two straight lines $f_1 = x + 1$, $f_2 = x - 1$, and one hyperbola

$$f_3 = \frac{a_{11}(a_{11} - 2)}{4a_{02}}x^2 + (a_{11} - 1)xy + a_{02}y^2 + \frac{a_{01}(a_{11} - 1)}{2a_{02}}x + a_{01}y + \frac{a_{01}^2 + 1}{4a_{02}} = 0.$$

Their cofactors are $K_1 = x - 1$, $K_2 = x + 1$ and $K_3 = (a_{11} + 1)x + 2a_{02}y + a_{01}$, respectively. Since $\sum_{i=1}^{3} \lambda_i K_i = -\mathrm{div}(P, Q)$ for $\lambda_1 = \lambda_2 = -1/2$, and $\lambda_3 = -1$, from Theorem 8.7(iv) it follows that $f_1^{\lambda_1} f_2^{\lambda_2} f_3^{\lambda_3}$ is a Darbouxian integrating factor. By computing its associated first integral we obtain

$$H = -2\,\mathrm{arctanh}\left[\frac{(a_{11} - 1)x + 2a_{02}y + a_{01}}{(x^2 - 1)^{1/2}}\right] - \ln\left|x + (x^2 - 1)^{1/2}\right|.$$

□

An interesting application of the Darboux theory of integrability allow us to present a shorter proof of the classification theorem of centers of quadratic systems. First we need some preliminary results.

Let L be a straight line, and let q be a point of L. We say that q is a *contact point* of the straight line L with a vector field X, if the vector $X(q)$ is parallel to L.

Lemma 8.12 *If X is a quadratic vector field and L is a straight line, then either L is invariant under X, or X has at most two contact points (including the singularities) along L.*

Proof. Let $X = P\partial/\partial x + Q\partial/\partial y$, and let $L = \{(x,y) : ax + by + c = 0\}$. Then the contact points of X and L must satisfy the system

$$aP(x,y) + bQ(x,y) = 0,$$
$$ax + by + c = 0.$$

From this system the lemma follows easily. □

Lemma 8.12 is used for proving the next result.

Proposition 8.13 *Let X be a quadratic system and let C be a closed orbit of X.*

(i) The interior of C is a convex set.
(ii) There is at most one singularity in the interior of C. This singularity has index 1, is non–degenerate and has complex conjugate eigenvalues.

Proof. Suppose that we have a segment with endpoints in the interior of C, but the segment is not completely contained in this interior. Then the straight line containing this segment will have at least three contact points, in contradiction to Lemma 8.12. Hence statement (i) is proved.

If C had at least two singularities in its interior, then the straight line through them would have at least three contact points with X, contradicting Lemma 8.12.

Without loss of generality we may assume that the unique singularity in the interior of C is at the origin. Its linear part cannot be zero, because a homogeneous quadratic system always has an invariant straight line through the origin.

Using the Poincaré–Hopf Theorem it follows immediately that the index of the unique singularity in the interior of C must be 1.

To prove that the linear part cannot be nilpotent, we use the Jordan normal form theorem and we write X as

$$\dot{x} = y + Bx^2 + Cxy + Dy^2, \quad \dot{y} = Ex^2 + Fxy + Gy^2.$$

Along $\{y = 0\}$ the y-component of this vector field is given by $\dot{y} = Ex^2$. Clearly this prevents the existence of closed orbits around the origin.

Since the index of the unique singularity in the interior of C is 1, and this singularity is elementary, if the eigenvalues of $DX(0,0)$ were not complex conjugate, we would be able to write X as

$$\dot{x} = \lambda x + Bx^2 + Cxy + Dy^2, \quad \dot{y} = \delta x + \mu y + Ex^2 + Fxy + Gy^2,$$

with $\delta = 0$, unless $\lambda = \mu$ in which case $\delta = 1$. Along $\{x = 0\}$ the x–component of this vector field is given by $\dot{x} = Dy^2$, which again prevents the existence of a closed orbit around the origin.   □

Proposition 8.13 allows us to prove the following lemma, which will be used for the characterization of quadratic centers.

Lemma 8.14 *If a quadratic vector field X has a center, then up to a translation, a linear transformation and a time rescaling, it can be written as*

$$\dot{x} = -y - bx^2 - Cxy - dy^2, \quad \dot{y} = x + ax^2 + Axy - ay^2. \tag{8.14}$$

Proof. We position the center at the origin. Since it is a center, by Proposition 8.13, we know that X can be written as

$$\dot{x} = -y + \alpha x^2 + \beta xy + \gamma y^2, \quad \dot{y} = x + \overline{\alpha} x^2 + \overline{\beta} xy + \overline{\gamma} y^2,$$

If $\overline{\alpha} + \overline{\gamma} = 0$, then the lemma follows. So we suppose that $\overline{\alpha} + \overline{\gamma} \neq 0$ and show that by a well chosen rotation

$$\begin{pmatrix} x \\ y \end{pmatrix} = \begin{pmatrix} \cos\theta & -\sin\theta \\ \sin\theta & \cos\theta \end{pmatrix} \begin{pmatrix} X \\ Y \end{pmatrix},$$

we obtain expression (8.14). We write $c = \cos\theta$, $s = \sin\theta$ and make a straightforward computation of the X^2 and Y^2 terms in the expression of $\dot{Y}$. The equality of these terms with opposite signs is given by the equation

$$(\overline{\alpha} + \overline{\gamma})c^3 - (\alpha + \gamma)c^2 s + (\overline{\alpha} + \overline{\gamma})cs^2 - (\alpha + \gamma)s^3 = 0.$$

Since $c^2 + s^2 = 1$, this is equivalent to $(\overline{\alpha} + \overline{\gamma})c - (\alpha + \gamma)s = 0$, which clearly provides two solutions for θ.   □

In the next theorem we characterize the quadratic centers.

Theorem 8.15 (Kapteyn–Bautin Theorem) *A quadratic system that has a center at the origin can be written in the form*

$$\dot{x} = -y - bx^2 - Cxy - dy^2, \qquad \dot{y} = x + ax^2 + Axy - ay^2. \tag{8.15}$$

This system has a center at the origin if and only if at least one of the following conditions holds

(i) $A - 2b = C + 2a = 0$,

(ii) $C = a = 0$,
(iii) $b + d = 0$,
(iv) $C + 2a = A + 3b + 5d = a^2 + bd + 2d^2 = 0$.

Proof. By Lemma 8.14 a quadratic system can have a center only if it can be placed in the form (8.15).

Using the algorithm for computing the Lyapunov constants described at the end of Chap. 4, we can compute them for the origin of system (8.15). We obtain

$$V_3 = c_3(2a + C)(b + d),$$
$$V_5 = c_5 a(A - 2b)(b + d)(A + 3b + 5d),$$
$$V_7 = c_7 a(A - 2b)(b + d)^2(a^2 + bd + 2d^2),$$

where c_3, c_5, and c_7 are nonzero real numbers. From Chap. 4 we know that it is necessary that $V_3 = V_5 = V_7 = 0$ in order that the origin of system (8.15) be a center. This shows that the conditions *(i)*–*(iv)* are necessary. Now we shall prove that they are sufficient.

Since system (8.15) has a linear center at the origin, to prove that system (8.15) satisfying one of the four conditions of the Kapteyn–Bautin Theorem has a center at the origin, it is sufficient to show that it has a first integral in a neighborhood of the origin.

Assume that system (8.15) satisfies condition *(i)*. Then it is easy to check that the system is Hamiltonian, i.e., $\dot{x} = -\partial H/\partial y$, $\dot{y} = \partial H/\partial x$ with $H = \frac{1}{2}(x^2 + y^2) + \frac{a}{3}x^3 + bx^2 y - axy^2 + \frac{d}{3}y^3$. Therefore H is a first integral defined in a neighborhood of the origin.

Suppose that system (8.15) satisfies condition *(ii)*. Then the system can be written in the form

$$\dot{x} = -y - bx^2 - dy^2, \qquad \dot{y} = x + Axy.$$

Since it is invariant under $(x, y, t) \rightarrow (-x, y, t)$, hence time-reversible, the origin necessarily is a center.

Assume that system (8.15) satisfies condition *(iii)*. The form of system (8.15) with $b + d = 0$ is preserved under a rotation about the origin. After performing a rotation of angle θ, the new coefficient a' of x^2 in the second equation of system (8.15) is of the form $a' = a\cos^3\theta + \alpha\cos^2\theta\sin\theta + \beta\cos\theta\sin^2\theta + d\sin^3\theta$. Therefore, if $a \neq 0$ we can find θ such that $a' = 0$. So we can assume that $a = 0$, and consequently $C \neq 0$; otherwise we would be under the assumptions of condition *(ii)*.

The system $\dot{x} = -y - bx^2 - Cxy + by^2$, $\dot{y} = x + Axy$, has the algebraic solutions $f_1 = 1 + Ay = 0$ if $A \neq 0$ with cofactor $K_1 = Ax$, and $f_2 = (1 - by)^2 + C(1 - by)x - b(A + b)x^2 = 0$ with cofactor $K_2 = -2bx - Cy$. Since the divergence of the system is equal to $K_1 + K_2$, by Darboux's Theorem *(iii)* we obtain that $f_1^{-1}f_2^{-1}$ is an integrating factor. Hence again the first integral associated to the integrating factor is defined at the origin, and consequently the origin is a center.

We remark that if $A = 0$ then f_1 is not an algebraic solution of the system, but then the divergence of the system is equal to K_2 and the integrating factor of the system is f_2^{-1}; using the same arguments we obtain that the origin is a center.

Suppose that system (8.15) satisfies condition *(iv)*. Then if $d \neq 0$ the system becomes

$$\dot{x} = -y + \frac{a^2 + 2d^2}{d}x^2 + 2axy - dy^2, \qquad \dot{y} = x + ax^2 + \frac{3a^2 + d^2}{d}xy - ay^2.$$

We note that if $d = 0$ then we are under the assumptions of condition *(ii)*, so we take $d \neq 0$. The system has the algebraic solution $f_1 = (a^2 + d^2)\left[(dy - ax)^2 + 2dy\right] + d^2 = 0$ with cofactor $K_1 = 2(a^2 + d^2)x/d$. Therefore the divergence of the system is equal to $\frac{5}{2}K_1$. Hence by Darboux's Theorem *(iii)* the function $f_1^{-5/2}$ is an integrating factor of the system. Since $d \neq 0$, its associated first integral is defined in a neighborhood of the origin, and consequently the origin is a center.   □

8.8 Prelle–Singer and Singer Results

In this last section we want to mention the excellent results of Prelle–Singer and Singer related with the Darboux method.

Roughly speaking an *elementary first integral* is a first integral expressible in terms of exponentials, logarithms and algebraic functions. The notion of elementary function of one variable is due to Liouville who, between 1833 and 1841, used it in the theory of integration. Elementary functions of two variables are defined by starting with the field of rational functions in two variables $\mathbb{C}(x, y)$ and using extension fields but with two commuting differentiations $\frac{\partial}{\partial x}$ and $\frac{\partial}{\partial y}$.

Theorem 8.16 *If a polynomial system has an elementary first integral, then it has an integrating factor of the form $f_1^{n_1} \ldots f_p^{n_p}$ with $f_i \in \mathbb{C}[x, y]$, $n_i \in \mathbb{Z}$, and each $f_i = 0$ is an invariant algebraic curve.*

We remark that this theorem says that if a polynomial system (8.1) has an elementary first integral, then this integral can be computed by using the invariant algebraic curves of the system.

Roughly speaking the *Liouvillian functions* are those functions which can be obtained "by quadratures" of elementary functions; see for instance Singer [150]. Of course the class of elementary functions is a subclass of the Liouvillian one.

Theorem 8.17 *If a polynomial system has a Liouvillian first integral, then the system has a Darbouxian integrating factor.*

Theorem 8.17 says that the method of Darboux finds all Liouvillian first integrals.

8.9 Exercises

Exercise 8.1 Show that the rational function

$$H = \frac{a_{00} + a_{10}x + a_{01}y + a_{20}x^2 + a_{11}xy + a_{02}y^2}{b_{00} + b_{10}x + b_{01}y + b_{20}x^2 + b_{11}xy + b_{02}y^2}$$

is a first integral of a quadratic polynomial differential system if and only if

$$a_{20}b_{11} - a_{11}b_{20} = a_{20}b_{02} - a_{02}b_{20} = a_{11}b_{02} - a_{02}b_{11} = 0.$$

Exercise 8.2 Consider the polynomial differential system

$$\dot{x} = x(-Bx + (C-1)y + 1),$$
$$\dot{y} = y((1-B)x - y + A).$$

Prove that

(i) $f = x - Cy + AC = 0$ is an invariant straight line if $ABC = -1$.
(ii) $f = A^2(Bx-1)^2 - 2A(Bx+1)y + y^2 = 0$ is an invariant conic if $ABC = 1$ and $A(1+C) = -1$.
(iii) For $A = -6$, $B = C = 1/2$ find an invariant algebraic curve of degree 3.
(iv) For $A = -10/3$, $B = 3$, $C = -7/10$ find an invariant algebraic curve of degree 6.

Exercise 8.3 Compute a first integral of the system

$$\dot{x} = cx,$$
$$\dot{y} = y(Ax + By + C).$$

Hint: Use the fact that $x = 0$ and $y = 0$ are invariant straight lines and that e^x is an exponential factor.

Exercise 8.4 Find an integrating factor for the system

$$\dot{x} = xy,$$
$$\dot{y} = y(x + By + 1).$$

Exercise 8.5 Find a first integral of the system

$$\dot{x} = x(ax + by + c),$$
$$\dot{y} = By^2.$$

using the fact that e^{-y} is an exponential factor.

Exercise 8.6 Show that the system

$$\dot{x} = 1,$$
$$\dot{y} = 2n + 2xy + y^2,$$

has the invariant algebraic curve

$$h(x, y) = H_n(x)y + 2nH_{n-1}(x) = 0,$$

where $H_n(x)$ is the Hermite polynomial of degree n, defined by

$$H_{n+1}(x) = 2xH_n(x) - 2nH_{n-1}(x),$$
$$H_0(x) = 1,$$
$$H_1(x) = 2x.$$

Hint: $H'_n(x) = 2nH_{n-1}(x)$.

8.10 Bibliographical Comments

In 1878 Darboux [46] showed how to construct the first integrals of polynomial systems possessing a sufficient number of invariant algebraic curves. In particular, he proved that if a polynomial system of degree m has at least $m(m + 1)/2$ invariant algebraic curves, then it has a first integral. The best improvements to Darboux's results in the context of planar polynomial systems are due to Jouanolou [92] in 1979, to Prelle and Singer [133] in 1983, and to Singer [150] in 1992. Jouanolou shows that if the number of invariant algebraic curves of a polynomial system of degree m is at least $[m(m + 1)/2] + 2$, then the system has a rational first integral, and consequently all its solutions are invariant algebraic curves. The proof presented here of statement (iii) of the Theorem on the Darboux Theory of integrability is due to Christopher and Llibre [39]. Prelle and Singer prove that if a polynomial system has an elementary first integral, then this integral can be computed by using the invariant algebraic curves of the system. Singer proves that if a polynomial system has a Liouvillian first integral, then it can be computed by using the invariant algebraic curves and the exponential factors of the system.

Recently there have appeared several partial expositions of the Darboux theory of integration for planar polynomial systems. One may find the results of Cairo, Christopher, Feix, Llibre, and Schlomiuk in [24, 25, 37–39, 141–144]. This chapter heavily uses the paper [39]. People interested in some applications of the Darboux theory that are not considered here should see [39].

In another context we must mention the excellent extensions of the Darboux method to dimension greater than 2 for differential polynomial systems on k^n, where k is a differential field of characteristic zero; see for instance the paper of Weil [163] and the references quoted there; or extensions to algebraic Pfaff equations done by Jouanolou [92]. For extensions to algebraic surfaces see the works of Gutierrez, Llibre, Rodríguez, and Zhang [79, 100, 101, 103].

The first part of the chapter is a survey on the Darboux integrability theory for planar complex and real polynomial systems. As far as we know, the problem of integrating a polynomial system by using its invariant algebraic curves was first considered by Darboux in [46]. The version that we present improves Darboux's result substantially because we also take into account exponential factors (see [38]), independent singular points (see [31]), and invariants (see [25]). Exponential factors appear when an invariant algebraic curve has geometric multiplicity greater than 1; see [37, 40]. For additional information about invariants; see [25].

One of the natural questions in this subject is whether or not a polynomial system (8.1) can have invariant algebraic curves. The answer is not easy. See the large section in Jouanolou's book [92] or the long paper of Moulin, Nowicki, and Strelcyn [112] devoted to showing that one particular polynomial system has no invariant algebraic solutions. Even for one of the most frequently studied limit cycles, the limit cycle of the van der Pol system, it was unknown until 1995 that it is not algebraic [117].

The original works of Kapteyn and Bautin are [93,94] and [14], respectively. The proof of the sufficient part of Theorem 8.15 comes from [24].

Other approaches and more information on the classification of quadratic centers can be found in [142] and in [170].

Polynomial Planar Phase Portraits

In this chapter we present a computer program based on the tools introduced in the previous chapters. This program is prepared to draw the phase portrait of any polynomial differential system on the compactified plane obtained given by Poincaré or Poincaré–Lyapunov compatification. Of course, there are always some computational limitations which we shall describe in this chapter and in the next one.

9.1 The Program P4

P4 [9] is a tool which can be used in the study of a polynomial planar differential system. Depending on the user's choice it draws the phase portraits on the Poincaré disk, or on a Poincaré–Lyapunov disk, near a singular point or on any rectangle in the finite plane, or at infinity in one of the four traditional charts used in the compactification process. The first version of P4 was partly written in C and partly written in REDUCE [81]. It ran only under a UNIX or LINUX system and its developer was mainly C. Herssens. The new version of P4 has changed the symbolic language from REDUCE to MAPLE, and can now be implemented more easily in any system, either WINDOWS, UNIX, or MACINTOSH OS-X, as long as MAPLE is available. When running it on UNIX, the user can toggle between MAPLE or REDUCE. The new version has been developed by De Maesschalck.

It is possible to work in numerical mode or in mixed mode, i.e., if possible, the calculations are done in algebraic mode. We shall now describe the structure and possibilities of P4.

First P4 checks whether or not the vector field has a continuous set of singular points in the plane, that is, whether or not the two polynomial components of the vector field have a common factor. If they have a common factor, we divide the vector field by this common factor and study the new vector field. Sometimes, if the vector field is too big or complex, the computer algebra package used (i.e., REDUCE or MAPLE) cannot find this common factor. In such cases P4 will work incorrectly. If the user knows the common factor (e.g., by means of another computer algebra package such as Mathematica, Axiom, ...), or simply because this comes from an already prepared system with a certain common factor, he can avoid this problem by giving this factor, together with the complete vector field (i.e., the vector field after division by the common factor), to P4.

In what follows let $X = P(x,y)\partial/\partial x + Q(x,y)\partial/\partial y$ with $\gcd(P,Q) = 1$. Now we will determine the finite isolated singular points. This can be done in algebraic or numeric mode. In both cases P4 will ask the symbolic language to solve the problem. If the degree of the vector field is high, determining these singularities can take a lot of time. In such cases it is better working numerically. In general, we recommend to work numerically when the expected finite singular points do not have rational coordinates or simple radical ones. For example, quadratic vector fields with rational coefficients can normally be studied in algebraic mode (with some exceptions), but higher degree systems may normally need numeric mode, unless they are very simple or many of the coefficients are zero.

For each singular point (x_0, y_0), P4 determines the local phase portrait in the following way. First it computes the Jacobian matrix at each singular point, i.e.,

$$dX_{(x_0,y_0)} = \begin{pmatrix} \dfrac{\partial P}{\partial x}(x_0,y_0) & \dfrac{\partial P}{\partial y}(x_0,y_0) \\[2mm] \dfrac{\partial Q}{\partial x}(x_0,y_0) & \dfrac{\partial Q}{\partial y}(x_0,y_0) \end{pmatrix},$$

and evaluates its eigenvalues λ_1 and λ_2. We have to distinguish different cases, depending on whether both eigenvalues are real, purely imaginary, or complex.

1. λ_1 *and* λ_2 *are real.* If λ_1 and λ_2 have the same sign then (x_0, y_0) is a stable (unstable) node and we are done. If they have different sign, then (x_0, y_0) is a saddle, and we compute a Taylor approximation of some order n, to be specified later, of the stable and unstable manifold as follows.

Consider the transformations

$$\bar{x} = x - x_0, \quad \bar{y} = y - y_0,$$

and

$$\bar{x} = w_{11}u + w_{21}v, \quad \bar{y} = w_{12}u + w_{22}v,$$

with (w_{11}, w_{12}) (respectively, (w_{21}, w_{22})) an eigenvector associated to the eigenvalue λ_1 (respectively, λ_2).

Use of these transformations yields the vector field

$$\begin{aligned} \dot{u} &= \lambda_1 u + p(u,v), \\ \dot{v} &= \lambda_2 v + q(u,v), \end{aligned} \tag{9.1}$$

with $\deg(p) \geq 2$ and $\deg(q) \geq 2$. Writing the invariant manifold as a graph $(u, f(u))$ and using the invariance of the flow, we have that

$$f(u) = \sum_{i=2}^{n} a_i u^i + o(u^n),$$

with

$$a_i = \frac{b_i}{(i\lambda_1 - \lambda_2)}, \quad i = 2, \ldots, n,$$

where b_i is the coefficient of u^i in the expression $q(u, f(u)) - f'(u)p(u, f(u))$. The manifold $(v, g(v))$ is computed in the same way.

If $\lambda_1 = 0$ and $\lambda_2 \neq 0$ then the singularity (x_0, y_0) is semi-hyperbolic. In this case there is a center manifold which is tangent to the line $v_2(x - x_0) - v_1(y - y_0) = 0$, where (v_1, v_2) is an eigenvector associated to the zero eigenvalue. To compute the center manifold, we simplify the vector field in the same way as in the saddle case. Hence the new vector field satisfies

$$\begin{aligned}
\dot{u} &= p(u, v), \\
\dot{v} &= \lambda_2 v + q(u, v),
\end{aligned} \tag{9.2}$$

with $\deg(p) \geq 2$ and $\deg(q) \geq 2$. Writing the center manifold as a graph $(u, f(u))$, and using the invariance of the flow, we have

$$f(u) = \sum_{i=2}^{n} a_i u^i + o(u^n),$$

where a_i is the coefficient of u^i in the expression

$$-[q(u, f(u)) - f'(u)p(u, f(u))]/\lambda_2.$$

This results in the behavior

$$\dot{u} = c_m u^m + o(u^m).$$

Using this information we find that the origin is:

1. A stable node if $c_m < 0, m$ odd, and $\lambda_2 < 0$
2. An unstable node if $c_m > 0, m$ odd, and $\lambda_2 > 0$
3. A saddle–node if m is even
4. A saddle if $c_m > 0, m$ odd, and $\lambda_2 < 0$ or $c_m < 0, m$ odd, and $\lambda_2 > 0$

If the singularity is a saddle–node or a saddle, then we also compute a Taylor approximation for the unstable or stable manifold.

If the two eigenvalues are zero, the point (x_0, y_0) is nonelementary. To study the vector field near the singularity, we desingularize the singularity by means of quasihomogeneous blow-up as described in Chap. 3.

2. If the eigenvalues are purely imaginary, then the point (x_0, y_0) is a weak focus or a center. The program first checks if the system is Hamiltonian or not and if it is, then it states that for certain the singular point is a center since a Hamiltonian system cannot have focus. If it is not Hamiltonian, then to determine its type, we compute the Lyapunov constants using the technique developed in Chap. 4. In the case of a quadratic vector field or a linear plus homogeneous cubic vector field, P4 is able to determine whether or not the point is a center, or an unstable or a stable weak focus of a certain order. In all other cases P4 evaluates by default the first four Lyapunov constants. If they are all zero we have an undetermined weak focus; in the other case we have a stable or an unstable weak focus. The algorithm is written in C and hence the computations are done numerically. Thus the Lyapunov constants are calculated up to a certain precision. By default we say that a Lyapunov constant V is zero if $|V| < 10^{-8}$, since this is the default value given to the variable *Precision* (which can be modified by the user as we will describe later).

3. If the eigenvalues are complex but not purely imaginary, the singular point (x_0, y_0) is a strong stable (respectively, unstable) focus if $\text{Tr}(\text{D}X_{(x_0,y_0)}) <$ /break0 (respectively, > 0).

Now we determine the singularities at infinity. By default we study the vector field on the Poincaré disk; see Chap. 5. First we transform the vector field using the transformation

$$x = \frac{1}{z_2}, \quad y = \frac{z_1}{z_2}.$$

This yields the vector field (after multiplying the result by z_2^{d-1})

$$\dot{z}_1 = z_2^d \left(-z_1 P\left(\frac{1}{z_2}, \frac{z_1}{z_2}\right) + Q\left(\frac{1}{z_2}, \frac{z_1}{z_2}\right) \right),$$

$$\dot{z}_2 = -z_2^{d+1} P\left(\frac{1}{z_2}, \frac{z_1}{z_2}\right),$$

with d the degree of the vector field. Suppose that $Q_d(1, z_1) - z_1 P_d(1, z_1) \not\equiv 0$. The points $(z_1, 0)$ which satisfy $Q_d(1, z_1) - z_1 P_d(1, z_1) = 0$ are the infinite singular points of X. These points are studied in the same way as the finite ones. In the case that $Q_d(1, z_1) - z_1 P_d(1, z_1) \equiv 0$, the line at infinity is a line of singularities. To study the behavior near infinity we divide the vector field by z_2, and study this vector field near the line $\{z_2 = 0\}$.

Second we transform the vector field using the transformation

$$x = \frac{z_1}{z_2}, \quad y = \frac{1}{z_2}.$$

This yields the vector field (after multiplying the result by z_2^{d-1})

$$\dot{z}_1 = z_2^d \left(P\left(\frac{z_1}{z_2}, \frac{1}{z_2}\right) - z_1 Q\left(\frac{z_1}{z_2}, \frac{1}{z_2}\right) \right),$$

$$\dot{z}_2 = -z_2^{d+1} Q\left(\frac{z_1}{z_2}, \frac{1}{z_2}\right).$$

We only have to determine whether or not the point $(0,0)$ is a singular point, since the others have been studied in the first chart.

If there is a singularity at infinity which is nonelementary, it is sometimes better to study the vector field on a Poincaré–Lyapunov disk (see Sect. 5.3) of some degree (α, β), i.e., we use a transformation of the form

$$x = \frac{\cos\theta}{r^\alpha}, \quad y = \frac{\sin\theta}{r^\beta},$$

for the study near infinity, which yields the vector field (after multiplying the result by r^c)

$$\dot{r} = -r^{c+1} \sum_{\delta \leq c} r^{-\delta} (\cos\theta P_\delta(\cos\theta, \sin\theta) + \sin\theta Q_\delta(\cos\theta, \sin\theta)),$$

$$\dot{\theta} = r^c \sum_{\delta \leq c} r^{-\delta} (-\beta \sin\theta P_\delta(\cos\theta, \sin\theta) + \alpha \cos\theta Q_\delta(\cos\theta, \sin\theta)),$$

$$(9.3)$$

with $P_\delta(x, y)\frac{\partial}{\partial x} + Q_\delta(x, y)\frac{\partial}{\partial y}$ the quasi homogeneous component of type (α, β) and quasi homogeneous degree δ; c is chosen to be the maximal δ.

With an appropriate choice of (α, β) we will often only encounter elementary singularities at infinity. To simplify the calculations we prefer to work with charts.

First we transform the vector field using the transformation

$$x = \frac{1}{z_2^\alpha}, \quad y = \frac{z_1}{z_2^\beta}.$$

This yields the vector field (after multiplying the result by αz_2^c)

$$\dot{z}_1 = z_2^c \sum_{\delta \leq c} z_2^{-\delta} (\alpha Q_\delta(1, z_1) - \beta z_1 P_\delta(1, z_1)),$$

$$\dot{z}_2 = -z_2^{c+1} \sum_{\delta \leq c} z_2^{-\delta} P_\delta(1, z_1).$$

If $\alpha Q_c(1, z_1) - \beta z_1 P_c(1, z_1) \neq 0$, then the singular points $(z_1, 0)$ that satisfy $\alpha Q_c(1, z_1) - \beta z_1 P_c(1, z_1) = 0$ are the infinite singular points of X. These points are studied in the same way as the finite ones. In those cases in which $\alpha Q_c(1, z_1) - \beta z_1 P_c(1, z_1) \equiv 0$, the line at infinity is a line of singularities. To study the behavior near infinity we divide the vector field by z_2 and study this vector field near the line $\{z_2 = 0\}$.

Next we transform the vector field using the transformation

$$x = \frac{-1}{z_2^\alpha}, \quad y = \frac{z_1}{z_2^\beta}.$$

This yields the vector field (after multiplying the result by αz_2^c)

$$\dot{z}_1 = z_2^c \sum_{\delta \leq c} z_2^{-\delta} (\alpha Q_\delta(-1, z_1) + \beta z_1 P_\delta(-1, z_1)),$$

$$\dot{z}_2 = z_2^{c+1} \sum_{\delta \leq c} z_2^{-\delta} P_\delta(-1, z_1).$$

This vector field can be studied in the same way as the previous one.

Finally we consider the two transformations

$$x = \frac{z_1}{z_2^\alpha}, \quad y = \frac{1}{z_2^\beta},$$

and

$$x = \frac{z_1}{z_2^\alpha}, \quad y = \frac{-1}{z_2^\beta}.$$

For these two vector fields we only have to determine whether or not the point $(0,0)$ is a singular point, since the others have been studied in the first two charts.

At this stage P4 is ready to draw a large part of the phase portrait of the vector field. First it draws the invariant separatrices, in the following way. In the case that the singularity is a saddle or a saddle–node, we use the Taylor approximation of the invariant manifold until it meets the boundary of a circle of radius ε, for a certain choice of $\varepsilon \geq 0$. From this point on we integrate the separatrices with the multistep Runge–Kutta method of orders 7 and 8. To prevent numeric overflow in the Taylor approximation, we normalize the vector fields (9.1) and (9.2), before we compute the Taylor approximation, as follows. Let a be the coefficient in the vector field that maximize $|a|$. We rescale the time such that this coefficient becomes equal to $1,000 \cdot \text{sign}(a)$. At the beginning of the numerical integration of the separatrices we have an error that comes from the Taylor approximation. By default we take $\varepsilon = 0.01$ and as order of approximation $n = 6$. So we have an error of order 10^{-14}. To make sure that this error is not too large, we do a test to decide whether or not the Taylor approximation "fits" the real invariant manifold. Let $f(t)$ be the Taylor approximation of the invariant manifold, which is tangent to the line $v = 0$. Suppose that $t_1^2 + f(t_1)^2 = \varepsilon^2$ and consider the points $(ih, f(ih))$, $i = 1, \ldots, 100$, with $h = t_1/100$. Consider the angles $\alpha_i = \arctan(f'(ih))$ and $\beta_i = \arctan(\dot{v}(ih, f(ih))/\dot{u}(ih, f(ih)))$, $i = 1, \ldots, 100$. If $|\alpha_i - \beta_i| < 10^{-8}$, for $i = 1, \ldots, 100$, we accept the Taylor approximation, otherwise we compute the Taylor approximation to the next higher order and do the test again. By default we take as maximum order $n = 20$. In this case the error is of order 10^{-42}. This test works very well for the stable and unstable manifolds of vector fields which are not close to structurally unstable systems, but for the center manifolds or stable and unstable manifolds of vector fields that are close to structurally unstable systems it sometimes fails, especially if the nonzero eigenvalue is large in absolute value, or both eigenvalues are of very different magnitudes.

If the singularity is nonelementary, we split the point into several singularities which are elementary. For each of these points we draw the invariant manifold (which correspond to a separatrix of the nonelementary singularity) as follows. First we use the Taylor approximation in the blow-up chart that corresponds to the elementary singularity, up to distance ε from the singularity. Then we extend the separatrix in this chart by numeric integration, up to distance 1 from the singularity. Next we extend by numeric integration in the real plane. The number of steps has to be decided in an interactive way by the user.

To prevent numerical overflow when integrating the vector field, we do not always integrate the vector field on the real plane and project it on the Poincaré sphere, but we use different charts which cover the Poincaré sphere as follows. Let (X, Y, Z) be a point on the Poincaré sphere with $Z > 0$, and let (θ, φ) be the sphere coordinates of the point, i.e., $X = \cos\theta \sin\varphi$, $Y = \sin\theta \sin\varphi$, and $Z = \cos\varphi$.

If $0 \le \varphi \le \pi/4$ we transform the point to the real plane, i.e., we consider the point $(X/Z, Y/Z)$ and integrate the original vector field. If $\varphi > \pi/4$ then we consider the following four cases.

1. If $-\pi/4 \le \theta \le \pi/4$, we consider the point $(z_1, z_2) = (Y/X, Z/X)$ and integrate the vector field

$$\dot{z}_1 = z_2^d \left(-z_1 P\left(\frac{1}{z_2}, \frac{z_1}{z_2} \right) + Q\left(\frac{1}{z_2}, \frac{z_1}{z_2} \right) \right),$$

$$\dot{z}_2 = -z_2^{d+1} P\left(\frac{1}{z_2}, \frac{z_1}{z_2} \right).$$

2. If $\pi/4 < \theta < 3\pi/4$, we consider the point $(z_1, z_2) = (X/Y, Z/Y)$ and integrate the vector field

$$\dot{z}_1 = z_2^d \left(P\left(\frac{z_1}{z_2}, \frac{1}{z_2} \right) - z_1 Q\left(\frac{z_1}{z_2}, \frac{1}{z_2} \right) \right),$$

$$\dot{z}_2 = -z_2^{d+1} Q\left(\frac{z_1}{z_2}, \frac{1}{z_2} \right).$$

3. If $3\pi/4 \le \theta \le 5\pi/4$, we consider the point $(z_1, z_2) = (Y/X, Z/X)$ and integrate the vector field

$$\dot{z}_1 = (-1)^{d-1} z_2^d \left(-z_1 P\left(\frac{1}{z_2}, \frac{z_1}{z_2} \right) + Q\left(\frac{1}{z_2}, \frac{z_1}{z_2} \right) \right),$$

$$\dot{z}_2 = (-1)^d z_2^{d+1} P\left(\frac{1}{z_2}, \frac{z_1}{z_2} \right).$$

4. If $5\pi/4 < \theta < 7\pi/4$, we consider the point $(z_1, z_2) = (X/Y, Z/Y)$ and integrate the vector field

$$\dot{z}_1 = (-1)^{d-1} z_2^d \left(P\left(\frac{z_1}{z_2}, \frac{1}{z_2} \right) - z_1 Q\left(\frac{z_1}{z_2}, \frac{1}{z_2} \right) \right),$$

$$\dot{z}_2 = (-1)^d z_2^{d+1} Q\left(\frac{z_1}{z_2}, \frac{1}{z_2} \right).$$

The pattern of singularities, the infinite as well as the finite ones, together with their separatrices constitute the separatrix skeleton and already give a very good idea of the global phase portrait; see Sect. 1.9. We of course do not see the exact number and location of the closed orbits, but we have confined the regions in which limit cycles or annuli of closed orbits can occur. If one has the impression that closed orbits and especially limit cycles will occur, one can ask P4 to find these limit cycles as follows. First one has to select two points x and y. The two points should be close to the region where one expects to find a limit cycle, and the segment L joining both points should cut the expected limit cycle transversely. P4 tries to determine the limit cycle as follows. First it divides the segment into segments $[p_i, p_{i+1}]$ of length h and starts integrating from one end of the subsegment L to the other. Every orbit close to the limit cycle is supposed to cut the segment L again. From this we detect the existence of the limit cycle when we find a change in the Poincaré Return Map. P4 detects such change as follows. Suppose that we start integrating from a point p_i on L, and that the orbit cuts the segment L again at a point q_i with $p_i < q_i$. P4 takes now the point p_j nearest to q_i with $p_j > q_i$ and starts integrating in the same direction. If this orbit cuts L at a point q_j with $q_j < p_j$, then there is a limit cycle between the points q_i and q_j. By default we take $h = 10^{-4}$. Of course in this way we can say only that in a region of length 10^{-4} there exists at least one limit cycle. Sometimes it is possible that P4 finds nonexistent limit cycles. The reason is that in these cases the Poincaré Return Map is very close to the identity. One can solve this by decreasing the value of h or increasing the precision with which we use the Runge–Kutta method, but these modifications will also increase the amount of time needed to check if there are closed orbits or not. The only way to reduce the time without losing precision is to reduce the length of the segment L, but then the user must be certain that he gives it on the region where limit cycles are expected.

In no case will P4 be able to detect semistable limit cycles. Indeed, detection of semistable limit cycles for a specific vector field seems to be beyond the capabilities of any numeric tool.

In case we study the vector field on a Poincaré–Lyapunov disk of degree (α, β), P4 draws the orbits of the vector field as follows; see Fig. 9.1.

Let $(x, y) \in \mathbb{R}^2$. If $x^2 + y^2 \leq 1$ then (x, y) will be plotted in the interior of the unit circle around the origin, by integrating the original vector field (of course making the detailed analysis of the finite singularities as presented in the case of Poincaré compactification). If $x^2 + y^2 > 1$, P4 makes a transformation of the form $x = \cos \theta / r^\alpha$ and $y = \sin \theta / r^\beta$ in order to plot in the annulus limited by the finite circle of radius 1 and the infinite one, integrating the vector field (9.3), to extend the information near the singularities. Unfortunately orbits crossing the circle of radius 1 give the impression of having a discontinuous derivative. This is due to the fact that we are using two different transformations which do not match in a differentiable way on the unit circle.

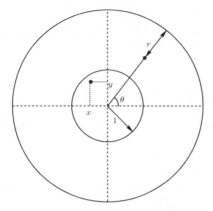

Fig. 9.1. Representation of the Poincaré–Lyapunov disk of degree (α, β)

9.2 Technical Overview

Following algorithms are written in symbolic language (REDUCE or MAPLE):

- The determination of the finite and infinite singularities.
- The local phase portrait of each singularity.
- The desingularization of a nonelementary singularity.
- The Taylor approximation of the invariant manifolds.
- Drawing the lines of singularities.

Following algorithms are written in C:

- The graphical interface. This interface is written using Trolltech's Qt (www.trolltech.com) library, which is platform independent, and hence works for WINDOWS, MacOSX, and most versions of UNIX.
- The test whether or not the Taylor approximation is a sufficiently good approximation of the real manifold.
- The calculations of the Lyapunov constants.
- The integration of the orbits and invariant separatrices. We use the Runge–Kutta 7/8 method for the integration [63].
- The search for limit cycles.

In order to run P4, one needs either a UNIX system with a C compiler, Xview 3.2 libraries and the computer algebra package REDUCE or MAPLE, or a WINDOWS system XP[1] with MAPLE. A compiler is not needed unless you need to recompile it. If so you will need a C++ compiler.

In Sect. 9.3 we will describe the graphical interface, and in the next chapter we will give a short guideline of the program based on examples. We will show

---

[1] Its performance has not been checked on other WINDOWS systems, on which it may run equally well.

and describe the windows as they appear in the WINDOWS system. The ones in UNIX are very similar. There are no exercises in this chapter but you will find them at the end of the next chapter.

9.3 Attributes of Interface Windows

9.3.1 The Planar Polynomial Phase Portraits Window

Function: The *Planar Polynomial Phase Portraits* window is the main control panel for the tool P4. We may call it also the *main* window.

Description: The *Planar Polynomial Phase Portraits* window is opened at start-up. The main function of this window is to introduce the system to be studied and to modify some of the working parameters.

Top section:

Quit button: Allows the user to stop the program. This will close all the related windows.

View menu button: Shows the description of the singular points of the system which the user is studying.

Finite... Gives information about the finite singular points.

Infinite... Gives information about the infinite singular points.

Plot button: Opens and brings to the foreground the *Phase Portrait* window.

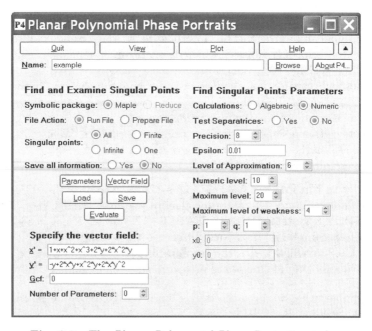

Fig. 9.2. The *Planar Polynomial Phase Portraits* window

Help button: Opens and brings to the foreground the *Help* window. The help files are written in HTML format.

Up arrow button: This toggles to a reduced version of the *Planar Polynomial Phase Portraits* window when the user has already introduced initial data. Clicking it again returns to the large window.

Name: Allows the user to enter the name of the file which contains a valid input for the polynomial vector field he wants to study, or the name of the file where he wants to store the system under examination. If loading a previous file, the user has to enter a name (e.g., *file1*) and immediately after click the *Load* button before pressing the *Evaluate*, the *View*, or the *Plot* button. By default all the input files have the extension *.inp*. If the loaded vector field has been already evaluated, you do not need to evaluate it again, and you can *View* or *Plot* it directly.

Browse: Allows the user to enter the WINDOWS system to look for a previously stored file.

About P4: Gives the logo of the program, the authors, and gives access to a window where one can change the main settings of the program.

Find and examine singular points section:

Symbolic package: In the case of a UNIX system, this allows the user to toggle between the two symbolic languages. In WINDOWS it is forced to MAPLE.

File action: Allows the user to choose between *Run File* or *Prepare File*. The default option is *Run File*.

Run file... Once the *Evaluate* button is clicked, the program will start studying the system. The singular points and the Taylor approximations of the invariant manifolds are determined and this information is stored in several files, namely *file1_fin.res*, *file1_inf.res*, *file1_vec.tab*, *file1_fin.tab* and *file1_inf.tab*.

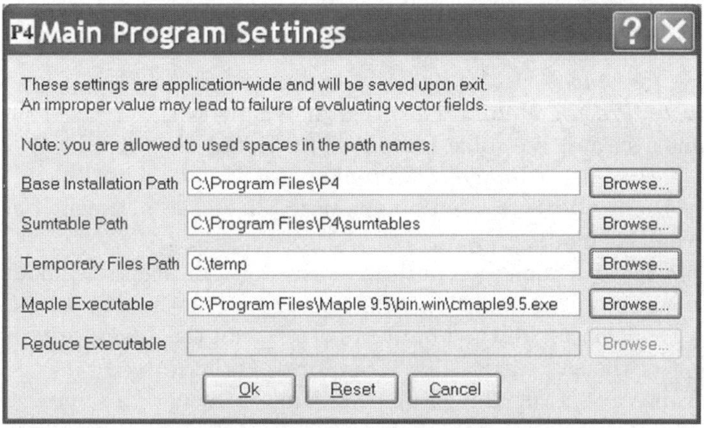

Fig. 9.3. The *Main settings* window

Prepare file ... If the *Prepare File* option is chosen, the *Evaluate* button is renamed to *Prepare* and by clicking it the program will generate a MAPLE or REDUCE file (*file1.txt*) or (*file1.red*) with the algebraic study of the system. The user can run this file directly with MAPLE/REDUCE for a further algebraic manipulation of the problem. He may be interested in this option if the amount of computations is very large and he prefers to run a MAPLE/REDUCE program in batch mode. The MAPLE file is in fact a text file with instructions in MAPLE that can be transferred to it.

Singular points: Allows the user to choose between the following options.

All... Determines all the finite and infinite singular points.

Finite... Determines the finite singular points.

Infinite... Determines the infinite singular points.

One... With this option the user can study the polynomial vector field near a singular point (x_0, y_0). The user has to enter the coordinates x_0 and y_0 in the *Find Singular Points Parameters section*. This option is useful if one wants to study the behavior near a nonelementary singularity.

Save all:

Yes... Gives an exhaustive description of every step executed by the program.

No... Reduces the amount of information that the user will get.

Parameters: The right side of the *Planar Polynomial Phase Portraits* window is closed. By default it appears and contains some parameters that may be changed by the user regarding the study of the problem. It can be reopened by clicking the same button again.

Vector field: The bottom side of the *Planar Polynomial Phase Portraits* window is closed. By default it appears and contains the polynomial vector field to be studied. It can be reopened by clicking the same button again.

Load button: Loads the file *file1.inp*.

Save button: Saves the polynomial vector field and the parameters to the file *file1.inp*.

Evaluate/Prepare button: The program will make the study of the system and you may see some computations in the *Output* window.

Find singular points parameters Section:

Calculations: With this option the user can toggle between *Algebraic* or *Numeric* mode.

Algebraic... Some computations are done in algebraic mode. These computations are the determination of all the singular points, the calculation of the lowest order terms of the Taylor approximation of the separatrices and the blow-up procedure.

Numeric... Everything is done in numeric mode. This option (default) is recommended if the degree of the polynomial vector field is high and if it has many coefficients.

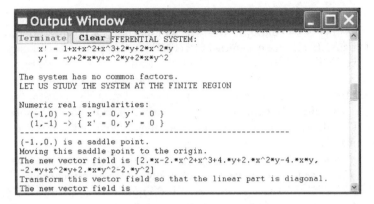

Fig. 9.4. The *Output* window

Test separatrices: With this option the user can decide whether or not P4
   has to test every Taylor approximation of the separatrices. In general
   it is recommended to set this option to *Yes*, but if the user has a
   specific system from which he knows it has a separatrix which is hard
   to deal with, he may deactivate this option.
Precision: Here the user needs to define a precision to avoid rounding
   errors. It tells the program to set a number a equal to zero if $|a| < 10^{-\delta}$
   where δ is the *Precision* parameter. Of course, this means that it
   is possible that a nonzero number be considered equal to zero. In
   such cases the *Precision* has to be modified. The program makes all
   computations with 16 digits, so it is recommended that the value for
   Precision be kept between 7 and 12.
Epsilon: In order to start integrating the separatrices we take an initial
   point at a certain distance *Epsilon* away from the singularity. This
   value is the default one we will use for every separatrix.
Level of approximation: Allows the user to set the order of the Taylor
   approximation for the separatrices. If the option *Test Separatrices* is
   activated, then P4 will test whether or not the Taylor approximation is
   a sufficiency good approximation of the real manifold (or separatrix).
   In case it is not, P4 computes the Taylor approximation one order
   higher and repeats the test again until the maximum degree is reached
   or until the Taylor approximation is a sufficiency good approximation
   of the real manifold.
Numeric level: If the option *Calculations* is set to *Algebraic*, then the computation of the coefficients of the Taylor approximation will be done
   in *Algebraic* mode until the value in *Numeric level* is reached. From
   this stage the computation will be done in *Numeric* mode.
Maximum level: Gives the maximum order of the Taylor approximation.
   If the test fails up to this level, then this will be explained in the
   report that P4 will produce.

Maximum level of weakness: Sets the number of Lyapunov constants that P4 has to calculate, in case the singularity is a non–degenerate weak focus. If all these values are zero then the program concludes that we have a center-focus (expect for Hamiltonian, quadratic systems, or linear plus homogeneous cubic systems). Sometimes the user is interested in getting a large number of Lyapunov constants, but he must realize that the time for computing them increases exponentially. The algorithm is written in C and hence the calculations are done in numeric mode.

p and q: Gives the degree of the Poincaré–Lyapunov compactification. If $(P, Q) = (1, 1)$, then we use the Poincaré compactification.

x_0 *and* y_0: Gives the coordinates of the singularity. The user will see this if he has selected *One* singular point in the *Singular Points* option and thus we will have to set it at will.

Vector field section:

x' *and* y': Defines the equation of the system in the variables x and y. The user can use the symbols +,-,*,/,^, (and) and any function that is valid in MAPLE/REDUCE, like sqrt(),sin(),cos(),.... Notice that if MAPLE is used, you must add a multiplicative * between numbers and letters. If REDUCE is used, this is not needed. The multiplicative * is always compulsory between letters (whether they are variables or parameters). On the contrary, REDUCE does not allow leaving any blank space inside the formulas, but MAPLE does. If one wants his files to be compatible with both systems, one should always insert the * denoting multiplication and leave no blank spaces.

Gcf: States the Greatest Common Factor between the two polynomials which define the system. If the user gives the greatest common factor, the program will believe the user and will skip computing it. It is also possible to ask the program to determine the GCF. In this case the value for GCF has to be set to zero. If the user says that there is no common factor (or the program cannot find it) when there is a nontrivial one, then the program will work incorrectly. To speed up the program when the user knows that there is no common factor, one can set this parameter to 1.

Number of parameters: Gives the number of parameters of the system. After the user has entered this number, the window will enlarge showing pairs of boxes to give room for them. Put in the left box the name of the parameter and in the right box the value. These names are not case sensitive in REDUCE, but they are case sensitive in MAPLE.

9.3.2 The Phase Portrait Window

Function: In this window the user will be able to draw the phase portrait of the polynomial vector field.

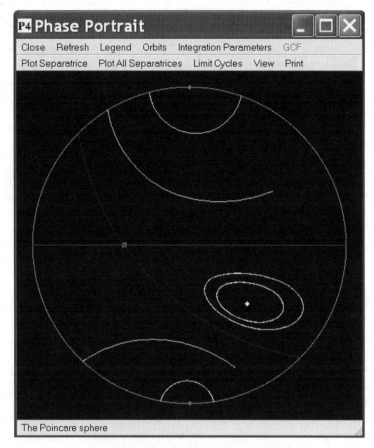

Fig. 9.5. The *Poincaré Disc* window

Description: This window is opened by selecting the *Plot* button in the *Planar Polynomial Phase Portraits* window. The user will see the window as in Fig. 9.5. In this window there is a circle representing infinity and some symbols representing the finite and infinite singular points of the system. If one presses the *Legend* button, the explanation of these symbols will appear.

If $(P, Q) \neq (1, 1)$, then the Poincaré–Lyapunov compactification is shown. In this case the user will see two circles. In the inner circle all the finite singular points with modulus less than one are plotted. If the modulus of a certain singular point is greater than one, then this point is plotted in the annulus limited by the circle of radius one and the circle at infinity. If the option *Singular Points* is set to *One*, then the user will see a planar representation of the neighborhood of such a point.

If the user moves the mouse in the drawing canvas, the current coordinates of the mouse position are displayed in the window's panel. If he studies the system on a Poincaré or Poincaré–Lyapunov disk, this region is blank when the mouse does not point to a region within the disk.

The user can always enlarge or reduce this window as usual and an *Aspect Ratio* value will be shown on the bottom line to remind the user if proportion holds (*Aspect Ratio*= 1) or not (*Aspect Ratio*≠ 1).

Mouse events in the main drawing canvas have the following effects:

- Clicking the LEFT button in the drawing canvas will select that point and opens the *Orbits* window.
- Clicking the LEFT button while holding down the SHIFT key will select the nearest singular point having separatrices (finite or infinite) and opens the *Plot Separatrices* window. You will see flashes around the selected singular point.
- Clicking the LEFT button while holding down the CONTROL key creates a rectangle used to make a zoom of a portion of the picture. You must move from one corner of the rectangular region to the opposite while holding both the LEFT button and the CONTROL KEY. Now the *Phase Portrait - Zoom* window will appear. At any time it is possible to cancel the zoom by clicking on the RIGHT button.

Panel items:

Close button: Closes this window and all the related windows.

Refresh button: Clears the window and redraws the drawing canvas. This is useful if all the separatrices are drawn, because the redraw will bring up the singularities which may have be shadowed by the lines.

Legend button: Opens and brings up to the foreground the *Legend* window.

Orbits button: Opens and brings up to the foreground the *Orbits* window.

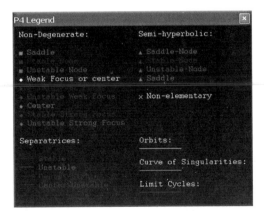

Fig. 9.6. The *Legend* window

Integration Parameters button: Opens and brings up to the foreground the *Parameters of Integration* window

GCF button: Opens and brings up to the foreground the *GCF* window. This button will be active if the system has a nontrivial greatest common factor.

Plot Separatrices button: Opens and brings up to the foreground the *Plot Separatrices* window.

Plot All Separatrices button: Will plot every separatrix. It is often possible that some separatrices are not completely plotted or even not plotted at all. In this case the user has to modify the *# Points* option in the *Parameters of Integration* window before he presses the button *Plot All Separatrices* again. Another possibility for dealing with these "slow" separatrices is to go to the *Plot Separatrices* window to plot them individually. We highly recommend this option.

Limit Cycles button: Opens and brings to the foreground the *Limit Cycles* window.

Print button: Opens and brings to the foreground the *Print* window.

View button: This button opens the *View Parameters* window that allows the user to change the way of viewing the phase portrait.

    Type of view: If you have done the study for all the singular points, you can toggle between the complete phase portrait on the Poincaré disk, or a reduced planar drawing on a certain rectangle. You may also choose to view the traditional U_1, U_2, V_1, and V_2 local charts which show the phase portrait from an infinite point of view. When using these local charts, take into account that the positive direction is always toward the inner disk, and the image we show maintains the usual standard of "right" and "up" being

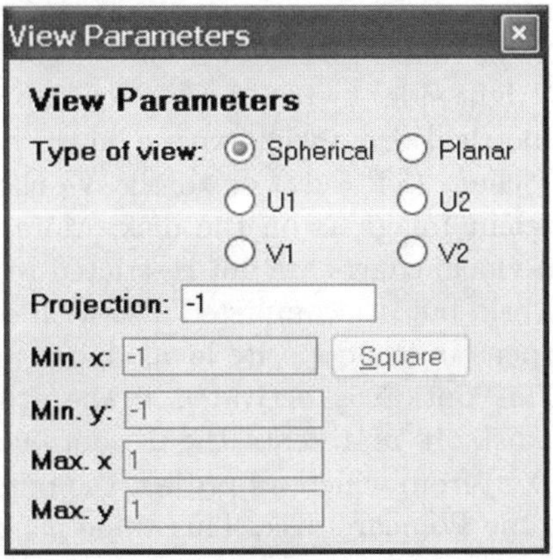

Fig. 9.7. The *View Parameters* window

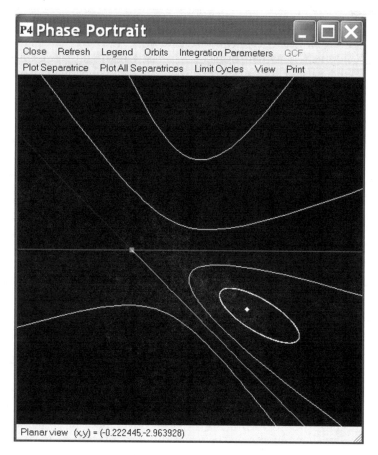

Fig. 9.8. A planar plot

positive while "left" and "down" are negative. So looking to the U_1 chart (right side) or to the U_2 chart (up side) may look different (symmetrical) from what we see on the disk. However, looking to the V_1 chart (left side) or to the V_2 chart (down side) gives the equivalent image as on the disk. Moreover, these images on the infinite local charts are not restricted to the upper half of the Poincaré disk, but are completed also with the lower part, that is, what happens in the opposite local chart.

Projection: This option is activated if the system is studied on a Poincaré disk. It represents the z coordinate of the projection point $(0, 0, z)$ from which we project the points from the Poincaré sphere to the Poincaré disk. This value has to be negative. If the user wants a parallel projection then he has to set this value to zero. The drawing canvas is refreshed after the user selects the *Refresh* button.

$x0$, $y0$, $x1$, $y1$, *and square:* This option is activated if the system is studied on a planar rectangle. One can either give the four coordinates of the lower-left (x_0, y_0) and upper-right corner (x_1, y_1), or set simply x_0 and *Square* to set the square centered at the origin of side $2x_0$.

9.3.3 The Plot Orbits Window

Function: Allows the user to draw any orbit of the system.

Description: The *Plot Orbits* window is opened by selecting the *Orbits* button in the *Phase Portrait* window or by selecting a point in the drawing canvas. In this window the user can integrate and delete orbits.

Panel items:

x_0 *and* y_0: Defines the starting point of the orbit to integrate. One can either do this by typing the precise point where integration must start, or by clicking on the canvas.

Select: If one has typed the specific coordinates of a point, it is not stored and ready for computation until this button is clicked.

Forwards button: Integrates the selected point in the positive direction. This button will be active when the user has selected a point in the drawing canvas or if the user has selected the *Select* button. After the integration the button will be inactive.

Continue button: Continues the integration in the current direction. This button will be active if the user has pressed the *Forwards* or *Backwards* button.

Backwards button: Integrates the selected point in the negative direction. This button will be active when the user has selected a point in the drawing canvas or if the user has selected the *Select* button. After the integration the button will be inactive.

Delete last orbit button: Erases the last orbit which has been drawn.

Delete all orbits button: Erases all the orbits which have been drawn.

Shortcuts: Once a point is selected, and the *Phase Portrait* window active, one can ask to integrate Forward, Backward, to Continue, to Delete last orbit, or to delete All orbits by pressing F, B, C, D, or A, respectively.

Fig. 9.9. The *Orbits* window

9.3.4 The Parameters of Integration Window

Function: Allows the user to modify the parameters which affect the integration of separatrices and orbits through the Runge–Kutta 7/8 method and the parameters which are used in the graphical representation.

Description: The *Parameters of Integration* window is opened by selecting the *Integration Parameters* button in the *Phase Portrait* window. In this window the user can change the parameters of integration. These parameters should be modified if the user is not satisfied with the results obtained. The default values are shown in Fig. 9.10.

Panel items:

Vector field: This option is activated if the system has a nontrivial greatest common factor or if the line at infinity is a line of singularities.

Original... Uses the original system for integration of the separatrices and orbits. Thus all trajectories stop when reaching the line of singularities.

Reduced... Uses the system which is obtained by dividing out the greatest common factor for integrating the separatrices and orbits. Thus, the trajectories seem to continue across the line of singularities.

Type: Sets the line style in which the separatrices and orbits are drawn.

Dots... Draws the separatrices and orbits as a sequence of dots.

Dashes... Connects the integration points of the separatrices and orbits with a line.

Step size: Defines the step size. This value is used if we start integrating a separatrix or orbit.

Current step size: Gives the current step size. This is just an output to show how integration is working.

Max step size: Defines the maximum step size that we allow in the Runge–Kutta 7/8 method.

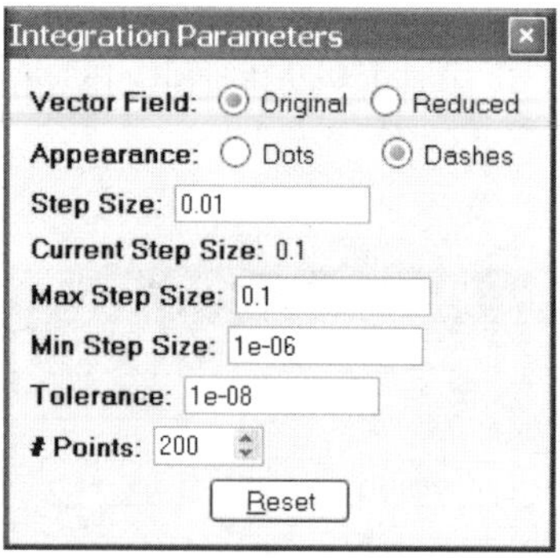

Fig. 9.10. The *Parameter of Integration* window

Min step size: Defines the minimum step size that we allow in the Runge–
Kutta 7/8 method.

Tolerance: Defines the required accuracy for the Runge–Kutta 7/8 method.

 # Points: This parameter indicates to the Runge–Kutta 7/8 method how
many steps it has to do each time we want to integrate a separatrix
or orbit.

Reset: This button resets all parameters in this window to their default
values. Useful when changing from one problem to another.

9.3.5 The Greatest Common Factor Window

Function: This window deals with the drawing of the greatest common factor
between the two polynomials which define the system.

Description: The *Greatest Common Factor* window is opened by selecting
the *GCF* button in the *Phase Portrait* window. In this window we call
MAPLE/REDUCE to plot the greatest common factor. Note that this
plot is a two-dimensional implicit plot which in the case of MAPLE has
needed some improvement, since it yielded very poor results when implic-
itly plotting reducible functions.

Panel items:

Appearance: Sets the line style in which the lines of singularities are
drawn. If the user already asked P4 to plot the lines and wants to
change the line style, he has to press the *Refresh* button in the *Phase
Portrait* window after he has changed this style.

Dots... Draws the lines as a sequence of dots.

Dashes... Connects the dots of each line.

Points: Denotes the number of unconditional data points. Note that a
high value may increase the computer time significantly. If the user
wants more information about this item, then he can check the "Reduce: Gnuplot interface Version 4" guide [109] or the help of MAPLE,
whichever applies.

Precision: Defines the maximum error which we will allow, expressed as
the negative exponent of a power of ten. This parameter is irrelevant when MAPLE is used. If the user wants more precision, simply
increase the number of points.

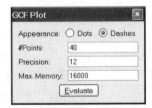

Fig. 9.11. The *Greatest Common Factor* window

Memory: Sets the maximum size of working space which we will allow to REDUCE. If the user increases the *Points* or *Precision* item, then he also has to increase this item. This parameter is irrelevant when MAPLE is used.

Evaluate button: Asks P4 to plot the lines of singularities. This may take some time, especially if the *Points* or *Precision* item is high.

9.3.6 The Plot Separatrices Window

Function: Allows the user to select the separatrices one by one.

Description: The *Plot Separatrices* window is opened by selecting the *Plot Separatrices* button in the *Phase Portrait* window or by selecting a singular point which has separatrices in the drawing canvas. If the user selects a point in the drawing canvas while holding down the SHIFT key, then P4 will select the closest singular point which has separatrices. A flash will show the selected point. The user will see in the *Plot Separatrices* window the type and the coordinates of this singularity. These coordinates are real in case the singularity is finite. If the user has selected a singularity at infinity, he gets the coordinates on the Poincaré sphere (i.e., $(X, Y, 0)$, where $X^2 + Y^2 = 1$), or on the Poincaré–Lyapunov sphere of degree (p, q) (i.e. $(0, \theta)$). If there are already some separatrices of this singular point drawn, the color of one of them is changed to gold. This is the first separatrix which will be studied.

Panel items:

Epsilon: This is the distance we move away from a singular point in order to start the integration of the separatrices. This value is equal to the one which is set in the *Find Singular Points Parameters* section of the *Planar Polynomial Phase Portraits* window. For some separatrices this value may be too small or too large. In this case the user has to modify this value, but should never allow it to be greater to 10^{-1}, to avoid

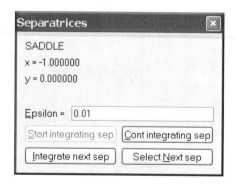

Fig. 9.12. The *Plot Separatrices* window

large errors when choosing the initial integration point. Do not forget to press the RETURN button after changing this value.

Start Integrating Sep button: Starts the integration of the chosen separatrix.

Cont Integrating Sep button: Continues the integration of the selected separatrix.

Integrate Next Sep button: Selects another separatrix of the same singular point and starts with the integration.

Select Next Sep button: It selects another separatrix of the same singular point.

9.3.7 The Limit Cycles Window

Function: Allows the user to search for non semistable limit cycles up to a certain degree of precision.

Description: The *Limit Cycles* window is opened by selecting the *Limit Cycles* button from the *Phase Portrait* window. In this window the user has to give two points forming a segment which he suspects is cut by at least one limit cycle.

Panel items:

x0, y0: Defines the first point of the line segment.

x1, y1: Defines the last point of the line segment. The user can select these two points by clicking the left button of the mouse on the first point, and while holding the button down, moving the mouse to the second point and releasing the button.

Grid: Determines the precision up to which the limit cycles will be determined. That is, if two consecutive limit cycles cut the selected segment in two points at distance greater than the *Grid* value, then P4 will detect them. Otherwise, it is possible that not both limit cycles are detected or that a continuum of limit cycles may be detected because of lack of precision.

 # Points: This parameter equals the number of steps the Runge–Kutta 7/8 method has to do each time we want to integrate an orbit with initial condition a point of the segment. If the orbit does not cross the

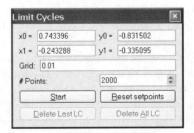

Fig. 9.13. The *Limit Cycles* window

Fig. 9.14. The *LC Progress* window

line defined by the segment at this time, the program will presume that the orbit does not cut the segment again. The user may note that this value is greater than the *# Points* value in the *Parameters of Integration* window. We suggest keeping it around the default value or even greater, since there may be very slow limit cycles which would remain undetected with low values. The user may get an approximate idea of which number he should enter by studying the integration of an orbit close to the limit cycle.

Start: This initiates the search for limit cycles. The window *LC Progress* showing a time bar is given, also including a button to abort the computation if something goes wrong, or becomes too time consuming. If so maybe a larger grid may help then.

Reset setpoints: Allows redefinition of the transverse section.

Delete Last LC: Deletes the last limit cycle found.

Delete All LC: Deletes all limit cycles found.

9.3.8 The Print Window

Function: Allows the user to output the phase portrait of the system to a file or a printing device.

Description: The *Print* window is opened by selecting the *Print* button in the *Phase Portrait* window.

Panel items:

EPS Image: Translates the picture into EPS format and saves it to a file.

XFIG Image: Translates the picture into FIG format and saves it to a file. This option is useful if the user wants to add arrows to the picture. One can use XFIG under UNIX to do it, or a JAVA tool for XFIG in WINDOWS.

JPEG image: Translates the picture into JPEG format and saves it to a file. The default saving name is always the name of the input file with the proper extension. The file is also saved by default in the same directory of the input file.

Black & White printing: Allows the user to choose between a full color image or a black and white image. In case of color printing, the black background

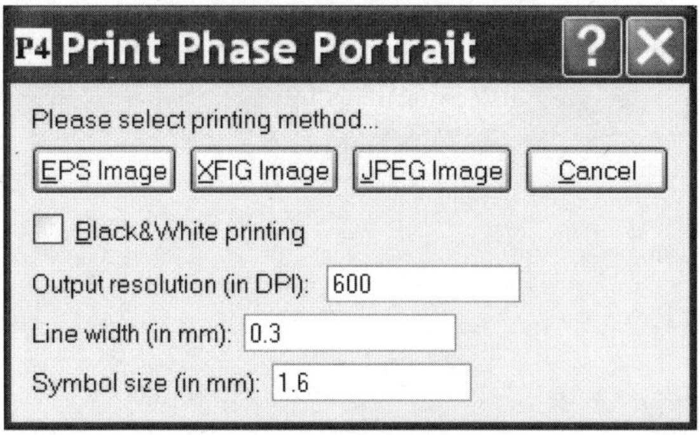

Fig. 9.15. The *Print* window

is turned into white, while the white line at infinity and the yellow orbits are turned into black and other colors are left unchanged. In case of Black & White, background is white and all drawing is in black.

Cancel button: Cancels everything and closes the window.

Output resolution (in DPI): All images are produced 15 cm wide, so if the user needs a bigger picture, it is suggested that he increases the *Output resolution* so as to maintain quality after rescaling the picture with some other tool.

Line with (in mm): The higher, the wider.

Symbol size (in mm): The higher, the larger.

10

Examples for Running P4

In this chapter we are going to provide several examples about the use of P4 [9], moving from easy examples to more complex ones, and trying to produce all the different situations and the most tricky problems that we have met up to now. Of course, we cannot pretend to cover all possible situations that may appear, since we are not aware of the complexity of the problem the reader may try to study, but at least we hope to give enough clues about how to solve them, or to show the impossibility of getting the complete study.

For these examples we will use the program running on a WINDOWS system, and using MAPLE as symbolic language. Where something different is expected for REDUCE's users, it will be described.

10.1 Some Basic Examples

Example 10.1 Consider the cubic vector field

$$
\begin{aligned}
\dot{x} &= y - x^3, \\
\dot{y} &= x + y + y^3.
\end{aligned}
\tag{10.1}
$$

The first thing to do after starting P4 is to introduce system (10.1) to the program. First you provide a name for the system such as, e.g., *example1*.

At the bottom of the window you see two fields where you can introduce the equation of the vector field. In the x' field you type y-x^3 and in the y' field x+y+y^3. Since there is no line of singularities, you can leave the *Gcf* field equal to 1. Now you are ready to study the system. Simply press the *Evaluate* button. The program now calls MAPLE (or REDUCE) which determines the singular points. You have to wait until you see in the *Output* window a message as in Fig. 10.1, or simply check that the button *Evaluate* has become active again.

Go to the top of the window and press the *View* menu button with the left button of the mouse. This will open a menu where you can choose to see

Fig. 10.1. The end of the calculations

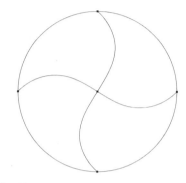

Fig. 10.2. Stable and unstable separatrices of system (10.1)

either information about the finite plane or about the points at infinity. Click for the finite plane. In this window you see that the system has only one finite singular point, namely $(0,0)$ which is a saddle. If you click again on the *View* menu button and choose for the infinite, you will be informed that the origin of each infinite local chart in the Poincaré compactification is also a singular point (in fact, each is a node).

Now you are ready to plot the phase portrait. Go to the top of the window and press the *Plot* button. This will open the *Phase Portrait* window. It may be necessary to press the *Refresh* button. In this window you see one green box, which represents the finite saddle. On the circle you see two blue boxes which represent the stable nodes at infinity, and two red boxes, which represent the unstable nodes at infinity. Pressing the *Legend* button will open the *Legend* window. In this window you get all the information about the symbols that you have in the drawing canvas. Now if you press the *Plot All Separatrices* button, some lines in red and blue will appear. These lines are the unstable and stable separatrices of the saddle; see Fig. 10.2. Red means unstable and blue means stable, and when referring to separatrices this means with respect to the point where you have started their integration. In any case, all curves are drawn in black in this book.

This vector field has only hyperbolic singular points and shows no complication. Thus, one single click to the *Plot All Separatrices* button has been enough to see everything interesting on it. But you may be interested in seeing

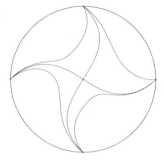

Fig. 10.3. Some more orbits of system (10.1)

how orbits flow and so you can left-click anywhere inside the Poincaré disk and the *Orbit* window will appear which will ask you if you want to integrate time-forwards or time-backwards from that point. You press the *Forward* button and you see a yellow orbit going to infinity. You may press the *Continue* button to make it longer if wished. Once you are satisfied with it, you may press also the *Backwards* button. You can also use the letters "F" for Forward, "B" for Backwards, and "C" for Continue after clicking the starting point, provided the *Phase Portrait* window is active.

You may also have noticed that there is a two-dimensional number in the bottom-left corner of the *Phase Portrait* window that changes while you move the mouse. This gives you the exact point in planar coordinates, i.e., with the Poincaré compactification undone, at which the mouse is located at that moment. This can be very useful for getting an approximate idea of the coordinate points in this compactification. If you need to draw an orbit passing through a certain point, there is no need to play with the mouse to look for it. You can always specify it exactly on the *Orbit* window. See Fig. 10.3.   □

Example 10.2 Consider the quadratic vector field

$$\dot{x} = x - x^2 - y^2,$$
$$\dot{y} = x^2 - 2*x*y + 2*y^2. \tag{10.2}$$

We will not repeat the explanation of steps already described but will concentrate on new options. So you introduce the vector field as described in Example 10.1, and evaluate it. Just notice that since you are using MAPLE now, you must introduce the system according to the conditions of this program, that is, you must include a multiplying symbol $*$ between letters, but also between letters and numbers. If you are using REDUCE on a LINUX system, then the second would not be needed (but is permissible). However, REDUCE does not allow any blank spaces within formulas, as MAPLE does. If you want your files to be compatible with both systems, then always put the $*$'s and leave no blanks.

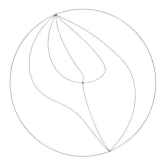

Fig. 10.4. Stable and unstable separatrices of system (10.2)

By pressing *View* button of the main window you see that the system has only one finite singular point, namely $(0,0)$, which is a semi-hyperbolic saddle-node and a pair of infinite singular points, which are nodes.

When plotting the phase portrait you see one purple triangle which represents the finite saddle-node. On the circle you see one blue box which represents the stable node at infinity, and a red box, which represents the unstable node at infinity. Now if you press the *Plot All Separatrices* button, there will appear two lines in red, the two unstable separatices which have one of the infinite singular points as their ω-limit. But there is one separatrix missing. If you look carefully, you may see a very small blue slash on the purple triangle. Nonelementary points usually have "slow" separatrices and the default number of iterations might not be enough to draw them satisfactorily. In this case you must again press the *Plot All Separatrices* button to begin to see the blue (stable) separatrix. To get this separatrix two clicks are enough to display it completely. Other cases may require more clicks or deeper study. You may also like to see one orbit in each connected component, for which it is enough to click once in each region and command the plot of the orbit through the point selected; see Fig. 10.4.   □

Example 10.3 Consider the quadratic vector field

$$\dot{x} = -y + l * x^2 + 5 * a * x * y + y^2,$$
$$\dot{y} = x + a * x^2 + (3 * l + 5) * x * y. \tag{10.3}$$

Introduce system (10.3) using the parameters and formulation given. As you can see, you may use parameters, which may help you to modify the vector field and test several vector fields belonging to a certain family. But these parameters must have a fixed value. Now you must click on the *Number of parameters* option and set how many parameters you are using. Fix it to just two parameters. You will see that two lines of boxes are opened below. Introduce the name of the parameter (i.e., a and l) in the left column and a fixed number in the second column, for example $a = 1$ and $l = -0.5$. The fixed number may be also a formula depending on other parameters, but you must avoid cyclic recursions. There is no limit to the number of parameters you may use. You can use parameters with several letters, or even parameters with letters and numbers, just remembering that they may not start with numbers. It is generally preferable to set the computing mode to numeric by switching the *Algebraic/Numeric* option in the *Parameters Find Singular Points* window if it is not. This must be done when the complexity of the vector field is high or when you detect computing problems which slow down the study in algebraic mode. Now make the program work with the *Evaluate* button. With the *View* button you see that the system has four finite singular points, which are three saddles and one weak focus. For this last point, the program also computes its Lyapunov constants and detects that the first two are zero (or simply smaller than the *Precision* you are allowing) and the third is different from zero (negative), so the point is a stable weak focus of third-order. There are also three pairs of infinite singular points, all of them nodes.

By pressing the *Plot* button of the main window you see three green boxes, which represent the finite saddles, and one dark blue diamond, which represents the stable weak focus. On the circle you see red and blue boxes for the nodes. Now if you press the *Plot All Separatrices* button, there will appear some lines in red and blue. These lines are the unstable and stable separatrices of the finite saddles; see Fig. 10.5.

For system (10.3) put now $a = 1$, $l = 0$ and evaluate again. The only interesting change is that now the singular point at $(0,0)$ has its three Lyapunov constants equal to zero. The vector field is quadratic and the program knows it. It knows that when a quadratic system has a singular point whose first three Lyapunov constants are zero, then the point must be a center. The program also knows that cubic systems with no quadratic part (and no constant part) have five independent Lyapunov constants that allow one to decide if the singular point is a weak focus or a center; in the latter case, the first five Lyapunov constants must be equal to zero. Finally, the program also knows that all finite antisaddles of a Hamiltonian vector field must be centers.

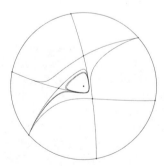

Fig. 10.5. Stable and unstable separatrices of system (10.3) for $a = 1$ and $l = -0.5$

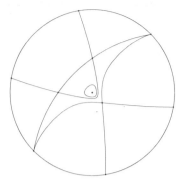

Fig. 10.6. Separatrix skeleton of the system (10.3) for $a = 1$ and $l = 0$

In this case $(0,0)$ is plotted now with a green diamond to show it is a center, and some separatrices from saddles coincide, so you may see the final result in red with some pixels in blue (or the opposite), and if you plot an orbit inside the canonical region where the center is situated, you will see it as periodic; see Fig. 10.6.   □

10.2 Modifying Parameters

Example 10.4 Consider the quadratic vector field

$$\dot{x} = (y - x) * (1/2 - y),$$
$$\dot{y} = (a - y) * (1 - a - x). \tag{10.4}$$

Introduce system (10.4) using the parameter and formulation given. As you can see, you are not limited to writing the vector field coefficient after coefficient. You may give it using symbolic formulas. Set the parameter $a = 0.6$. Now make the program work with the *Evaluate* button. With the *View* button you see that the system has three finite singular points which are one saddle, one unstable node, and one strong stable focus. There is also one pair of infinite singular points, which are saddle–nodes.

By pressing the *Plot* button of the main window you see one green box, which represents the finite saddle, one red box, which represents the unstable node and one blue diamond for the stable focus. On the circle you see two purple triangles, which represent a pair of saddle–nodes at infinity. Now if you press the *Plot All Separatrices* button, there will appear some lines in red and blue. These lines are the unstable and stable separatrices of the finite saddle; see Fig. 10.7. Some additional clicks may be needed to prolong the separatrices.

Your study is not yet complete since one separatrix is still missing. Near the point $(1,0,0)$ (that is, the saddle–node at infinity) you see a small dark

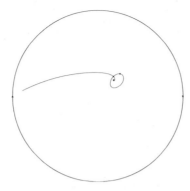

Fig. 10.7. Stable and unstable separatrices of system (10.4)

blue line. This line represents the center stable separatrix of that point. You can use the same method as described in Example 10.2 of pressing the *Plot All Separatrices* button but you must consider that this will integrate again all other separatrices, and this may take longer than if you only integrate the separatrix you need. You can concentrate your efforts on this separatrix by selecting this point. Go with the mouse near that point and press the left button while holding down the shift key. This opens the *Plot Separatrices* window. Notice also that a flash can be seen around the saddle–node, which confirms that you have selected it. By pressing the *Start Integrating Sep* button the center separatrix will be integrated. You see that this separatrix is very slow, so you have to press the *Cont Integrating Sep* button several times. Another possibility is changing the default number of integrations done for each click. You can press the *Integration Parameters* button in the *Phase Portrait* window and change the *# Points* option to a higher number, for example 2,000. This value affects all integrations that you do from now on, so it can slow down the procedure if you press the *Plot All Separatrices* button in a system with many separatrices, but it can be very useful when studying one single slow separatrix. You may also do the same with the finite saddle since one of its separatrices has not yet reached its ω-limit. By clicking close to it (with the shift key pressed) you select the closest singular point with separatrices, in your case, the finite saddle. You will see as one of the separatrices turns into orange color. This is the currently selected separatrix. You may move from one to another by clicking the *Select Next Sep* button until you get the one you want to prolong; see Fig. 10.8.   □

Example 10.5 Consider the quadratic vector field

$$
\dot{x} = \frac{1}{25} - \frac{9}{100} * x + \frac{3}{10} * y + \frac{9}{2} * x^2,
$$
$$
\dot{y} = -\frac{3}{125} + x - \frac{9}{50} * y + \frac{15}{2} * x * y.
$$

(10.5)

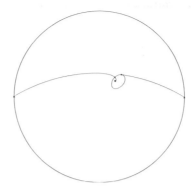

Fig. 10.8. Separatrix skeleton of the system (10.4)

Enter this system into P4 and ask the program to determine all the singularities. If you now press the *Plot* button then you see near the origin some singularities that are very close to each other. In fact there are three points, namely the points $(3/125, -2527/18750), (1/50, -2/15)$, and $(0, -2/15)$. So you have to zoom in to see these three singularities. In order to zoom in you must left-click in a corner of a rectangle you want to amplify while you hold down the Control key. Do not release either the Control key or the left button of the mouse while you move the mouse. You will see a white rectangle which changes as you move the mouse. You then release the left button at the opposite corner of the region to magnify and you will get a new window which shows only that rectangle. Be aware that the unit of measure for the coordinates x and y may have changed during the zoom if you have selected a very narrow or a very flat rectangle. The program will try to make a rectangular box that approximates as closely as possible your selection. However, because the plane has been compactified to form the Poincaré disk, you might have specified very large coordinate corners, in which case, the narrowness of the box would imply an almost invisible window. After you zoom in and press the *Plot All Separatrices* button, you get in the zoom window a plot as in Fig. 10.9.

As you see there are strange lines in the picture. This is because the *epsilon* value with which P4 has started to integrate the separatrices of the saddle point is too large. The *epsilon* value corresponds to the distance that you move away from the separatrix along its Taylor approximation when starting to integrate. Therefore, you have to change the *epsilon* parameter. Select this saddle point by pressing the left button of the mouse while holding down the shift key. Now the *Plot separatrices* window appears. In this window you change the value of *epsilon* to 0.001 and press the return key. Now you select the *Refresh* button in the *Zoom* window and press the *Start Integrating Sep* button followed by the *Cont Integrating Sep* button several times. You do the same for the other separatrices. After drawing all the separatrices you get

Fig. 10.9. *Epsilon* value too great

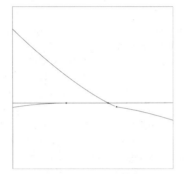

Fig. 10.10. Good *epsilon* value

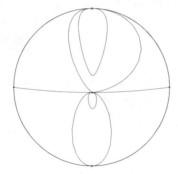

Fig. 10.11. Phase portrait of system (10.5)

the picture of Fig. 10.10. To obtain a global vision of the phase portrait more orbits have to be drawn; see Fig. 10.11.

There is also a global *epsilon* value which can be set in the *Find Singular Points Parameters* section of the main window and that is used by default in all singular points with separatrices. This number is set by default to 0.01 since lower values are needed only for very special cases and they would delay most other systems. Also, sometimes, and for some specific cases, it may be necessary to set *epsilon* to higher values, such as 0.05 or even 0.1, to help the program and start integrating further from the singular point, avoiding the low start of some central manifolds.   □

Example 10.6 Consider the cubic vector field

$$\dot{x} = y,$$
$$\dot{y} = -x^2 * (x + 1) + d * (a + b * x + x^2) * y. \tag{10.6}$$

Introduce system (10.6) using the parameters and formulation given. You test first the case $d = 0.1$, $a = b = 0$ and evaluate it in *Numeric* mode. With the *View* button you see that the system has two finite singular points, one strong unstable focus and one degenerate point at $(0,0)$. This point is studied in detail and you see that the program informs you that it has a stable separatrix arriving in the direction $(x(t), y(t)) = (-t^2, 0.8165t^3 + \ldots)$ and an unstable separatrix departing in the direction $(x(t), y(t)) = (-t^2, -0.8165t^3 + \ldots)$. It also tells you that the sector between the first and second separatrix is hyperbolic, and the sector between the last (second) and first separatrix is also hyperbolic. It also gives you the index of the singular point, which is 0. Thus the point is a cusp. There are also two pairs of infinite singular points, one a semi-hyperbolic saddle and the other a degenerate point. For this last point, the program has found four sectors but it happens that they are all parabolic. So the four separatrices described by the program are just orbits which arrive to that point but are not separatrices, even though it says so. In fact, the degenerate point is nothing more than a degenerate stable node. Only separatrices which coincide with separatrices of hyperbolic sectors are real ones. Those which limit with parabolic or elliptic sectors are just local separatrices in the blowing up system which may not maintain their character in the global system.

By pressing the *Plot* button of the main window you see one red diamond, which represent the finite focus, and one cross which represents the cusp (in general any degenerate singular point). On the circle you see two green triangles and two more crosses, which represent a pair of semi-hyperbolic saddles and the degenerate singular points at infinity, respectively. Now if you press the *Plot All Separatrices* button, there will appear some lines in red and blue starting at the cusp. Two more clicks on the same button give you a better idea of what those separatrices do; see Fig. 10.12.

Your study is not yet complete since two separatrices are still missing. The infinite semi-hyperbolic saddles must have one separatrix each moving to the finite plane, but the most that you can see are some unclear blue dashes at the top and the bottom. If you zoom in on the topmost region as described in Example 10.4, you will realize that the blue dash does not correspond to the saddle, but it is one of the orbits of the degenerate node in which you are not interested at all. So you must select the semi-hyperbolic saddle by left-clicking close to it while holding the shift key. This time, you must modify the

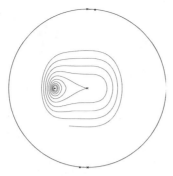

Fig. 10.12.   Stable and unstable separatrices of system (10.6) with $d = 0.1$ and $a = b = 0$

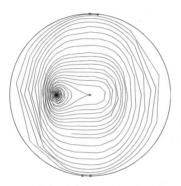

Fig. 10.13.   Stepping too fast

epsilon to a greater value (like 0.1), but you must also modify the maximum integration step size allowed in *Phase Portrait–Integration Parameters* window given by *Max Step Size*. Try setting it to 10 and also setting *# Points* to 2,000. After pressing the *Cont Integrating Sep* two times you see it drawing, but what appears does not seem to be a differentiable curve, but a dashed one, which becomes even worse if you continue integrating; see Fig. 10.13. The integration speed needed to get away from the singular point is too fast when you are far from it.

You need to play with the parameters in order to get a nice phase portrait. For example, click on the *Epsilon* box of the *Separatrices* window and press return to redraw this orbit, make two integrations with the previous conditions (with 2,000 integration steps) and then move back to 200. After nine integrations more you will see the separatrix already departing from the infinite singular point. At that moment, again change the maximum time to 0.1 and *# Points* to 2,000 , and continue integrating. Now it appears differentiable, but when moving close to the other semi-hyperbolic saddle, it slows down again. As it takes too long to cross this region, it is convenient to set again the maximum integration step to 1 and once the separatrix starts leaving the

Fig. 10.14. Separatrix skeleton of the system (10.6) with $d = 0.1$ and $a = b = 0$

slow region to change it again to 0.1 to avoid the dashes. New crosses close to both semi-hyperbolic points will mean new slow downs, but since new crosses will take place further and further away, the effect is not so severe. You may do the same with the semi-hyperbolic saddle on the bottom of the screen, but it will not add much information.

You will finally see that both separatrices from the semi-hyperbolic saddles have the same α-limit as the stable separatrix of the cusp; see Fig. 10.14.

Now if you set the parameters $d = a = b = 0.1$ and evaluate it in *Numeric* mode you see that the system now has a saddle–node at $(0, 0)$. You may notice that the program takes a bit of time when computing and that it often shows a line saying `test failed for i=xx`. The reason is that the separatrices of some singular point twist very fast just upon leaving the point and thus it becomes quite complicated to find the correct Taylor approximation for the separatrices. It may happen sometimes that a program collapses during that procedure, in which case you are advised to disconnect the *Test Separatrices* option in the *Parameters Find Singular Points* section of the main window. Be aware then that you cannot completely trust the separatrices drawn, since tests have not been passed. The infinite configuration has not changed.

By pressing the *Plot* button of the main window you see one red diamond, which represents the finite focus, and one purple triangle which represents the saddle-node. On the circle you see two green triangles and two more crosses. Now if you press the *Plot All Separatrices* button, there will appear some lines in red and blue starting at the saddle-node. Two more clicks on the same button give you a better idea of what those separatrices do; see Fig. 10.15. However, all that you see greatly resembles the previous case and the behavior of the saddle-node seems to be a cusp. In fact, what happens is that there are two unstable separatrices and a stable one (central manifold). But both unstable separatrices are very close to each other and even though they start in different directions from the singular point, one of them twists very fast and is impossible to distinguish one from the other. You can nevertheless check that there are two by integrating the separatrices of this point one by one

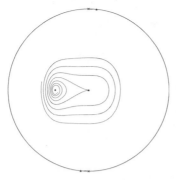

Fig. 10.15. Stable and unstable separatrices of system (10.6) with $d = a = b = 0.1$

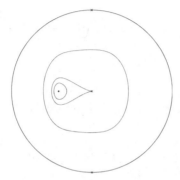

Fig. 10.16. Stable and unstable separatrices of system (10.6) with $d = 0$

with the *Plot Separatrices* window, a very small *Epsilon* value (try it with 10^{-4}, 10^{-5}, and up to 10^{-6}), and a very deep zoom.

Now if you set the parameter $d = 0$ (a and b become irrelevant) and evaluate it in any mode you prefer, you see that the system has a cusp at $(0,0)$ and a center at $(-1,0)$. The program knows that it is a center because the system is Hamiltonian. The infinite configuration has only one degenerate point which has only two separatrices (the ones corresponding to infinity), and two hyperbolic sectors. Thus, this infinite singular point has a kind of box flow around it. If you integrate all separatrices, you see that the separatrices from the cusp seem to coincide. Since the antisaddle is a center, you can be sure that they coincide. Any other orbit you integrate will resemble a periodic orbit, which is what in fact they are, although the program cannot assure it; see Fig. 10.16.   □

Example 10.7 Consider the cubic vector field

$$\dot{x} = (x - 3) * (-x + x^2 - 3 * x * y + 2 * y^2),$$
$$\dot{y} = (x - 3) * (y + 4 * x^2 - 5 * x * y - 2 * y^2). \qquad (10.7)$$

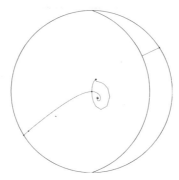

Fig. 10.17.   Stable and unstable separatrices of system (10.7)

Introduce system (10.7). It is clear that this vector field has a common factor and thus it will have an infinite number of finite singular points. So the first thing to do is set the *Gcf* option to 0. Now you are saying to the computer that you are not sure if there is a common factor and that it must look for it. When you wrote 1 for this option in all previous examples, you were saying to the program that there was no greatest common factor between the equations, and that it did not need to spend its time looking for it. In fact, this task can be quite time-consuming and it is advisable to avoid it if possible. Another possibility is to say to the computer that the greatest common factor is $x - 3$ and that taking that, it may skip its computation also. Of course, if you cheat, the program will fail.

Now the study of the system says that it has a common factor, and that it will study the reduced system which has some singular points. When plotting the phase portrait you will see a button called *GCF* which is normally inactive. By clicking it you enter a window where the *Evaluate* button will draw for you the curve defined by the common factor. The parameters there do not normally need to be changed, but since the complexity of the common factor can be great, they may help in some cases.

Now you see a green line in the Poincaré disk. Although it seems a curve it corresponds to a straight line due to the compactification. You may plot all separatrices (one of them may be plotted better with a smaller time step since dashes appear); see Fig. 10.17.

You are interested in what happens close to the line of singularities.

You may notice that the separatrix coming from the infinite saddle on the right-hand side has stopped on that line. Also, if you click on a point close to the upper end of the line of singularities, and ask to integrate it *Backwards* you will see that it stops on the line of singularities. But you may wish to see those separatices and orbits continued across the line of singularities. You can do that by changing the *Original/Reduced* option in the *Integration Parameters* window. Now when you ask to integrate all separatrices, you will see that the separatrix coming from the right infinite saddle crosses the line of singularities,

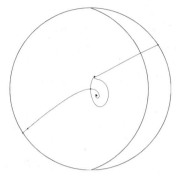

Fig. 10.18.   Portrait in the reduced mode of system (10.7)

and moreover, it changes its color when doing so. The reason is that the right part of it is unstable with respect to the infinite saddle, but the other part must be seen as stable with respect to that singular point. Notice also that the speed of the flow has changed when integrating the reduced system and that even the separatrix that looked dashed in the previous sample now looks nicer; see Fig. 10.18.   □

10.3 Systems with Weak Foci or Limit Cycles

The presence of limit cycles (or even just their possible presence) greatly complicates the realization of the phase portrait of a vector field. Most times you will just be able to assure the existence of an odd number of limit cycles (most probably one); an example is when you see a stable orbit (stable from the point of view of the saddle or more degenerate point from which it comes) that spirals around a stable focus (or exactly the opposite situation). In some examples you will see a fast transition from a focus to a limit cycle which will clearly detect the limit cycle, but other systems will either have an infinitesimal limit cycle or a very slow transition which will make it almost impossible to detect the cycle numerically. In cases having more than one limit cycle, these problems may become worse. Nevertheless, with a bit of patience and experience, you may be able to detect several limit cycles in some examples.

Also the detection of a weak focus is important since once you have it, by small perturbations you can produce as many limit cycles as the order of weakness of the focus.

P4 is ready to detect any order of weakness of a focus, but by default this order is set to 4, that is, if the proposed system has a weak focus it will start checking its order of weakness until 4th degree. If it is even weaker, that is, if all previous Lyapunov constants are zero, you may adjust a parameter in the main window. Be careful that the amount of time needed to compute each

Lyapunov constant grows exponentially, and thus it is not advisable to set this parameter to high values without good reason.

You must also remember that when studying a quadratic system, if a singular point has its first three Lyapunov constants equal to zero, then it is a center. The same happens when in a linear plus cubic system a singular point has its first five Lyapunov constants equal to zero. The program knows that and in those cases it will note that fact. Also if a system is Hamiltonian it will know that an elementary point that is not a saddle must be a center. But for any other polynomial system not included in these cases, the program cannot know that a point is a center when even all computed Lyapunov constants are zero.

Example 10.8 Consider the quadratic vector field

$$\dot{x} = -y + (b - v)/3 * x^2 + (2 * a + l) * x * y + n * y^2,$$
$$\dot{y} = x + a * x^2 + (2 * v + b)/3 * x * y - a * y^2. \tag{10.8}$$

Introduce system (10.8) using the parameters and formulation given. You first test the case $b = a = l = v = n = 1$ and evaluate it in *Numeric* mode. With the *View* button you see that the system has four finite singular points which are three saddles and one weak focus. The computer automatically evaluates the first Lyapunov constant, detects that it is 0.25, clearly different from zero, and determines that it is an unstable weak focus of first-order.

By pressing the *Plot* button of the main window you see one dark red diamond, which represents the finite focus, and three green squares for the saddles. Now if you press the *Plot All Separatrices* button, there will appear some lines in red and blue. You see one blue separatrix coming from one of the saddles spiralling around the weak red focus. That is, a stable separatrix which has the saddle as ω-limit is spiralling out from an unstable focus. This is consistent with either the complete absence of limit cycles, the presence of an even number of hyperbolic limit cycles, or the presence of any number of semistable limit cycles. If you want to prolong this separatrix, you may click several more times on the *Plot All Separatrices* button or use the *Plot Separatrix* button specifically for this particular separatrix alone, and you will see that the blue separatrix seems to actually spiral out from the focus, which indicates that either there are no limit cycles, or that all limit cycles that exist are infinitesimal at this scale; see Fig. 10.19. You could use the zoom techniques explained before to improve the picture.

Now change the parameter l to $l = 0$ and evaluate the system again. The first Lyapunov constant of the singularity at the origin is now zero, and since by default P4 computes up to four Lyapunov constants automatically, it determines both this and the fact that the second Lyapunov constant is 0.25. Thus it determines that the origin is now an unstable weak focus of second-order. The phase portrait is very similar to the previous one.

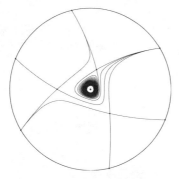

Fig. 10.19. Phase portrait of system (10.8) with $a = b = l = n = v = 1$

Now set also $b = -5$ and evaluate again. The first two Lyapunov constants are zero and the computer finds the third one to be -1.5625; thus you have a stable weak focus of third-order.

Finally set $b = -10$ and $v = -1$ and evaluate again. Now you see that the first two Lyapunov constants are zero, so the third is required. The computed value is 1.818989×10^{-12} (this number may vary slightly depending on the CPU used, due to numerical rounding). Since we set the precision parameter to 10^{-8} by default, the program will take this number as zero, assuming that some numerical perturbations have occurred. Since the system is quadratic and the first three Lyapunov constants of the singularity are zero, the program concludes that the singularity is a center.

If you set $b = 0$ and $v = 3$ the program does not even start computing Lyapunov constants since it detects from the beginning that the system is Hamiltonian, and automatically determines that all elementary points that are not saddles are centers.   □

Example 10.9 You consider the quadratic vector field

$$\dot{x} = -x^2 - 0.4 * x * y + 1.5 * x + 0.27 * y^2 - 0.88 * y - 0.89,$$
$$\dot{y} = 1.2 * x^2 + 0.15 * x * y - 1.5 * x + 0.12 * y^2 + 1.6 * y + 0.15. \qquad (10.9)$$

Introduce system (10.9) and evaluate it in *Numeric* mode. With the *View* button you see that the system has two finite singular points which are one saddle and one strong stable focus. Now if you press the *Plot All Separatrices* button a couple of times you will see that one of the stable separatrices of the saddle spirals out from a neighborhood of the strong focus which is also stable; see Fig. 10.20. This is conclusive proof of the existence of at least one limit cycle. You may easily remember that if a red separatrix spirals around a red point (or blue around blue) you have a limit cycle.

You can confirm the relative position of the limit cycle by clicking inside the region apparently enclosed by the separatrix and asking the program to

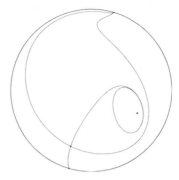

Fig. 10.20. Phase portrait of system (10.9)

Fig. 10.21. One orbit inside the limit cycle

integrate one orbit backward. After several more clicks on the *Continue* button you can determine that the limit cycle must be very close to where the blue separatrix has stopped its advance toward the focus; see Fig. 10.21.

But you can also detect the limit cycle using the *Limit Cycles* button. Click on it and a new window will open. Once it is opened you must simply left-click on the *Phase Portrait* window and move the mouse a little while holding the left button down to generate a transverse section where the limit cycle is supposed to be, that is, close to the region where the blue separatrix is spiralling. We suggest that you simply make a short segment which crosses it. Once you have specified the section just press the *Start* button in the *Limit Cycles* window and in a few seconds you will see a purple closed orbit portrayed. That is the limit cycle; see Fig. 10.22. A *Searching for limit cycles* window appears with a time bar which should show the time left for computing but whose most useful application is to stop searching, since it may easily delay a lot before or after finding a limit cycle.

You may change the parameter *Grid* at will. The smaller you set it, the more precisely the limit cycle will be found (and more time will be needed).

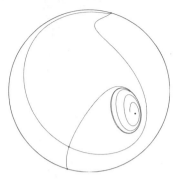

Fig. 10.22. The limit cycle

The greater you set it, the faster you will find the limit cycle (with some loss of precision). □

Example 10.10 Consider the quadratic vector field

$$\dot{x} = 1 + x * y,$$
$$\dot{y} = a00 + a10 * x + a01 * y + a20 * x^2 + a11 * x * y + a02 * y^2. \tag{10.10}$$

Introduce system (10.10) using the parameters and formulation given. Set $a00 = a01 + a11 - a10 - a20 - a02$. You test the case $a10 = 15.28$, $a01 = 8.4$, $a20 = -12$, $a11 = -1.398$, and $a02 = 3$. Evaluate it in *Numeric* mode. With the *View* button you see that the system has two finite singular points which are a node and a strong focus, both unstable. Now if you press the *Plot All Separatrices* button you will immediately see that the default parameters for integration are not the proper ones, especially in reference to the unstable infinite separatrix that spirals around the strong focus, since it goes too fast and appears in dashes. Nevertheless, it is enough to make you infer of the existence of at least one limit cycle.

Press the *Plot* button in the main window again, and before clicking the *Plot All Separatrices* button modify the *Max Step Size* to 0.01 in the *Integration Parameters* window. Now you may again click the *Plot All Separatrices* button. You must do it several times since the integration is much slower, and you get the main phase portrait; see Fig. 10.23.

But in fact, this is an example of quadratic system with three limit cycles, luckily of visible size, which were found in [32]. Set the *# Points* in the *Integration Parameters* windows to 2,000 to compensate for the smaller integration step. Click in the *Phase Portrait* window approximately on the point with coordinates $(3 \pm 0.1, -3 \pm 0.1)$ (or set it directly on the *Plot Orbits* window). You do not need strict precision in clicking. It is just to be sure that you follow the example as it is described. Integrate from this point in the forward sense and continue it a bit. You will see a yellow orbit which is clearly increasing turn after turn. Since the red separatrix was also moving in

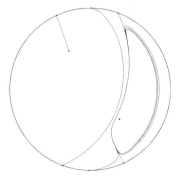

Fig. 10.23. Phase portrait of system (10.10) with given conditions

Fig. 10.24. Outer limit cycle

positive time, you already have a limit cycle located in the apparent annulus formed by this separatrix and the orbit drawn. You may use the *Limit cycles* button to try to draw it. We suggest checking that the *Grid* parameter is set to 0.01 before starting since the Poincaré return map for this system is quite close to the identity and it may take a lot of time to compute the limit cycle with a smaller grid. Once the grid is set, the section across the annulus is specified and the procedure is started, you will see the limit cycle portrayed in purple; see Fig. 10.24.

Now click on the *Phase Portrait* window approximately on the point with coordinates $(1.5\pm0.1, -1\pm0.1)$ and integrate forward. It may need some clicks on the *Continue* button before you can observe that the orbit is moving inside; see Fig. 10.25. You need to pay attention to the pixels which turn yellow in the inner border of the orbit already drawn. This fact, together with all you know already of the phase portrait, confirms the existence of two more limit cycles, one in the apparent annulus generated by both orbits drawn, and another around the strong focus and inside the inner orbit.

In short, the program has helped you to prove numerically the existence of at least 3 limit cycles. However, trying to draw the two inner limit cycles as you have done with the outer one is not so easy. The reason is that the

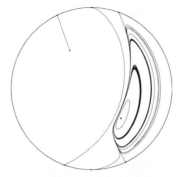

Fig. 10.25. Looking for more limit cycles

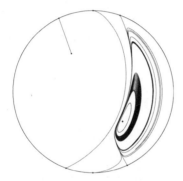

Fig. 10.26. When limit cycles are hard to find

Poincaré return map near these limit cycles is too close to the identity. If you set the grid to 0.01 as before what you will get will be that the program will detect a continuum of periodic orbits; see Fig. 10.26. So you might need a bit of patience, work on a zoomed screen, and use a lower grid with very precise transverse sections to detect the other two limit cycles. Once done, you can clear all drawn orbits and even separatrices to have a clean image of the limit cycles. A *Refresh* click may be needed.   □

10.4 Exercises

Exercise 10.1 Construct the phase portraits of all topologically different linear phase portraits on the Poincaré disk. Determine which of them are topologically equivalent on the plane but nonequivalent on the disk.

(i) $\dot{x} = x,$ $\dot{y} = y,$
(ii) $\dot{x} = x,$ $\dot{y} = 2y,$
(iii) $\dot{x} = x,$ $\dot{y} = -y,$

(iv) $\dot{x} = x + y,$ $\dot{y} = -3x + 2y,$
(v) $\dot{x} = -y,$ $\dot{y} = x,$
(vi) $\dot{x} = y,$ $\dot{y} = 1,$
(vii) $\dot{x} = x,$ $\dot{y} = 1,$
(viii) $\dot{x} = x,$ $\dot{y} = 0,$
(ix) $\dot{x} = y,$ $\dot{y} = 0.$
(x) $\dot{x} = x + y,$ $\dot{y} = y.$

Exercise 10.2 Prove that the following system forms a rotated vector field family with rotation parameter α, and has a semistable limit cycle $x^2 + y^2 - 1$ of multiplicity two for $\alpha = 0$, using the tools given in Chap. 7.

$$\dot{x} = -y + x(x^2 + y^2 - 1)^2,$$
$$\dot{y} = x + y(x^2 + y^2 - 1)^2 + \alpha(-y + x(x^2 + y^2 - 1)^2).$$

Also prove for which sign of α (either positive or negative) you have two hyperbolic limit cycles, and check it numerically with P4.

Exercise 10.3 Construct the phase portrait of system

$$\dot{x} = y + x^2 - x^2 y + 2y^3,$$
$$\dot{y} = x + y^2 - x^3 - xy^2,$$

check that there is a finite limit cycle, and show that the circle at infinity (the equator of the Poincaré sphere) behaves like a semistable limit cycle. Now you perturb the infinite limit cycle by adding some higher order terms, trying to produce a finite limit cycle. Determine a value of ε for which there are two finite limit cycles for

$$\dot{x} = y + x^2 - x^2 y + 2y^3 + \varepsilon(x^3 y + y^4),$$
$$\dot{y} = x + y^2 - x^3 - xy^2 + \varepsilon x^2 y^2.$$

Then use a rotated vector field family

$$\dot{x} = y + x^2 - x^2 y + 2y^3 + \varepsilon(x^3 y + y^4),$$
$$\dot{y} = x + y^2 - x^3 - xy^2 + \varepsilon x^2 y^2 + \alpha(y + x^2 - x^2 y + 2y^3 + \varepsilon(x^3 y + y^4)),$$

so as to reduce the distance that separates the limit cycles, and get an approximation of the value of the rotation parameter for which a finite semistable limit cycle occurs. You will see that it is impossible to be certain of the value for which the semistable limit cycle exists. At most, you can be sure that for a certain value of α there are two limit cycle, and for another, there is none.

Hint: It may be necessary to play with the *Projection* parameter a little since there may be interesting information to view close to infinity which is not clear with the default value.

Exercise 10.4 Construct the phase portrait of

$$\dot{x} = x + 4x^4y - 12x^2y^3,$$
$$\dot{y} = y + 12x^3y^2 - 4xy^4,$$

and describe it.

Hint: This time, the separatrices do not seem conclusive enough to determine a complete qualitative and quantitative phase portrait since the existence of several infinite singular points may raise the doubt that some important separatrices are still undrawn. Depending on which orbits the user draws to see it more clearly, the draft of the picture may even look erroneous; the user needs to extract detailed information about the infinite singular points in order to understand it.

Exercise 10.5 Change the linear part of the previous exercise. Construct the phase portrait of

$$\dot{x} = -y + 4x^4y - 12x^2y^3,$$
$$\dot{y} = x + 12x^3y^2 - 4xy^4.$$

It shows a nice set of apparently periodic orbits but the program is not conclusive regarding the center-focus problem. Anyway, it induces to search for an algebraic proof of it. Which?

Exercise 10.6 Construct the phase portrait of

$$\dot{x} = (x^2 - 1)(x^2 - (2k-1)^2)(x + \sqrt{5}y),$$
$$\dot{y} = (y^2 - 1)(y^2 - (2k-1)^2)(\sqrt{5}x + y),$$

for $k = (\sqrt{5} - 1)/2$ and check graphically how many invariant straight lines appear.

Hint: This a good time to change the picture from Poincaré disk to the plane.

Exercise 10.7 Construct the phase portrait of

$$\dot{x} = 2x(x^2 - 3)(4x^2 - 3)(x^2 + 21y^2 - 12),$$
$$\dot{y} = y(-216 + 378x^2 + 378y^2 - 315x^4 - 189y^4$$
$$+ 35x^6 + 105x^4y^2 - 63x^2y^4 + 27y^6),$$

and check graphically how many invariant straight lines appear.

Exercise 10.8 Prove numerically that the system

$$\dot{x} = 1 + xy,$$
$$\dot{y} = a_{00} + a_{10}x + a_{01}y + a_{20}x^2 + a_{11}xy + a_{02}y^2,$$
$$a_{00} = a_{01} + a_{11} - a_{10} - a_{20} - a_{02}$$
$$a_{10} = -26.5$$
$$a_{01} = 67/220$$
$$a_{20} = -12$$
$$a_{11} = 2.1502$$
$$a_{02} = 8/11$$

has a $(3, 1)$ limit cycle configuration and determine the annuli which contain those limit cycles one by one.

Exercise 10.9 Determine the behavior of the degenerate singular point of the system

$$\dot{x} = (x^2 - y^2)x + 2xy^2 + x^4 - 6x^2y^2 + y^4,$$
$$\dot{y} = 2x^2y - (x^2 - y^2)y - 4x^3y + 4xy^3.$$

How many invariant straight lines does it seem to have? Check it algebraically.

Exercise 10.10 Construct the phase portrait of the system

$$\dot{x} = 1 + x + x^2 + x^3 + 2y + 2x^2y,$$
$$\dot{y} = -y + 2xy + x^2y + 2xy^2.$$

Check the curious behavior of the orbits when they are drawn using the reduced option in the *Parameters of Integration* window. Find out why the program cannot be sure of the center-focus problem regarding the finite linear center, and is certain that the reduced system has a center at infinity.

Make a small perturbation of the linear coefficient of y in the first equation and enjoy the curious new behavior around the infinite singular point in the reduced system.

How can one force the program to draw orbits that do not seem to end before reaching infinity?

Exercise 10.11 Study the system

$$\dot{x} = -70 - 100x + 70x^2 + 100x^3 - 200y + 200x^2y,$$
$$\dot{y} = 146x + 100y + 140xy + 100x^2y + 200xy^2,$$

using different values for the *Precision* parameter and observe the numerical instability of the Lyapunov constants.

Exercise 10.12 Study the system

$$\dot{x} = y - 3 + (x^2 - 1)^2(x + 3) - ax,$$
$$\dot{y} = -0.01x.$$

for values of $a \in [0, 0.2]$.

   Hint: This system has a very fast transition from a phase portrait with two limit cycles to one without any cycles, which makes it impossible to determine numerically the value of a for which a semistable limit cycle occurs. It also shows an odd behavior of the orbits with fast regions followed by terribly slow ones. You will need to play with the integration parameters a little.

Exercise 10.13 Study the system

$$\dot{x} = -y + axy(x^2 - y^2),$$
$$\dot{y} = x + bxy(x^2 - y^2),$$

for parameters $ab \neq 0$. We suggest you to set them to random numbers between -5 and 5 and not close to zero. Would you say that the origin is a center?

   Hint: Try plotting an orbit far from the origin. Is it a periodic orbit? Then try to compute more than the default number of Lyapunov constants, but do not set a high number directly. It is better to test the speed of your computer by increasing the number of constants by two at a time.

Exercise 10.14 Study the system

$$\dot{x} = y,$$
$$\dot{y} = x^5 - xy.$$

First use Poincaré compactification. Then try it with a Poincaré–Lyapunov compactification, first looking for a couple of well chosen powers.

Exercise 10.15 Construct the phase portrait of the homogeneous systems:

$$\dot{x} = -42x^7 + 68x^6y - 23x^5y^2 + 86x^4y^3 + 39x^3y^4 + 10x^2y^5 + 20xy^6 - 8y^7,$$
$$\dot{y} = y(1110x^6 - 220x^5y - 1591x^4y^2 + 478x^3y^3 + 487x^2y^4 - 102xy^5 - 12y^6),$$

$$\dot{x} = 315x^7 + 477x^6y - 113x^5y^2 + 301x^4y^3 - 300x^3y^4 -$$
$$\quad - 192x^2y^5 + 128xy^6 - 16y^7,$$
$$\dot{y} = y(2619x^6 - 99x^5y - 3249x^4y^2 + 1085x^3y^3 + 596x^2y^4 - 416xy^5 + 64y^6),$$

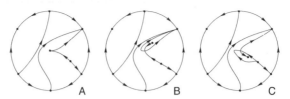

Fig. 10.27. Exercise 10.16

$$\dot{x} = 6x^7 + 4x^6y - 15x^5y^2 - 10x^4y^3 - 33x^3y^4 - 22x^2y^5 - 12xy^6 - 8y^7,$$
$$\dot{y} = y(-90x^6 - 28x^5y + 201x^4y^2 + 62x^3y^3 - 177x^2y^4 - 70xy^5 + 12y^6),$$

$$\dot{x} = -45x^7 - 9x^6y + 155x^5y^2 + 31x^4y^3 + 120x^3y^4 +$$
$$+ 24x^2y^5 - 80xy6 - 16y^7,$$
$$\dot{y} = y(2619x^6 - 99x^5y - 3249x^4y^2 + 1085x^3y^3 + 596x^2y^4 - 416xy^5 + 64y^6),$$

$$\dot{x} = y^2(25x^5 + 20x^4y + 18x^3y^2 + 18x^2y^3 - 7xy^4 - 2y^5),$$
$$\dot{y} = y(288x^6 - 72x^5y - 367x^4y^2 + 118x^3y^3 + 130x^2y^4 - 10xy^5 - 15y^6).$$

Exercise 10.16 Consider the system

$$x' = y + 4x(x + y), \qquad y' = x^2 + 4xy + 3y^2. \tag{10.11}$$

The finite singularities of system (10.11) are $P_0(0,0)$ which is a second-order cusp and $P_1(1/8, -1/24)$, an unstable hyperbolic node. At infinity there is a stable node at $(1,0)$ of local chart U_1, a saddle at the origin of U_2 and a saddle–node coming from the collision of a finite singularity with an infinite one located at $(-1,0)$ of U_1. The phase portrait of system (10.11) is given in Fig. 10.27a.

Make some linear and constant small perturbations to system (10.11) such that you split the cusp into an elementary saddle and an elementary stable or unstable node, in a way that yields the phase portraits given in Fig. 10.27b, c.

Hint: The perturbations may be very small and thus the quantitative image that you get with P4 at normal zoom appear different, or simply difficult to grasp.

Bibliography

1. M.I. Al'mukhamedov. On the construction of a differential equation having given curves as limit cycles. *Izv. Vyss. Ucebn. Zaved Matematika (in Russian)*, 44:3–6, 1963.
2. A.F. Andreev. Investigation of the behaviour of the integral curves of a system of two differential equations in the neighbourhood of a singular point. *Trans. Amer. Math. Soc.* 8:183–207, 1958.
3. A.A. Andronov. Les cycles limites de Poincaré et la théorie des oscillations auto–entretenues. *C.R. Acad. Sci. Paris*, 189:559–561, 1929.
4. A.A. Andronov, E.A. Leontovich, I.I. Gordon, and A.G. Maier. *Qualitative theory of second-order dynamic systems*. Israel Program for Scientific Translations. Halsted Press (A division of Wiley), New York, 1973.
5. V.I. Arnol'd. *Geometrical methods in the theory of ordinary differential equations*. Fundamental Principles of Mathematical Science. Springer, Berlin Heidelberg New York, 1983.
6. V.I. Arnol'd and Y. Il'Yashenko. *Ordinary differential equations*, pages 1–148. Encyclopaedia Math. Sci., 1, Dynamical systems, I. Springer, Berlin Heidelberg New York, 1988.
7. V.I. Arnol'd, A. Varchenko, and S. Goussien-Zade. *Singularités des applications différentiables*. Israel Program for Scientific Translations. Mir, Moscow, 1982.
8. D.K. Arrowsmith. The singularity $x\,\partial/\partial y$. *J. Differ. Equat.*, 34:153–166, 1979.
9. J.C. Artés, F. Dumortier, C. Herssens, J. Llibre, and P. De Maesschalck. Computer program P4 to study Phase Portraits of Planar Polynomial differential equations. http://mat.uab.es/~artes/p4/p4.htm, 2005.
10. J.C. Artés, R. Kooij, and J. Llibre. Sructurally stable quadratic vector fields. *Mem. Am. Math. Soc.* 134(639), 1998.
11. J.C. Artés, J. Llibre, and J. C. Medrano. Nonexistence of limit cycles for a class of structurally stable quadratic vector fields. *Preprint*, 2005.
12. F. Balibrea and V. Jimenez. A characterization of the ω-limit sets of planar continuous dynamical systems. *J. Differ. Equat.* 145:469–488, 1998.
13. R. Bamon. Quadratic vector fields in the plane have a finite number of limit cycles. *Int. Hautes Études Sci. Publ. Math.* 64:111–142, 1986.
14. N.N. Bautin. On the number of limit cycles which appear with the variation of the coefficients from an equilibrium position of focus or center type. *Math. USSR–Sb.* 100:397–413, 1954.

15. I. Bendixson. Sur les courbes définies par des équations différentielles. *Acta Math.* 24:1–88, 1901.

16. A.N. Berlinskii. On the number of elliptic domains adherent to a singularity. *Soviet Math. Dokl.* 9:169–173, 1968.

17. A.N. Berlinskii. On the structure of the neighborhood of a singular point of a two-dimensional autonomous system. *Soviet Math. Dokl.* 10:882–885, 1969.

18. E. Bierstone and P.D. Milman. Semi-analytic and subanalytic sets. *Inst. Hautes tudes Sci. Publ. Math.* 67:5–42, 1988.

19. Th. Bröcker. Differentiable germs and catastrophes. *London Math. Soc. Lect. Note Ser.* 17, 1975.

20. H. Broer. Formal normal forms theorems for vector fields and some consequences for bifurcations in the volume preserving case. *Springer Lecture Notes in Maths.* 898:54–74, 1981.

21. I.U. Bronstein and A.Y. Kopanskii. *Smooth invariant manifolds and normal forms. World Scientific Series on Nonlinear Science*, volume 7 of *Series A: Monographs and Treatises*. World Scientific, Singapore, 1994.

22. M. Brunella and M. Miari. Topological equivalence of a plane vector field with its principal part defined through Newton polyhedra. *J. Differ. Equat.* 85:338–366, 1990.

23. A.D. Bruno. *Local methods in non–linear differential equations*. Springer Series Soviet Math. Springer, Berlin Heidelberg New York, 1989.

24. L. Cairó, M.R. Feix, and J. Llibre. Integrability and algebraic solutions for planar polynomial differential systems with emphasis on the quadratic systems. *Resenhas*, 4:127–161, 1999.

25. L. Cairó and J. Llibre. Darbouxian first integrals and invariants for real quadratic systems having an invariant conic. *J. Phys. A: Math. Gen.* 35:589–608, 2002.

26. J. Carr. *Applications of centre manifold theory.* Applied Mathematical Sciences. Springer, Berlin Heidelberg New York, 1981.

27. M. Caubergh and F. Dumortier. Hopf–Takens bifurcations and centres. *J. Differ. Equat.*, 202:1–31, 2004.

28. J. Chavarriga and J. Giné. Integrability of a linear center perturbed by fourth degree homogeneous polynomial. *Publicacions Matemàtiques*, 40:21–39, 1996.

29. J. Chavarriga and J. Giné. Integrability of a linear center perturbed by fifth degree homogeneous polynomial. *Publicacions Matemàtiques*, 41:335–356, 1997.

30. J. Chavarriga and J. Giné. Integrability of cubic with degenerate infinity. *Differ. Equat. Dyn. Syst.* 6:425–438, 1998.

31. J. Chavarriga, J. Llibre, and J. Sotomayor. Invariant algebraic solutions for polynomial systems with emphasis in the quadratic case. *Expositiones Math.* 15:161–173, 1997.

32. L.A. Cherkas, J.C. Artés, and J. Llibre. Quadratic systems with limit cycles of normal size. *Bull. Acad. Ştiinţe Repub. Mold. Mat.*, 1:31–46, 2003.

33. C. Chicone and F. Dumortier. Finiteness for critical points of planar analytic vector fields. *Nonlinear Anal. Theor. Methods Appl.* pages 315–335, 1993.

34. C. Chicone and W. Liu. Asymptotic phase revisited. *J. Differ. Equat.* 204:227–246, 2004.

35. S.N. Chow and J.K. Hale. *Methods of bifurcation theory*, volume 251 of *Grundlehren der Mathematischen Wissenschaften [Fundamental Principles of Mathematical Science]*. Springer, Berlin Heidelberg New York, 1982.

36. S.N. Chow, C.Z. Li, and D. Wang. *Normal forms and bifurcation of planar vector fields.* Grundlehren der Mathematischen Wissenschaften [Fundamental Principles of Mathematical Science]. Cambridge University Press, Cambridge, 1994.

37. C.J. Christopher. Invariant algebraic curves and conditions for a center. *Proc. Roy. Soc. Edinburgh*, 124A:1209–1229, 1994.

38. C.J. Christopher and J. Llibre. Algebraic aspects of integrability for polynomial systems. *Qualit. Theor. Dyn. Syst.* 1:71–95, 1999.

39. C.J. Christopher and J. Llibre. Integrability via invariant algebraic curves for planar polynomial differential systems. *Ann. Differ. Equat.* 16:5–19, 2000.

40. C.J. Christopher, J. Llibre, and J.V. Pereira. Multiplicity of invariant algebraic curves in polynomial vector fields. *Preprint*, 2004.

41. A. Cima, A. Gasull, V. Mañosa, and F. Mañosas. Algebraic properties of the Lyapunov and period constants. *Rocky Mountain J. Math.* 27:471–501, 1997.

42. A. Cima, A. Gasull, and J. Torregrosa. On the relation between index and multiplicity. *J. London Math. Soc.* 57:757–768, 1998.

43. J.L. Coolidge. *A Treatise on algebraic plane curves.* Dover, New York, 1959.

44. W.A. Coppel. *Stability and asymptotic behavior of differential equations.* Heath, Boston, 1965.

45. W.A. Coppel. A survey of quadratic systems. *J. Differ. Equat.* 2:293–304, 1966.

46. G. Darboux. Mémoire sur les équations différentielles algébriques du premier ordre et du premier degré (mélanges). *Bull. Sci. Math.* 124A:60–96, 123–144, 151–200, 1878.

47. J. Devlin. Word problems related to periodic solutions of a non–autonomous system. *Math. Proc. Cambridge Philos. Soc.* 108:127–151, 1990.

48. J. Devlin. Word problems related to derivatives of the displacement map. *Math. Proc. Cambridge Philos. Soc.* 110:569–579, 1991.

49. G.T. dos Santos. *Classification of generic quadratic vector fields without limit cycles*, volume 597 of *Lect. Notes in Math*, pages 605–640. Springer, Berlin Heidelberg New York, 1977.

50. G.F.D. Duff. Limit cycles and rotated vector fields. *Ann. Math.* 57:15–31, 1953.

51. H. Dulac. Sur les cycles limites. *Bull. Soc. Math. Fr.* 51:45–188, 1953.

52. F. Dumortier. Singularities of vector fields on the plane. *J. Differ. Equat.* 23:53–106, 1977.

53. F. Dumortier. *Singularities of vector fields*, volume 32 of *Monografias de Matemática*. IMPA, Rio de Janeiro, 1978.

54. F. Dumortier and P. Fiddelaers. Quadratic models for generic local 3–parameter bifurcations on the plane. *Trans. Am. Math. Soc.* 326:101–126, 1991.

55. F. Dumortier and C. Herssens. Polynomial Liénard equations near infinity. *J. Differ. Equat.* 153:1–29, 1999.

56. F. Dumortier and C. Li. Quadratic Liénard equations with quadratic damping. *J. Differ. Equat.*, 139:41–59, 1997.

57. F. Dumortier, P. Rodrigues, and R. Roussarie. *Germs of diffeomorphisms on the plane*, volume 902 of *Lectures Notes in Mathematics*. Springer, Berlin Heidelberg New York, 1981.

58. F. Dumortier and C. Rousseau. Cubic Liénard equations with linear damping. *Nonlinearity*, 3:1015–1039, 1990.

59. J. Écalle. *Introduction aux fonctions analysables et preuve constructive de la conjecture de Dulac*. Actualités Math. Hermann, Paris, 1992.

60. D. Eisenbud and H. Levine. An algebraic formula for the degree of a C^∞ map germ. *Ann. Math.* 106:19–44, 1977.

61. C. Elphick, E. Tirapegui, M.E. Brachet, P. Coullet, and G. Iooss. A simple global characterization for normal forms of singular vector fields. *Phys. D*, 29:95–127, 1987.

62. W.W. Farr, Chengzhi Li, I.S. Labouriau, and W.F. Langford. Degenerate Hopf bifurcation formulas and Hilbert's 16th problem. *SIAM J. Math. Anal.* 20:13–29, 1989.

63. E. Fehlberg. *Classical fifth–, sixth–, seventh–, and eight–order Runge–Kutta formulas with step size control*, volume 287. NASA Technical Report, 1968.

64. N. Fenichel. Persistence and smoothness of invariant manifolds for flows. *Indiana Univ. Math. J.* 21:193–226, 1971.

65. J.P. Françoise. Successive derivatives of a first return map, application to the study of quadratic vector fields. *Ergodic Theory Dyn. Syst.* 16:87–96, 1996.

66. J.P. Françoise and R. Pons. Computer algebra methods and the stability of differential systems. *Random Comput. Dyn.* 3:265–287, 1995.

67. J.P. Françoise and Y. Yomdin. Bernstein inequalities and applications to analytic geometry and differential equations. *J. Funct. Anal.* 146:185–205, 1997.

68. E. Freire, A. Gasull, and A. Guillamon. Limit cycles and lie symmetries. *Preprint*, 2004.

69. A. Gasull, A. Guillamon, and V. Mañosa. An explicit expression of the first Lyapunov and period constants with applications. *J. Math. Anal. Appl.* 211:190–202, 1997.

70. A. Gasull, A. Guillamon, and V. Mañosa. An analytic–numerical method of computation of the Lyapunov and period constants from their algebraic structure. *SIAM J. Numer. Anal.* 36:1030–1043, 1999.

71. A. Gasull and J. Torregrosa. A new approach to the computation of the Lyapunov constants. *Comput. Appl. Math.* 20:149–177, 2001.

72. H. Giacomini, J. Llibre, and M. Viano. On the nonexistence, existence and uniqueness of limit cycles. *Nonlinearity*, 9:501–516, 1996.

73. M.G. Golitsina. Nonproper polycycles of quadratic vector fields on the plane. *Selected translations, Selecta Math. Soviet.*, 10:143–155, 1991. translation of Methods of the qualitative theory of differential equations (Russian), 51–67, Gor'kov. Gos. University, Gorki, 1987.

74. M. Golubitsky and D.G. Schaeffer. *Singularities and groups in bifurcation theory. Vol. I*, volume 51 of *Applied Mathematical Sciences*. Springer, Berlin Heidelberg New York, 1985.

75. M. Golubitsky and D.G. Schaeffer. *Singularities and groups in bifurcation theory. Vol. II*, volume 69 of *Applied Mathematical Sciences*. Springer, Berlin Heidelberg New York, 1988.

76. E.A.V. Gonzales. Generic properties of polynomial vector fields at infinity. *Trans. Amer. Math. Soc.* 143:201–222, 1969.

77. J. Guckenheimer and P. Holmes. *Nonlinear oscillations, dynamical systems, and bifurcations of vector fields*, volume 42 of *Applied Mathematical Sciences*. Springer, Berlin Heidelberg New York, 1983.

78. A. Guillamon. *Estudi d'alguns problemes de teoria qualitativa al pla, amb aplicacions als sistemes depredador–presa*. PhD thesis, Universitat Autònoma de Barcelona, July 1995.

79. C. Gutierrez and J. Llibre. Darboux integrability for polynomial vector fields on the 2 dimensional sphere. *Extracta Math.* 17:289–301, 2002.

80. P. Hartman. *Ordinary differential equations.* Wiley New York London Sydney, 1964.

81. A.C. Hearn and J.P. Fitch. *Reduce 3.6, user's manual.* Konrad–Zuse–Zentrum, Berlin, 1995.

82. D. Hilbert. Mathematische probleme. In Nachr. Ges. Wiss., editor, *Second Internat. Congress Math. Paris, 1900,* pages 253–297. Göttingen Math.–Phys. Kl., 1900.

83. M. Hirsch, C.C. Pugh, and M. Shub. *Invariant manifolds,* volume 583 of *Lectures Notes in Mathematics.* Springer, Berlin Heidelberg New York, 1977.

84. W. Hirsch and S. Smale. *Differential equations, dynamical systems, and linear algebra,* volume 60 of *Pure and Applied Mathematics.* Academic Press, New York London, 1974.

85. H. Hopf. *Differential geometry in the large,* volume 1000 of *Lecture Notes in Math.* Springer, Berlin Heidelberg New York, 1989.

86. I.D. Iliev. On second order bifurcations of limit cycles. *J. London Math. Soc.* 58:353–366, 1998.

87. I.D. Iliev and L.M. Perko. Higher order bifurcations of limit cycles. *J. Differ. Equat.* 154:339–363, 1999.

88. Y. Il'Yashenko. *Finiteness theorem for limit cycles,* volume 94 of *Trans. of Math. Monogr.* Am. Math. Soc. 1991.

89. Y. Il'Yashenko. Centennial history of Hilbert's 16th problem. *Bull. Am. Math. Soc.* 39:301–354, 2002.

90. Y. Il'Yashenko and S. Yakovenko. *Lectures on analytic differential equations.* to be published.

91. Y. Il'Yashenko and S. Yakovenko. Finitely smooth normal forms of local families of diffeomorphisms and vector fields. *Uspekhi Mat. Nauk,* 46:3–39, 1991. translation in Russian Math. Surveys 46 (1991), no. 1, 1–43.

92. J.P. Jouanolou. *Equations de Pfaff algébriques,* volume 708 of *Lectures Notes in Mathematics.* Springer, Berlin Heidelberg New York, 1979.

93. W. Kapteyn. On the midpoints of integral curves of differential equations of the first degree. *Nederl. Akad. Wetensch. Verslag. Afd. Natuurk. Konikl. Nederland,* pages 1446–1457, 1911.

94. W. Kapteyn. New investigations on the midpoints of integrals of differential equations of the first degree. *Nederl. Akad. Wetensch. Verslag. Afd. Natuurk. Konikl. Nederland,* 20:1354–1365, 1912. Also **21**, 27–33 (Dutch).

95. A. Katok and B. Hasselblatt. *Introduction to the modern theory of dynamical systems,* volume 54 of *Encyclopedia of Mathematics and its Applications.* Cambridge University Press, Cambridge, 1995.

96. A. Kelley. The stable, center-stable, center, center-unstable, unstable manifolds. *J. Differ. Equat.* 3:546–570, 1967.

97. A.Y. Kotova. Finiteness theorem for limit cycles of quadratic systems. *Selected translations, Selecta Math. Soviet.* 10:131–142, 1991. translation of Methods of the qualitative theory of differential equations (Russian), 74–89, Gor'kov. Gos. University, Gorki, 1987.

98. S. Lefschetz. *Differential equations: geometric theory.* Pure and Applied Mathematics. Interscience Publishers, New York London, 1963.

99. A. Liénard. Etude des oscillations entretenues. *Rev. Générale de l'Electricité,* 23:335–357, 1928.

100. J. Llibre and G. Rodríguez. Darboux integrability of polynomial vector fields on 2 dimensional surfaces. *Int. J. Bifurcation Chaos*, 12:2821–2833, 2002.

101. J. Llibre and G. Rodríguez. Invariant hyperplanes and Darboux integrability for d dimensional polynomial differential systems. *Bull. Sci. Math.* 124:1–21, 2002.

102. J. Llibre and G. Rodríguez. Configurations of limit cycles and planar polynomial vector fields. *J. Differ. Equat.* 198:374–380, 2004.

103. J. Llibre and X. Zhang. Darboux integrability of real polynomial vector fields on regular algebraic hypersurfaces. *Rendiconti del circolo matematico di Palermo, Serie II*, 51:109–126, 2002.

104. N.G. Lloyd, C.J. Christopher, J.Devlin, J.M. Pearson, and N. Yasmin. Quadratic like cubic systems. *Differ. Equat. Dyn. Syst.* 5:329–345, 1997.

105. A.M. Lyapunov. Issledovanie odnogo iz osobenych slučaevzadači ob ustoičivosti dviženiya. *Matematičeskiĭ Sbornik*, 17:253–333, 1893.

106. A.M. Lyapunov. *Stability of motion*, volume 30 of *Mathematics in Science and Engineering*. Academic, New York–London, 1966.

107. L. Markus. Global structure of ordinary differential equations in the plane. *Trans. Am. Math. Soc.* 76:127–148, 1954.

108. L. Markus. Quadratic differential equations and non–associative algebras. *Ann. Math. Stud.* 45:185–213, 1960.

109. H. Melenk. *Reduce: gnuplot interface version 4*. Konrad–Zuse–Zentrum, Berlin, 1995.

110. K.R. Meyer. *The implicit function theorem and analytic differential equations*, volume 468 of *Lecture Notes in Math*, pages 191–208. Springer, Berlin Heidelberg New York, 1974.

111. J.W. Milnor. *Topology from the differentiable viewpoint*. Princenton Landmarks in Math. Princenton University Press, New Jersey, 1997.

112. J. Moulin-Ollagnier, A. Nowicki, and J.M. Strelcyn. On the non–existence of constants of derivations: the proof of a theorem of jouanolou and its development. *Bull. Sci. Math.* 119:195–233, 1995.

113. R. Moussu. Le problème de la finitude du nombre de cycles limites [the problem of finiteness of the number of limit cycles]. *Astérisque*, 145–146:89–101, 1987. Séminaire Bourbaki, Vol. 1985/86.

114. J. Murdock. *Normal forms and unfoldings for local dynamical systems*. Springer Monographs in Mathematics. Springer, Berlin Heidelberg New York, 2003.

115. R. Narasimhan. *Analysis on real and complex manifolds*. North-Holland, Amsterdam, 1968.

116. D. Neumann. Classification of continuous flows on 2–manifolds. *Proc. Am. Math. Soc.* 48:73–81, 1975.

117. K. Odani. The limit cycle of the van der Pol equation is not algebraic. *J. Differ. Equat.* 115:146–152, 1995.

118. J. Palis and W. de Melo. *Geometric theory of dynamical systems*. Springer, Berlin Heidelberg New York, 1982.

119. J.M. Pearson, N.G. Lloyd, and C.J. Christopher. Algorithmic derivation of centre condition. *SIAM Rev.* 38:619–636, 1996.

120. M.C. Peixoto and M.M. Peixoto. Structural stability in the plane with enlarged boundary conditions. *An. Acad. Brasil. Ci.*, 31:135–160, 1959.

121. M.M. Peixoto. On structural stability. *Ann. Math.* 69:199–222, 1959.

122. M.M. Peixoto. Structural stability on two-dimensional manifolds. *Topology*, 1:101–120, 1962.

123. M.M. Peixoto. On the classification of flows on 2–manifolds. Academic, New York, pages 389–419, 1973. Dynamical systems (Proc. Sympos., Univ. Bahia, Salvador,1971).

124. M. Pelletier. *Contribution a l'étude de quelques singularités de systèmes non linéaires*. PhD thesis, Université de Bourgogne, Dijon, 1994.

125. M. Pelletier. Eclatements quasi homogènes [Quasi homogeneous blow-ups]. *Ann. Fac. Sci. Toulouse Math.* 4:879–937, 1995.

126. C. Perelló. A note on analytic structural stability in compact m^2. *Bol. Soc. Math. Mexicana*, 15:40–41, 1970.

127. L. Perko. Rotated vector fields and the global behavior of limit cycles for a class of quadratic systems in the plane. *J. Differ. Equat.* 18:63–86, 1975.

128. L. Perko. *Differential equations and dynamical systems*, volume 7 of *Texts in applied mathematics*. Springer, Berlin Heidelberg New York, 2nd edition, 1996.

129. Zhang Pingguang. Quadratic systems with a 3rd–order (or 2nd–order) weak focus. *Ann. Differ. Equat.* 17:287–294, 2001.

130. H. Poincaré. Mémoire sur les courbes définies par une équation différentielle. *J. Maths. Pures Appl.* 7:375–422, 1881. Ouvre (1880–1890), Gauthier–Villar, Paris.

131. H. Poincaré. Mémoire sur les courbes définies par une équation différentielle. *J. Math. Pures Appl.* 8:251–296, 1882.

132. H. Poincaré. Sur l'intégration des équations différentielles du premier ordre et du premier degré I. *Rendiconti del circolo matematico di Palermo*, 5:161–191, 1891.

133. M.J. Prelle and M.F. Singer. Elementary first integrals of differential equations. *Trans. Am. Math. Soc.* 279:215–229, 1983.

134. C.C. Pugh. *Hilbert 16th problem: Limit cycles of polynomial vector fields on the plane*, volume 468 of *Lect. Notes in Math*, pages 55–57. Springer, Berlin Heidelberg New York, 1975.

135. Ye Yan Quian. *Theory of limit cycles*, volume 66 of *Trans. of Mathematical Monographs*. Am. Math. Soc. Providence, RI, 2nd edition, 1984.

136. C. Robinson. *Dynamical systems, stability, symbolic dynamics and chaos*. Studies in Advanced Math., CRC Press, Boca Raton, FL, 1999. 2nd edition.

137. R. Roussarie. *Bifurcation of planar vector fields and Hilbert's sixteenth problem*. Birkhäuser Verlag, Basel, 1998.

138. M.E. Sagalovich. Topological structure of the neigborhood of a critical point of a differential equation. *Differ. Equat.* 11:1498–1503, 1975.

139. M.E. Sagalovich. Classes of local topological structures of an equilibrium state. *Differ. Equat.* 15:253–255, 1979.

140. S. Schecter and F. Singer. A class of vector fields on S^2 that are topologically equivalent to polynomial vector fields. *J. Differ. Equat.* 57:406–435, 1985.

141. D. Schlomiuk. Algebraic and geometric aspects of the theory of polynomial vector fields. In D. Schlomiuk, editor, *Bifurcations and periodic orbits of vector fields*, volume 408 of *NATO ASI Series C: Mathematical and Physical Sciences*, pages 429–467. Kluwer, Dordrecht, 1993.

142. D. Schlomiuk. Algebraic particular integrals, integrability and the problem of the center. *Trans. Am. Math. Soc.* 338:799–841, 1993.

143. D. Schlomiuk. Elementary first integrals and algebraic invariant curves of differential equations. *Expositiones Math.* 11:433–454, 1993.

144. D. Schlomiuk. Basic algebro–geometric concepts in the study of planar polynomial vector fields. *Publ. Matemàtiques*, 41:269–295, 1997.

145. G. Seifert. A rotated vector approach to the problem of stability of solutions of pendulum–type equations. contributions to the theory of nonlinear oscillations. *Annals of Math. Studies*, 3:1–16, 1956.

146. D.S. Shafer. Structural stability and generic properties of planar polynomial vector fields. *Rev. Math. Iberoamericana*, 3:337–355, 1987.

147. K.S. Sibirskii. On the number of limit cycles on the neigboorhood of a singular point. *Differ. Equat.* 1:36–47, 1965.

148. C.L. Siegel and J.K. Moser. *Lectures on celestial mechanics*. Classics in Mathematics. Springer, Berlin Heidelberg New York, 1995. Translated from the German by C. I. Kalme. Reprint of the 1971 translation.

149. J. Sijbrand. Properties of center manifolds. *Trans. Am. Math. Soc.* 289:431–469, 1985.

150. M.F. Singer. Liouvillian first integrals of differential equations. *Trans. Am. Math. Soc.* 333:673–688, 1992.

151. J. Sotomayor. *Curves definides per equaçoes diferenciais no plano*. Instituto de Matemática Pura e Aplicada, Rio de Janeiro, 1979.

152. J. Sotomayor. *Liçoes de equaçoes diferenciais ordinárias*, volume 11 of *[Euclid Project*. Instituto de Matemática Pura e Aplicada, Rio de Janeiro, 1979.

153. S. Sternberg. On the structure of local homeomorphisms of euclidean n-space. ii. *Am. J. Math.* 80:623–631, 1958.

154. D. Stowe. Linearization in two-dimensions. *J. Differ. Equat.* 63:183–226, 1986.

155. R. Sverdlove. Inverse problems for dynamical systems. *J. Differ. Equat.* 42:72–105, 1981.

156. F. Takens. Normal forms for certain singularities of vector fields. *Ann. Inst. Fourier (Grenoble)*, 23:163–195, 1973.

157. J. Torregrosa. *Punts singulars i òrbites periòdiques per a camps vectorials*. PhD thesis, Universitat Autònoma de Barcelona, Barcelona, 1998.

158. R.T. Valeeva. Construction of a differential equation whith given limit cycles and critical points. *Volz Math. Sb.*, 5:83–85, 1966. (in Russian).

159. A. van den Essen. *Reduction of singularities of the differential equation $Ady = Bdx$. Équations différentielles et systèmes de Pfaff dans le champ complexe (Sem., Inst. Rech. Math. Avancé, Strasbourg, 1975)*, volume 712 of *Lecture Notes in Math.*, pages 44–59. Springer, Berlin Heidelberg New York, 1979.

160. B. van der Pol. On relaxation–oscillations. *Philos. Mag.* 2:978–992, 1926.

161. S.J. van Strien. Center manifolds are not C^∞. *Math. Z.* 166:143–145, 1979.

162. A. Vanderbauwhede. Centre manifolds, normal forms and elementary bifurcations. *Dynamics reported*, 2:89–169, 1989. *Dynam. Report. Ser. Dynam. Syst. Appl.* 2, Wiley, Chichester.

163. J.A. Weil. *Constant et polynômes de Darboux en algèbre différentielle: applications aux systèmes différentiels linéaires*. PhD thesis, École Polytechnique, 1995.

164. S. Wiggins. *Global bifurcations and chaos. Analytical methods*, volume 73 of *Applied Mathematical Sciences*. Gauthier Villars, Paris, 1988.

165. Chen Xiang-Yan. Applications of the theory of rotated vector fields i. *Nanjing Daxue Xuebao (Math.)*, 1:19–25, 1963. Chinese.

166. Chen Xiang-Yan. Applications of the theory of rotated vector fields ii. *Nanjing Daxue Xuebao (Math.)*, 2:43–50, 1963. Chinese.

167. Chen Xiang-Yan. On generalized rotated vector fields. *Nanjing Daxue Xuebao (Nat. Sci.)*, 1:100–108, 1975. Chinese.

168. S. Yakovenko. A geometric proof of the Bautin theorem, concerning the Hilbert's 16th problem. *Trans. Am. Math. Soc. Ser. 2*, 165:203–219, 1995. Providence.

169. Zhi Fen Zhang, Tong Ren Ding, Wen Zao Huang, and Zhen Xi Dong. *Qualitative theory of differential equations*, volume 101 of *Trans. of Mathematical Monographs*. Am. Math. Soc. Providence, RI, 1992.

170. H. Żołądek. Quadratic systems with center and their perturbations. *J. Differ. Equat.* 109:223–273, 1994.

171. H. Żołądek. On algebraic solutions of algebraic Pfaff equations. *Studia Math.* 114:117–126, 1995.

Index

Universitext

Aguilar, M.; Gitler, S.; Prieto, C.: Algebraic Topology from a Homotopical Viewpoint

Aksoy, A.; Khamsi, M. A.: Methods in Fixed Point Theory

Alevras, D.; Padberg M. W.: Linear Optimization and Extensions

Andersson, M.: Topics in Complex Analysis

Aoki, M.: State Space Modeling of Time Series

Arnold, V. I.: Lectures on Partial Differential Equations

Arnold, V. I.: Ordinary Differential Equations

Audin, M.: Geometry

Aupetit, B.: A Primer on Spectral Theory

Bachem, A.; Kern, W.: Linear Programming Duality

Bachmann, G.; Narici, L.; Beckenstein, E.: Fourier and Wavelet Analysis

Badescu, L.: Algebraic Surfaces

Balakrishnan, R.; Ranganathan, K.: A Textbook of Graph Theory

Balser, W.: Formal Power Series and Linear Systems of Meromorphic Ordinary Differential Equations

Bapat, R.B.: Linear Algebra and Linear Models

Benedetti, R.; Petronio, C.: Lectures on Hyperbolic Geometry

Benth, F. E.: Option Theory with Stochastic Analysis

Berberian, S. K.: Fundamentals of Real Analysis

Berger, M.: Geometry I, and II

Bliedtner, J.; Hansen, W.: Potential Theory

Blowey, J. F.; Coleman, J. P.; Craig, A. W. (Eds.): Theory and Numerics of Differential Equations

Blowey, J.; Craig, A.: Frontiers in Numerical Analysis. Durham 2004

Blyth, T. S.: Lattices and Ordered Algebraic Structures

Börger, E.; Grädel, E.; Gurevich, Y.: The Classical Decision Problem

Böttcher, A; Silbermann, B.: Introduction to Large Truncated Toeplitz Matrices

Boltyanski, V.; Martini, H.; Soltan, P. S.: Excursions into Combinatorial Geometry

Boltyanskii, V. G.; Efremovich, V. A.: Intuitive Combinatorial Topology

Bonnans, J. F.; Gilbert, J. C.; Lemaréchal, C.; Sagastizábal, C. A.: Numerical Optimization

Booss, B.; Bleecker, D. D.: Topology and Analysis

Borkar, V. S.: Probability Theory

Brunt B. van: The Calculus of Variations

Bühlmann, H.; Gisler, A.: A Course in Credibility Theory and Its Applications

Carleson, L.; Gamelin, T. W.: Complex Dynamics

Cecil, T. E.: Lie Sphere Geometry: With Applications of Submanifolds

Chae, S. B.: Lebesgue Integration

Chandrasekharan, K.: Classical Fourier Transform

Charlap, L. S.: Bieberbach Groups and Flat Manifolds

Chern, S.: Complex Manifolds without Potential Theory

Chorin, A. J.; Marsden, J. E.: Mathematical Introduction to Fluid Mechanics

Cohn, H.: A Classical Invitation to Algebraic Numbers and Class Fields

Curtis, M. L.: Abstract Linear Algebra

Curtis, M. L.: Matrix Groups

Cyganowski, S.; Kloeden, P.; Ombach, J.: From Elementary Probability to Stochastic Differential Equations with MAPLE

Dalen, D. van: Logic and Structure

Da Prato, G.: An Introduction to Infinite-Dimensional Analysis

Das, A.: The Special Theory of Relativity: A Mathematical Exposition

Debarre, O.: Higher-Dimensional Algebraic Geometry

Deitmar, A.: A First Course in Harmonic Analysis

Demazure, M.: Bifurcations and Catastrophes

Devlin, K. J.: Fundamentals of Contemporary Set Theory

DiBenedetto, E.: Degenerate Parabolic Equations

Diener, F.; Diener, M.(Eds.): Nonstandard Analysis in Practice

Dimca, A.: Sheaves in Topology

Dimca, A.: Singularities and Topology of Hypersurfaces

DoCarmo, M. P.: Differential Forms and Applications

Duistermaat, J. J.; Kolk, J. A. C.: Lie Groups

Dumortier. F.; Llibre, J.; Artés J. C.: Qualitative Theory of Planar Differential Systems

Edwards, R. E.: A Formal Background to Higher Mathematics Ia, and Ib

Edwards, R. E.: A Formal Background to Higher Mathematics IIa, and IIb

Emery, M.: Stochastic Calculus in Manifolds

Emmanouil, I.: Idempotent Matrices over Complex Group Algebras

Endler, O.: Valuation Theory

Erez, B.: Galois Modules in Arithmetic

Everest, G.; Ward, T.: Heights of Polynomials and Entropy in Algebraic Dynamics

Farenick, D. R.: Algebras of Linear Transformations

Foulds, L. R.: Graph Theory Applications

Franke, J.; Härdle, W.; Hafner, C. M.: Statistics of Financial Markets: An Introduction

Frauenthal, J. C.: Mathematical Modeling in Epidemiology

Freitag, E.; Busam, R.: Complex Analysis

Friedman, R.: Algebraic Surfaces and Holomorphic Vector Bundles

Fuks, D. B.; Rokhlin, V. A.: Beginner's Course in Topology

Fuhrmann, P. A.: A Polynomial Approach to Linear Algebra

Gallot, S.; Hulin, D.; Lafontaine, J.: Riemannian Geometry

Gardiner, C. F.: A First Course in Group Theory

Gårding, L.; Tambour, T.: Algebra for Computer Science

Godbillon, C.: Dynamical Systems on Surfaces

Godement, R.: Analysis I, and II

Goldblatt, R.: Orthogonality and Spacetime Geometry

Gouvêa, F. Q.: p-Adic Numbers

Gross, M. et al.: Calabi-Yau Manifolds and Related Geometries

Gustafson, K. E.; Rao, D. K. M.: Numerical Range. The Field of Values of Linear Operators and Matrices

Gustafson, S. J.; Sigal, I. M.: Mathematical Concepts of Quantum Mechanics

Hahn, A. J.: Quadratic Algebras, Clifford Algebras, and Arithmetic Witt Groups

Hájek, P.; Havránek, T.: Mechanizing Hypothesis Formation

Heinonen, J.: Lectures on Analysis on Metric Spaces

Hlawka, E.; Schoißengeier, J.; Taschner, R.: Geometric and Analytic Number Theory

Holmgren, R. A.: A First Course in Discrete Dynamical Systems

Howe, R., Tan, E. Ch.: Non-Abelian Harmonic Analysis

Howes, N. R.: Modern Analysis and Topology

Hsieh, P.-F.; Sibuya, Y. (Eds.): Basic Theory of Ordinary Differential Equations

Humi, M., Miller, W.: Second Course in Ordinary Differential Equations for Scientists and Engineers

Hurwitz, A.; Kritikos, N.: Lectures on Number Theory

Huybrechts, D.: Complex Geometry: An Introduction

Isaev, A.: Introduction to Mathematical Methods in Bioinformatics

Istas, J.: Mathematical Modeling for the Life Sciences

Iversen, B.: Cohomology of Sheaves

Jacod, J.; Protter, P.: Probability Essentials

Jennings, G. A.: Modern Geometry with Applications

Jones, A.; Morris, S. A.; Pearson, K. R.: Abstract Algebra and Famous Inpossibilities

Rubel, L. A.: Entire and Meromorphic Functions

Ruiz-Tolosa, J. R.; Castillo E.: From Vectors to Tensors

Runde, V.: A Taste of Topology

Rybakowski, K. P.: The Homotopy Index and Partial Differential Equations

Sagan, H.: Space-Filling Curves

Samelson, H.: Notes on Lie Algebras

Schiff, J. L.: Normal Families

Sengupta, J. K.: Optimal Decisions under Uncertainty

Séroul, R.: Programming for Mathematicians

Seydel, R.: Tools for Computational Finance

Shafarevich, I. R.: Discourses on Algebra

Shapiro, J. H.: Composition Operators and Classical Function Theory

Simonnet, M.: Measures and Probabilities

Smith, K. E.; Kahanpää, L.; Kekäläinen, P.; Traves, W.: An Invitation to Algebraic Geometry

Smith, K. T.: Power Series from a Computational Point of View

Smoryński, C.: Logical Number Theory I. An Introduction

Stichtenoth, H.: Algebraic Function Fields and Codes

Stillwell, J.: Geometry of Surfaces

Stroock, D. W.: An Introduction to the Theory of Large Deviations

Sunder, V. S.: An Invitation to von Neumann Algebras

Tamme, G.: Introduction to Étale Cohomology

Tondeur, P.: Foliations on Riemannian Manifolds

Toth, G.: Finite Möbius Groups, Minimal Immersions of Spheres, and Moduli

Verhulst, F.: Nonlinear Differential Equations and Dynamical Systems

Weintraub, S.H.: Galois Theory

Wong, M. W.: Weyl Transforms

Xambó-Descamps, S.: Block Error-Correcting Codes

Zaanen, A.C.: Continuity, Integration and Fourier Theory

Zhang, F.: Matrix Theory

Zong, C.: Sphere Packings

Zong, C.: Strange Phenomena in Convex and Discrete Geometry

Zorich, V. A.: Mathematical Analysis I

Zorich, V. A.: Mathematical Analysis II

Printing: Krips bv, Meppel
Binding: Stürtz, Würzburg